Water Productivity

Principles and Practices

The Editors

Dr M R Umesh, Assistant Professor, University of Agricultural Sciences, Raichur (UAS), Karnataka. He earned MS and Ph.D. from UAS, Bangalore, specialized in Agronomy and later Post–Doctoral Fellow in New Mexico State University, NM, US. He has nine years of experience at various levels on maize agronomy, water management and cropping system research. He has published more than 40 research articles, 20 book chapters, 10 research bulletins and extension literature for the benefit of farming community. Based on his significant contribution, many professional societies and organizations conferred awards including young scientist award from National Environmental Science Academy, New Delhi, IPNI best photo award, Department of Science and Technology fast track young scientist, UAS Raichur incentive award, ISA best poster *etc*. He is an active referee of five national and international journals, life member of five professional societies. He is involved in teaching at BS, MS and Ph.D. students and research guidance for the last five years. He has visited USA, England, and the Philippines for various academic purposes.

Dr B M Chittapur has earned his B.Sc. (Agri.), M.Sc.(Agri.) and Ph.D. degrees from UAS, Dharwad, Karnataka. Presently he is working as Director of Research, UAS, Raichur. He also served as Professor and Head of Agronomy and has three decades experience in teaching, research and farm production. Also worked as Dean (Agri.), College of Agriculture, Bheemarayanagudi, UAS, Raichur for three and half years and at College of Agriculture, Raichur for one and half years. He worked on soil solarization for his Ph.D. and is a recipient of Gold medal for his Masters degree and has meritorious academic carrier throughout. He implemented many innovative ideas in research after taking the charge of Director of Research. He guided 15 Masters and 6 Doctoral students, among them one Doctoral student theses has been awarded as "Best Thesis" from the Indian Society of Agronomy, New Delhi. He has published hundreds of research papers in national and international scientifically peer reviewed journals and has half a dozen of books besides useful technologies for farming community.

Water Productivity

Principles and Practices

– Editors –

M R Umesh

Assistant Professor

B M Chittapur

Director of Research

University of Agricultural Sciences,
Raichur, Karnataka

2018

Daya Publishing House®

A Division of

Astral International Pvt. Ltd.

New Delhi–110 002

© 2018 EDITORS

ISBN: 9789388173766 (Int. Edn.)

Published by   :   **Daya Publishing House®**
                             A Division of
                             Astral International Pvt. Ltd.
                             – ISO 9001:2015 Certified Company –
                             4736/23, Ansari Road, Darya Ganj
                             New Delhi-110 002
                             Ph. 011-43549197, 23278134
                             E-mail: info@astralint.com
                             Website: www.astralint.com

Digitally Printed at   :   **Replika Press Pvt. Ltd.**

Foreword

In the current agricultural scenario of the country, the crop productivity has to climb demographic geometric ladder under dwindling resource base. Over the last few years net cultivated area has remained constant around 142± 2 million hectare. There is no option in a situation of horizontal expansion restrictions, except enhanced efforts for sustained and higher reap per drop of water. In agriculture, water holds the key and its availability and utilization by plants through precise application is going to be very vital in the future. Further, in rainfed agriculture, water use efficiency is relatively higher but crop productivity is very low. The primary concern is to enhance soil moisture status and increase length of growing period. On the other side, most of the irrigation commands in India have very low irrigation efficiency (<30 per cent) due to cumulative losses along the course and at farm level. Further, flooding of water courses including rivers and untimely rains and hailstorms have also affected crops. Water,therefore, is considered to be both boon and bane for crop growth. It is a fact that even when all the resources are supplied except water nothing will ensure the plant system to survive.

Scientific use of the precious water resource through improved methods, proper scheduling of when and how much to use, tailor made soil management options, effective drainage, good quality water use *etc.*, will go a long way in attaining higher use efficiency. There is a need to have a refined conceptual background on water management practices so that the potential which exists in the agriculture production system can be fully utilized. Increasing uncertainty in the distribution and quantum of rainfall is affecting cropping pattern even in the irrigation commands. For instance, during rabi/summer 2015-16, paddy cultivation has been withheld in the TungaBhadra Command of Karnataka due to water shortage in the reservoir mainly to reserve water for drinking purpose during critical summer months.

Therefore, it becomes necessary to suggest alternate productive crops to paddy. The story remains same in the Upper Krishna Project of Karnataka. Alternate crops to paddy not only save water but also restore soil health. Nevertheless, efforts have to be made in development of water efficient production technologies for large scale adoption, besides working on conservation agriculture system which is an emerging practice in the North.

Considering all these issues the fundamental and applied aspects of water use have been given due importance in this publication, **'Water Productivity: Principles and Practices.** It will also prove to be a very useful reference material for decision makers, scientists, students, extension personnel and all others interested in water productivity. I take this opportunity to congratulate Dr. Umesh, M.R. and Dr. B.M. Chittapur, authors of this book, for their splendid efforts in compiling, editing and making it available to the scientific and extension fraternity.

P.M. Salimath

Vice-Chancellor

UAS, Raichur

Karnataka, India

Preface

Over the last few decades degradation of natural resources, indiscriminate use of precious water and spatial and temporal variation in rainfall due to climate change have resulted in low agricultural productivity. Indian agriculture, as aptly said, is a gamble with monsoon and this insecurity is increasing with climate change witnessed these days. Therefore, conservation and maintenance of natural resources is a prime concern and responsibility of everyone for sustainable agriculture production. In that, there is a need to restore and/improve our water resources and use it appropriately based on crop water need in commensuration with soil, climate and other critical resources. This requires proper knowledge of different aspects of water management and its conservation for future generation.

Irrigation commands which are thought as the solution to the nation's food problem are also suffering increasingly from low productivity and water use efficiency because of unscientific water use and land management. This calls for efforts to educate every stake holder in the efficient use of this scarce resource. In this effort a Model Training Course on water productivity enhancement through alternate production techniques in command areas was organized at UAS, Raichur December, 2016 with the financial assistance from Department of Agriculture Extension, Ministry of Agriculture and Farmers Welfare, Government of India, New Delhi. The University is grateful to DAC, Govt. of India, New Delhi for their assistance to this cause. Thinking not to waste efforts of good compilation of information by eminent scientists the compendium is now brought out in this book form.

The book deals with general aspects of water management, water efficient cropping systems, alternate cropping systems for command area, substitution of paddy in commands, conservation agriculture practices to enhance water use productivity, micro irrigation, herbigation, precision water management, management of problematic soils, specific water management practices in field and horticulture crops, crop residue management, climate change and global warming and so on. A sincere effort has been made to give a broader understanding of water management practices and other related issues which are of regional and global concern as well.

We express our gratitude to the University of Agricultural Sciences, Raichur for giving us an opportunity to come out with this book containing contributions from scientists across India and the globe. We are highly thankful to them who are too numerous to name here. Special thanks are due to Dr. P.M. Salimath, Vice Chancellor, who has encouraged us besides guiding during the course of training and for providing illuminating Foreword. We are also thankful to Dr. A.S. Halepyati, Dean (Agri.), UAS, Raichur, Dr. Anand, N., Dr. M.Y. Ajaykumar and Dr. Basavaraj Hulgur Assistant Professors, UAS, Raichur for coordinating the course. The editors are highly indebted to all the staff members of Department of Agronomy and Directorate of Research, UAS, Raichur for their direct and indirect support. This book is expected to be useful to teachers, research workers, scientists, scholars, policy makers and others involved in agriculture development particularly in irrigation commands of the country.

M R Umesh

B M Chittapur

Contents

List of Contributors

Dr. S. Bhaskar
Assistant Director General (AAF and CC)
Krishi Anusandhan Bhavan-II
Pusa, New Delhi

Dr. B.M. Chittapur
Director of Research
University of Agricultural Sciences, Raichur, Karnataka

Dr. M.R. Umesh
Assistant Professor (Agronomy)
Directorate of Research
UAS, Raichur

Dr. Ajaykumar, M.Y.
Assistant Professor (Agronomy)
AICRP on Cotton, MARS
UAS, Raichur

Dr. Djanaguiraman Maduraimuthu
Assistant Professor
Department of Plant Pathology
College of Agriculture
Tamil Nadu Agriculture University
Coimbatore, Tamil Nadu

Dr. D.M. Chandargi
Registrar
UAS, Raichur, Karnataka

Dr. C.A. Srinivasamurthy
Director of Research
Central Agriculture University
Imphal, Manipur

Dr. Suresh S Patil
Dean (Agri.)
College of Agriculture
Bheemarayanagudi, UAs, Raichur

Mr. R.H. Rajkumar
Assistant Professor (Soil and Water Engg.)
AICRP on salt affected soil, Gangavati
UAS, Raichur

Dr. S.R. Anand
Assistant Professor (Agronomy)
AICRP on Salt Affected Soils
ARS,Gangavati, UAS, Raichur

Dr. Mukesh Kumar Meena
Assistant Professor (Crop Physiology)
Directorate of Research
UAS, Raichur

Dr. S. G. Patil
Retd. Director of Education
UAS, Raichur

Mr. Chandra Naik M
Assistant Professor (Crop Physiology)
Directorate of Extension, UAS, Raichur

Dr. P.S. Kanannavar
Assistant Professor (Soil and Water Engg.)
College of Agriculture Engineering, Raichur
UAS, Raichur

Dr. Ananda, N.
Assistant Professor (Agronomy)
AICRP on Weed Management
UAS, Raichur

Dr. Vidyavathi G Yadahalli
Assistant Professor (Soil Science)
KVK, Raichur UAS, Raichur

Dr. Ravi, MV
Assistant Professor (SS and AC)
College of Agriculture, UAS, Raichur

Dr. Premalatha, B.R.
Assistant Professor (Agronomy)
College of Horticulture, Kolar
University of Horticultural Sciences
Bagalkot, Karnataka

Dr. Latha, HS
Assistant Professor (Agronomy)
College of Agriculture, Raichur
UAS, Raichur

Dr. BN Maruthi Prasad
Assistant Professor
College of Horticulture
Kolar, UHS Bagalkot

Dr. Shubha, G.V.
Assistant Professor
College of Agriculture, Raichur
UAS, Raichur

Dr. Manjunatha, N.
Assistant Professor (Agronomy)
Krishi Vignana Kendra, Kalaburagi
UAS, Raichur

Dr. G.V. Srinivasa Reddy
Assistant Professor (Soil and Water Engg.)
College of Agriculture Engineering, Raichur
UAS, Raichur

Dr. Yogeshkumar Singh
Assustant, CIMMYT- India
Karnataka

Dr. G.S. Yadahalli
Assistant Professor (Agronomy)
MARS, Raichur, UAS, Raichur

Dr. B.G. Masthana Reddy
Professor and Head
AICRP on Rice, Gangavati
UAS, Raichur

Dr. M S Dinesha
AICRP on Underutilized Crops, Hebbal
UAS, GKVK, Bangalore

Dr. P.H. Kuchanur
Associate Professor
College of Agriculture
Bheemarayanagudi
UAS, Raichur

Dr. Manjunatha Bhanuvally
Assistant Professor
Agriculture Research Station, Dhadesugur
UAS, Raichur

Dr. Ramesha, Y.M.
Assistant Professor (Agronomy)
Agriculture Research Station, Dhadesugur
UAS, Raichur

Dr. B.K. Desai
Professor
Department of Agronomy
College of Agriculture, UAS, Raichur

Er.Karegoudar AV
Assistant Professor (Soil and Water Engg.)
AICRP on Salt Affected Soils
ARS, Gangavati, UAS, Raichur

Dr. U.K. Shanwad
Assistant Professor (Agronomy)
AICRP on Sunflower, MARS, Raichur
UAS, Raichur

Dr. Rajesh N.L.
Assistant Professor
Department of Soil Science
College of Agriculture, UAS, Raichur

Dr. Satyanarayana Rao
Professor and Head
Organic Farming Research Institute
MARS, UAS, Raichur

Dr. Vishwanath Jowkin
Professor and Head
AICRP on Salt Affected soils, Gangavti
UAS, Raichur

Dr. P Balakrishnan
Dean (Agri. Engineering)
College of Agriculture Engineering
Raichur

Dr. Rekha D.
Department of Plant Pathology
Indian Institute of Horeticulture Research
Hesargatta, Bangalore

Dr. K.R. Sreenivas
Assistant Professor (Plant Pathology)
Krishi Vignana Kendra
Tiptur, UAS, Bangalore

Dr. M.N. Thimmegowda
Assosicate Professor (Agronomy)
AICRP on Dryland Agriculture
GKVK, UAS, Bangalore

Dr. B.K. Ramachandrappa
Professor and Head
AICRP on Dryland Agriculture
GKVK, UAs, Bangalore

Dr. GB Lokesh
Assistant Professor,
Dept. of Agril. Economics
College of Agriculture
UAS, Raihcur

Dr. Sushilendra
Assistant Professor
Dept. of Farm Machinery and Power Engg.
College of Agricultrure Engineering
UAS, Raichur

Dr. Praksh KV
Assistant Professor
AICRP on Farm machinery and Power Engg.
College of Agricultrure Engineering
UAS, Raichur

Dr. M. Anantachar
Professor and Head
AICRP on Farm Machinery and Power Engg.
College of Agricultrure Engineering
UAS, Raichur

Dr. Amrutha T Joshi
Professor, Department of Agril. Economics
College of Agriculture
UAS, Raichur

1

Water Productivity: Why and How?

B.M. Chittapur and M.R. Umesh

The blue planet, the earth, with nearly 70 per cent of its surface being covered with water still thirsts for water for agriculture and 'water crisis' is the major issue discussed in every forum these days. Water is one most critical input that enables a higher productive potential from the land, and significant production response from associated use of high yielding varieties, fertilizers, and others. For this reason, agriculture in India as elsewhere in Asia has major share of potable twater. However, climate change and consequent rainfall aberrations are affecting water supply and many river basins are likely to witness physical water scarcity by 2050. And, in the country, inter-state water sharing would be a real political problem now on. Fischer *et al.* (2007) estimated an increase in irrigation water requirements of 50 per cent in developing countries and 16 per cent per cent in developed region between 2000 and 2080. Therefore, enhancing water productivity, either by producing more with given water or producing the same quantity with lesser water, is now a priority through techniques that are cost-effective and eco-friendly unlike traditional ways of irrigation.

Water Resources of India

The source for all the surface and ground water is precipitation. The rainfall in the country is highly irregular, variable and undependable. Its distribution varies from 100 mm in western Rajasthan to about 11, 000 mm in Meghalaya. Not only the rainfall, but the potential evapotranspiration from the country's surface also varies widely due to wide variations in topography, temperature, solar radiation, wind speed and relative humidity. Average annual rainfall in the country is about 1,170 mm, which corresponds to an annual precipitation (including snowfall) of 4,000 billion cubic meters (BCM). Nearly 75 per cent of this (3000 BCM) occurs during the monsoon season, confined generally to 3-4 months (June to September) of a year. According to the Planning Commission, India has so far created a total of about 225 BCM of surface storage capacity. However, per capita storage capacity in India at 190 cubic meters is very less compared to the USA (5,961 m^3), Australia (4,717 m^3), Brazil (3,388 m^3) and China (2,486 m^3). This necessitates creation of large storage facilities for maximum utilization of the run-off (Amritha Patel, 2013). This warrants concerted efforts from governments as well as individuals. Further, for sustained agriculture production and high water productivity, it is necessary to manage the use of water resources such that i). both surface and ground water supplies are maintained at desired level, and ii). The quality of land and water resources does not deteriorate with time (Sitaram Singh *et al.*, 1998).

Table 1.1: Irrigated Area in India

	Utilization (m ha)	Capacity (m ha)	Ultimate Irrigation Potential (m ha)
Major and medium irrigation project	28.02	32.69	58.50
Ground water	42.50	45.73	64.05
Surface	10.12	10.89	17.38
Total	**80.54**	**89.31**	**139.9**

Gulati *et al.* (2005).

Climate change is threatening our ecosystems; water scarcity is becoming a way of life and pollution is a growing threat to our health and habitat. Rivers all over India are still being degraded. Not only per capita availability of water in the country is already low but also there is enormous wastage, growing pollution and contamination of surface as well as groundwater. Ground water pollution becomes major threat for human beings and agriculture crops besides to ecosystem. Some of the irrigation commands in the country have been reported to have high nitrate level. In some districts of Assam (Barpota, Darrang, Kamrup, Sonipni) and Orissa (Balasore, Cuttack, Puri) ground water have high iron content ranging from 1 to 10 mg l^{-1}.Thirteen states in India have been identified as endemic to fluorosis due to abundance in naturally occurring fluoride bearing minerals. There are nearly half million people in India suffering from ailment due to excess of fluoride in drinking water. Arsenic in ground water has been reported in a range of 0.05-3.2 mglp in shallow aquifers from 61 block in 8 districts of West Bengal namely Malda, Mushirbad, Nadia, North and South 24 Pargana, Bardhawan, Howrah and Hugli.

Furthermore, lowering of benefit cost ratio, lower productivity per unit area and time, rising costs of inputs, compaction of soil structure due to indiscriminate use of water, unfavourable weather conditions, nutrient deficiencies, toxicities and plant protection problems have shaken the trust of farmers in some cropping systems. It may result in lower irrigation efficiency in irrigation commands. The intended service from water in agriculture is the creation of favourable water regime for crop, fish and animal production. Hence, agricultural water management emphasizes those methods, systems and techniques of water conservation, remediation, development, application, use and removal that provide a socially and environmentally favourable level of water regime to agricultural production system at the least economic cost (Sitaram Singh *et al.*, 1998). Therefore, the efforts on generation of new knowledge and empowerment of stakeholders should focus on i. effective development, management, and conservation of on-farm water resources for its sustained availability, ii. Significant reduction in the use of irrigation water per unit irrigated area, iii. Removal of excess water from agricultural lands, iv. Development of sustainable cropping system in relation to the availability of water, v. Devising multiple uses of water in agricultural production programmes to enhance water productivity, vi. Reuse of poor quality water municipal, industrial, agricultural and other waste water, and vii. Avoiding/reversing the contaminations and further degradation of soil and water resources (Sitaram Singh *et al.*, 1998). Since the subject considered is vast in the following text a few aspects like water management in rice, pressurized (micro) irrigation, alternate ways of economising irrigation water under scarcity, use of poor quality water, field drainage, and water fortification and amelioration of poor quality water are considered.

Strategies of Higher Water Productivity

1. Strategies to Save Water in Rice Production

Among various crops rice considered to be largest consumer of water in different ecosystems. But rice is inevitable. In Asia rice is grown in large area and it is the largest consumer. According to Food and Agriculture Organization (FAO), the global rice requirement in 2025 will be of the order of 800 m t. At the moment, the production is less than 600 m t and hence an additional 200 m t needed will have to be produced by increasing productivity per unit area against the diminishing resources. In India, rice is commonly grown by transplanting seedlings into puddled soil. However, in addition to adverse effects of puddling on soil physical properties, puddling and transplanting require large amount of water and labour, both of which are becoming increasingly scarce and expensive, making rice production less profitable and sustainable. Therefore, there is shift from transplanting to other rice establishment methods namely alternate wetting and drying (AWD), direct seeded rice (DSR), system of rice intensification (SRI) and aerobic rice with differential water requirements and managerial expertise.

Table 1.2: Water Productivity as Influenced by Conventional vs. SRI Method

Water Measurement	Method	Irrigation (cu.m)	Increase (per cent)
	Conventional	149.33	38.00
Water Productivity	SRI-Org-ino-	91.89	
kg/cu.M	Conventional	1.18	
	SRI-Org-ino-	2.23	46.00
Rainfall(cu.M)		203.952	
Total water prodc.	Conventional	0.48	
kg/cu.M	SRI-Org-ino-	0.68	29.00

In case of wet land rice, percolation is the major problem. Compared to unpuddled control, compaction reduced relative water expense to 87.1 per cent and saved 40 cm water. Puddling with rotovator and disk harrow reduced the relative water expense to 67.74 and 73.87 per cent, respectively and saved 100 and 81 cm water (Sitaram Singh *et al.*, 1998). However, mere compaction is less effective while puddling with a cage-wheeled tractor driven cultivator resulted in minimum relative loss of 68.5 per cent and maximum grain yield. Again, puddling is less effective in light-textured sandy loam soils than in heavy soils. Generally, the farmers have a tendency to apply high quantities of irrigation water (about 10 to 15 cm) to partially overcome the effect of uncertainty in the availability of water, and to reduce the labour requirement. But higher depth of ponded water increases the rate of percolation owing to high hydraulic head and reduces tiller formation by way of submergence of tiller forming plant parts. Nevertheless, experiments revealed that comparable rice yields were obtained at shallow (5cm) and deep (10 cm) submergence with tendency of decreased yields at higher depths, besides there was considerable increase in irrigation water requirement with increase in ponding depth.

The practice of rice cultivation under continuous submergence requires large quantity of water for its growth. In order to reduce the irrigation requirement of rice, an approach of intermittent ponding (Alternate wetting and drying, AWD) in areas with shallow ground water in place of continuous submergence has been studied throughout the country. In this practice, irrigation was applied to rice fields to achieve 5 to 7 cm ponding depth (Sitaram Singh *et al.*, 1998). After cessation of irrigation, the ponded water infiltrates into the soil till it disappears from the land surface, and thereafter irrigation is withheld for one to several days. During the non-ponding period, there is continuous desaturation of the root zone owing to the downward movement of water and the evapotranspiation losses. The period of desaturation should be so chosen that crop yield is not adversely affected. For most of the locations where the study was carried, a three-day period appears to be permissible duration for desaturation, after which irrigation should be applied to rice. This schedule of irrigation, relative to continuous submergence, entailed substantial saving in irrigation water (23 to 65 per cent) while yields were comparable. Significantly lower yields were obtained with the desaturation period exceeding 3 days except that under some situations such as high-water table conditions, a longer period (5 days) could be allowed in case of Modipurum (Bihar). At some locations having relatively low water retentive soils, desaturation perod of 1 day was not found desirable in view of the significant yield reduction. But, even with 1 day desaturation a considerable saving in irrigation could be obtained.

A practical way to implement AWD safely is by using a 'field water tube (*pani pipe*)' to monitor the water depth in the field (Mahenderkumar *et al.*, 2016). After irrigation, the water depth will gradually decrease. When the water level has dropped to about 15cm below the surface of the soil, irrigation should be applied to re-flood the field to a depth of about 5 cm. From one week before to a week after flowering, the field should be kept flooded, topping up to a depth of 5 cm as needed. After flowering, during grain filling and ripening, the ware level can be allowed to drop again to 15 cm below the soil surface before irrigation. AWD can be started a few weeks (1- 2 weeks) after transplanting. When many weeds are present, AWD should be postponed for 2 -3 weeks to assist suppression of the weeds by the ponded water and improve the efficiency of the herbicide. Similar report was made from in the Philippines and Bangladesh (Bouman and Troug, 2001).

As it entails considerable saving of water, it makes possible to irrigate a large area with limited water supply. It also enables more effective use of rainfall in the rice fields. In the tubewell command, there is considerable reduction in pumping hours resulting in saving in energy and increased life of pumping equipment. The transplanting of rice should be planned in such a way that the cropping period matches with the monsoon period to effectively utilize the rain in crop production and reduce irrigation. For instance at Ludhiana, delaying transplanting from 16 May to 31 may and skipping a period of high evaporative demand, there was a saving of 21 to 40 cm irrigation water (Sitaram Singh *et al.*, 1998).

One more thing is farmers generally continue irrigation in rice till about 4 – 7 days before harvest. Studies revealed that suspension of irrigation 14 – 21 days before harvest caused more uniform ripening of crop and economized on water by about 16 cm (Sandhu *et al.*, 1982). It was also found that termination of water three weeks before harvest though caused a reduction of 200 kg ha^{-1} yield it saved 38 cm of water over irrigation cut off 7 days before harvest.

DSR though not new to farmers in many areas of Karnataka *e.g.* In Western Ghats (Belgaum, Dharwad and Karwar districts), DSR is common in uplands. During recent years with the joint efforts of CIMMYT and University of Agricultural Sciences, Raichur, development and validation of DSR technology had shown promise for its out-scaling through innovative strategies in the areas where water supplies are limited and farmers do not get sufficient water at right time and constrained with ON-OFF canal water supply (Chittapur *et al.*, 2016). Moreover, insufficient water in barrages, delayed, erratic and untimely canal supplies are leading to delayed transplanting (beyond August) and consequently lower yields. Therefore, a participatory intervention on DSR in the Upper Krishna Project (UKP) and Tunga Bhadra Project (TBP) in addressing such predicaments was initiated (GoK- CGIAR Project Progress Report (2013)-CIMMYT). Response to early dry seeding or taking advantage of early rains received before canal supplies was met with imminent success with farmers. Success of Kasabe camp, Raichur today has spread over 90,000 acres in TBP and UKP commands. At Siruguppa falling under TBP in Karnataka, a farmer by adopting DSR and AWD has obtained yields exceeding 42 q ac^{-1} over 60 acres of land. He even overcame an irrigation break of 20 days due to non release of water during crop growth using a foliar spray of KNO$_3$ @ 1.0 per cent at that time (Basavanneppa, 2016 personal communication). These days studies on drip irrigation and subsurface irrigation are being carried out in rice and drip at 80 cm interval found to save nearly 40 per cent irrigation water in TBP irrigation command of Karnataka. In other words, farmers need to be proactive in situations of water scarcity. In addition to increase in net income, timely sowing, reduced seed rate by half, reduced fuel consumption by 40-50 l/ha, reduced water use by 25-35 per cent, reduced emission of GHGs, and increased NUE are the other benefits (the details are covered in future chapters).

2. Pressurised Irrigation

A majority of farmers in the irrigated areas use traditional surface (flood) irrigation method where most of the water is wasted as deep drainage or result in waterlogging and related salinization apart from health issues. To overcome these situations and even to use bore wells with limited output, pressurized irrigation system is the best option (Kukal *et al.*, 2014). The main types available are sprinkler irrigation and drip irrigation (surface and subsurface). Drip and sprinkler systems are much more water efficient than conventional basin irrigation practices. These have a conveyance efficiency of 100 per cent and an application efficiency of 70 to 90 per cent, while the corresponding figures for basin irrigation are 40 – 70 per cent and 60 – 70 per cent respectively (Von *et al.*, 2004; Narayanamoorthy, 2006). Sprinkler irrigation method has relatively lower water saving (up to 70 per cent efficiency) than drip irrigation, since it supplies water over entire field of the crop (INCD, 1998; Kulkarni, 2005).

The pressurized irrigation systems have the potential to increase irrigation water productivity by providing water to match crop requirements, reducing deep water drainage losses, and generally keeping soil drier, thereby reducing soil evaporation (Camp, 1998). However, Jenson (1984) remarked that in row crops planted in moderately wide rows like sugarcane and cotton, drip irrigation did not help reduce ET. Only drippers located in the subsoil reduced ET by 10 – 15 per cent. Drip irrigation not only results in savings in water usage, but also increases the yield (Tiwari *et al.*, 2003; Yuan, 2003; Dhawan, 2002). Micro irrigation technologies developed for irrigation of horticultural and plantation crops have been reported to save 30 -50 per cent of water and increase crop yields by 20 to 50 per cent (Hanson *et al.*, 1997; Fekadu and Teshome, 1998). In fact drip system has made commercial horticulture possible in most arid areas with too little and too deep ground water. Kolar district in Karnataka is the best example while the story is true in Maharashtra and areas adjoining Karnataka also. The use of sprinkler and drips led to doubling of yield in sugarcane, vegetable and horticultural crops. It is reported that adoption of drip irrigation for suitable crops in the potential areas may lead to reduction in crop water requirements to the level of 44. 46 BCM in India (Sharma *et al.*, 2009).

Sprinkler irrigation system is especially suitable to shallow sandy soils of uneven topography where levelling is not possible, and to the regions where both labour and water are scarce, it has been reported that water productivity is substantially improved in wheat, maize, sorghum, sugarcane, and cotton under sprinkler irrigation system compared to basin irrigation (Ali *et al.*, 2012; Verma and Shrivastava, 1992; Home *et al.*, 2002; Pawar *et al.*, 2002). Sprinklers are advantageous compared to the surface methods as water can be delivered at a desired and controlled rate, thereby ensuring a uniform distribution of water and hence high WUE. In tomato, with at same irrigation depth ridges and furrow required seven irrigation, while in sprinkler with six irrigation only yield was 18.6 per cent higher. Application of 3.5 cm water with sprinkler gave as much as yield as 6 cm application by surface method, thereby saving about 34 per cent water. In groundnut yields were 24 and 15 per cent higher under sprinkler method compared to border and check-basin methods. The sprinkler saved about 31 per cent water over border and check basin methods, which required 90 cm irrigation water. Moreover, sprinkler irrigation can prove to be important intervention in mitigating the adverse impact of climate change on crops, as it may alter the micro climatic conditions of the field crop (Kukal *et al.*, 2014).

Similarly, substantial improvement in yield and water productivity were reported in a number crops such as cotton, sugarcane, soybean, maize and wheat in drip system over conventional method (Kumar and Singh, 2002; Ahuja *et al.*, 2007; Zhangg *et al.*, 2008; Chandrashekhar, 2009). In transplanted pigeonpea yields tripled under drip irrigation system (Vanishree, 2016). Drip irrigation system can reduce irrigation requirements from 20 to 72 per cent while increasing crop yields by 20 –90 per cent compared with surface irrigation (Postel *et al.*, 2001; Howell, 2011). At Rahuri, studies in sugarcane revealed a saving of more than 60 per cent with drip irrigation over furrow method of irrigation. With fertigation besides higher yields, up to 50 per cent of NPK fertilizers and 39 per cent of water could be saved. They also observed 57 to 80 per cent less weed infestation under drip fertigation compared to surface irrigation method. In Banana at Bhubaneshwar application of 24 l of water on alternate days with drip produced higher yields (31 t ha⁻¹) at IW/CPE = 1 and saved 50 per cent water in comparison to surface irrigation. Similar was the trend at Parbhani with a saving of 22 per cent irrigation water.

Of late drip system is being tried in paddy crops with substantial benefits. With laterals 80 cm apart found to save nearly 40 per cent irrigation water in TBP irrigation command of Karnataka. Drip method of irrigation in rice also revealed higher WUE (98.14 kg hamm⁻¹) compared to semi irrigated paddy (64.21 kg ha mm⁻¹) and KRH-2 variety had higher grain and straw yield (56.8 q ha⁻¹ and 62.4 q ha⁻¹, respectively) than Rasi (54.0 q ha⁻¹ and 61.7 q ha⁻¹, respectively) in a study in Karnataka (Puspa, 2006). KRH-2 variety specifically developed for southern dry land region has more adaptability than Rasi variety and there is no wastage of water either through seepage or by percolation in drip irrigation, hence, water productivity was high.

At Patna, combination of system of rice intensification (SRI) and micro irrigation (low energy water application (LEWA) and micro-sprinkler) significantly increased water productivity by 153 and 156 per cent respectively in comparison to check basin under farmers (0.34 kg m⁻³) practices of rice establishment (Ajay Kumar *et al.*, 2015). There was a mean saving of 27 and 39 per cent water observed in LEWA and micro sprinkler, respectively compared to check basin irrigation (340 mm) in rice crop in a rice-wheat cropping system.

In an attempt of lower operational cost of drip system, planting pattern (normal, paired row and pit method) could be adjusted without sacrificing yield. For instance, Ahuja *et al.* (2005) demonstrated that when same quantity of water was applied through drip irrigation seed cotton yield increased by 32 per cent and water productivity increased by 26 per cent compared to normal sowing. Furthermore, under paired sowing a yield increase of 20 per cent was achieved alongwith the saving of 50 per cent of water and cost of laterals. Thind *et al.* (2008) reported that the water productivity in dense planting of cotton under drip irrigation increased from 0.98 to 1.65 under check basin method to 1.79 – 2.13 kg hamm[-1]. When the quantity of water through drip irrigation was reduced to 75 per cent, the increase in seed cotton yield was 12 per cent over check basin, thus dense planting under drip irrigation increased seed cotton yield and water productivity over normal sowing and also saved 25 per cent irrigation water as well as cost of laterals. Therefore, in India pressurised irrigation systems are being promoted starting in 1980s with subsidies 50 to 90 per cent, and with 'more crop per drop' slogan further impetus is given to tap full potential by the government. And, it is not irrigation alone but intelligent use of fertilizers could increase agricultural production.

3. Deficit Irrigation

Deficit irrigation is a water-saving strategy used in many parts of the world in which irrigation water is applied at amounts less than full crop water requirements (*i.e.* ET) thereby increasing WUE (Morrison *et al.*, 2008). Deficit irrigation, in fact, promotes the use of profile-stored water by encouraging deeper rooting in crops. This practice was reported to save two out of six irrigations applied in wheat at fixed growth stages without any adverse effect on crop yield (Prihar *et al.*, 1976). In deficit irrigation water is applied during drought –sensitive growth stages of a crop to maximize the productivity of water by allowing the crops to sustain some degree of water deficit and yield reduction (Zhangg and Oweis, 1999; Pereira *et al.*, 2002, Fereres and Soriano, 2007). For deficit irrigation, critical growth stages of crops for irrigation are to be identified and water is not missed at these stages. For instance, in wheat crown root initiation and booting to heading are the two most important sensitive stages and drought stress should be avoided during these stages.

The deficit irrigation has been widely used in irrigation of fruit and vineyards (Collins *et al.*, 2010) and annual crops (Kirda *et al.*, 2007; Wang *et al.*, 2101). For example, a study conducted at the International Centre for Agricultural Research in the Dry Areas revealed that application of just 50 per cent of full IW requirement caused a yield reduction of only 10 – 15 per cent (Ahang and Oweis, 1999). Deficit irrigation can increase the water productivity of wheat, maize and rice by 10 -42 per cent (Kang *et al.*, 2010a; Li *et al.*, 2010; Soundharajan and Sudheer, 2009) without causing sever yield reductions (Greets and Raes, 2009). The deficit irrigation can also save water by reducing irrigation water depth by watering only the plant root zone and increasing the interval between successive irrigation.

It has been suggested that yields and water productivity could increase even more if deficit irrigation is used in combination with water conservation practices such as mulching or rain water harvesting techniques (Ali and Talukder, 2008; Oweis and Hachum, 2006). On this line another farm of deficit irrigation system named partial root zone drying (PRD) has been developed. Sepaskhah and Ahmadi (2010) opined that partial root deficit is a successful alternative irrigation technique compared to full irrigation that can save irrigation water up to approximately 50 per cent without significant yield loss. In partial root deficit, the roots sense the soil drying and induce abscisic acid (ABA) that reduce leaf expansion and stomatal conductance, and simultaneously the roots in wet soil absorb sufficient water to maintain a high water status in shoot (Liu *et al.*, 2006; Ahmadi *et al.*, 2010a). A small narrowing of the stomatal opening way reduces water loss substantially with little effect on the photosynthesis. Secondly, part of the root system in a dying soil can respond to the drying by sending a root sourced signal to the shoot where stomatal loss may be inhibited so that water loss is reduced.

For instance, partial rootzone drying in grape wines improved the water use efficiency (by up to 50 per cent) without significant yield reduction. The technique was developed on the basis of knowledge of the mechanisms controlling transpiration and requires that approximately half of the root system is always maintained in a dry or drying state while the remainder of the root system is irrigated. The wetted and dried sides of the root system are alternated on a 10–14d cycle. Abscisic acid (ABA) concentration in the drying roots increases 10fold, but ABA concentration in leaves of grapevines under PRD only increased by 60 per cent compared with a fully irrigated control. Stomatal conductance of vines under PRD irrigation was significantly reduced when compared with vines receiving water to the entire root system. Grapevines from which water was withheld from the entire root system, on the other hand, show a similar reduction in stomatal conductance, but leaf ABA increased 5-fold compared with the fully irrigated control. PRD results in increased xylem sap ABA concentration and increased xylem sap pH, both of which are likely to result in a reduction in stomatal conductance. In addition, there was a reduction in zeatin and zeatin-riboside concentrations in roots, shoot tips and buds of 60, 50 and 70 per cent, respectively, and this may contribute to the reduction in shoot growth and intensified apical dominance of vines under PRD irrigation. There is a nocturnal net flux of water from wetter roots to the roots in dry soil and this may assist in the distribution of chemical signals necessary to sustain the PRD effect. It was concluded that a major effect of PRD is the production of chemical signals in drying roots that are transported to the leaves where they bring about a reduction in stomatal conductance. However, prolonged exposure to drying soil may cause anatomical changes in the roots, such as suberization of epidermis, collapse of the cortex, and loss of succulence. Therefore, it is necessary to alternatively irrigate the two sides of the root system to keep root in dry soil alive and fully functional and sustain the supply of root signals (Kang *et al.*, 2004).

Alternate or every other furrow irrigation is also considered as partial root deficit irrigation. Alternate furrow irrigation and alternate furrow irrigation alternately were widely tried and were found particularly useful in crops like cotton on heavy soils where lack of aeration is the limitation under irrigation. Alternate furrow irrigation was successfully used as water saving irrigation (Grimes *et al.*, 1968). Alternate furrow irrigation in different crops has resulted in higher water productivity (Sepaskhah and Ahmadi, 2010). Kang *et al.* (2000b) reported that alternate furrow irrigation maintained higher maizae grain yields coupled with a 50 per cent reduction in the amount of irrigation compared with the fixed furrow irrigation. A study conducted in warm semi arid region of Iran (Shani-Dashtgol *et al.*, 2006) revealed that using variable alternate furrow irrigation, compared to ordinary every furrow irrigation, applied irrigation water was reduced by 26 per cent with even 10 per cent higher cane production. Interestingly studies also showed that crops under partial root deficit yielded better than under deficit irrigation when the same amount of water is applied, resulting in higher water productivity (Kirda *et al.*, 2004, Liu *et al.*, 2006; Du *et al.*, 2008; Sadras, 2009).

4. Water Fortification and Amelioration of Poor Quality Water

These days, water fortification either with organic or inorganic nutrient sources is very common particularly under drip irrigation in conventional and organic farming which not only enhances nutrient use efficiency but also improves water productivity. Fertilizer (solid/liquid mineral, single or multiple) application through the drip irrigation/micro-irrigation system is fertigation (Srinivas, 2004). It is the most advanced and efficient practice of fertilization. Fertigation combines the application of water and nutrient required for plant growth and development and it allows an accurate and uniform application of nutrients to the wetted area in the root zone, where the active roots are concentrated. Therefore, it is possible to precisely adjust the nutrients quantity and concentration to their demand throughout the growing season of the crop. To produce high yield and quality fruits and vegetables the right combination of water and nutrients is the key. Consequently, recommendations were developed for the most suitable fertilizer formulation (including the basic nutrients NPK and microelements) according to the type of soil, physiological stage, climate and other factors. Special attention should be given to the pH and NO_3/NH_4 ratio, nutrient mobility in soil and salinity conditions. The efficiency of conventional fertilizers application is very low. Fertigation has an edge with regard to nutrient use efficiency (NUE) in comparison to conventional application. Thomson *et al.* (2003) observed the NUE of 90 and 81 per cent with 250 and 350 kg N/ha in broccoli. Fertigation can save fertilizers by 50 per cent and may increase the crop yield by 20-30 per cent.

For effective fertigation farmer should have knowledge of plant growth behaviour including nutrient requirements and rooting patterns, soil chemistry such as solubility and mobility of the nutrients, fertilizers chemistry (mixing compatibility, precipitation, clogging and corrosion) and quality of irrigation water like pH, salt and sodium hazards, and toxic ions. The fertilizer requirements will depend on soil and leaf analysis. The choice of fertilizers will also depend on availability of product, price and the quality of the irrigation water. The fertilizer products which can be used are limited to those that are readily soluble. Many of these formulations prepared for specific crops, or combinations can be used depending on the crop cycle.

These are stable and highly soluble, dissolving rapidly and providing a balance of nutrients. Generally liquid or 100 per cent water soluble fertilizers are more expensive per unit of nutrient when compared to standard fertilizers. While selecting a fertilizer for fertigation one should have criteria of solubility, compatibility with other fertilizers, convenience and cost (Rajput and Patel 2002). Sources of fertilizers have different effects on the pH of irrigation water. The pH of irrigation water should be near to neutral. Higher pH values (>7.5) of irrigation waters are undesirable. Any water source having Ca, Mg and bicarbonates can have interaction with fertilizers and can cause diverse problems, as formation of precipitation in fertilizer tank and clogging of drip system. In water having high Ca and carbonates may cause the precipitation of $CaSO_4$ if sulphate fertilizer is used, urea may precipitate as $CaCO_3$ because urea increase pH. In case of high concentration of Ca and Mg and high pH value may lead the precipitation of Ca and Mg phosphate. When Ca and Mg is high in irrigation water the Phosphoric acid or Mono-ammonium phosphate fertilizer is used for P. Similarly, low pH values (<5) of irrigation water can have detrimental effect on plant roots. The ideal pH of irrigation used for fertigation should be between 5-7.5.

In organic farming alongwith drip irrigation water liquid nutrient sources like cow urine, biodigester solution, Jeevamrit, dashaparni, vermivash *etc.* are supplied to crops which have been found to have tremendous effect on quality and self life besides reduction in pest and disease incidence. Humic acid is another organic found favour with cultivators these days which can be applied with irrigation water. At Coimbatore, Tamil Nadu, Jayakumar *et al.* (2014) observed that application of 150 per cent RDF as drip fertigation combined with bio-fertigation of liquid formulation of azophosmet @ 250 ml ha^{-1}(10 12 cells/ml) significantly increased seed cotton yield and a progressive increase in seed yield was noticed with increasing levels of NPK fertilizer application. Sabreen *et al.* (2015) observed highest sesame yield (533 kg/fed),water use efficiency (0.307 kg seeds/m^3 of irrigation water at 75 per cent ET irrigation levels) with liquid poultry manure (4 t ha^{-1}). Thus, organic bio-stimulants hold great promise for the future of agriculture in general and organic farming in particular.

When canal-water supplies are either unassured or in short supply, farmers are forced to pump saline ground or drainage waters to meet the crop water requirements. These waters from the two sources can be applied either separately or mixed. Mixing water to acceptable quality for crops also results in improving the stream size and thus the uniformity in irrigation especially for the surface method practiced on sandy soils. Allocation of two waters separately, if available on demand, can be done either to different fields, seasons or crop-growth stages such that higher-salinity water is not applied to sensitive crops or growth stages (Minhas, 1998). For instance, brackish ground water cannot be used for normal irrigation. However, when applied sequentially with good quality surface water, the former can be partially used and will reduce the chance of rise of water table. Two irrigations in a total of six to wheat can be by poor quality ground water with a marginal reduction in yield and with no long term adverse effect on the soil (Bhattacharya, 1998) (Table 1.3).

Conjunctive use of poor quality water revealed that use of saline water up to 4 dS/m in direct mode had no adverse effect on cotton yield in Tunga Bhadra irrigation command of Karnataka (Vishwanath, 2015). Use of saline water (4-6 dS/m) during canal lean period and then switching over to good quality water wherever available conclusively established that early establishment (June) with available saline water (with 4 irrigations) and later switching over to canal (August) water produced highest kapas yield (22.1q/ha) compared to a crop receiving good water but sown during August (12.6 q/ha). The salt balance remained favourable and did not raise any concern (Vishwanath, 2015). Drip irrigation can be advantageous in using poor quality waters high in salt contents (up to 8-10 dSm⁻¹) without affecting the yield.

Table 1.3: Conjunctive Use of Poor Quality Groundwater (TW) and Fresh Canal Water for Irrigation in South West Punjab

Irrigation Sequence	Soil Chemical Parameters (av. of 0–120 cm)				Wheat Yield (t ha⁻¹)
	EC (dS m⁻¹)	pH	Na (meq l⁻¹)	SAR (meq l⁻¹)⁻¹/²	
Before experiment	2.1	10.2	19.8	31.2	< 1.5
After 4 years of experiment					
CW only	0.5	8.3	6.7	6.7	3.85
2 CW 1 TW	0.8	9.0	7.1	7.1	3.59
1CW 1TW	0.9	9,2	7.5	9.3	2.85
1CW 2TW	0.9	9.4	8.0	9.0	1.95

CW – pH 7.9, EC 0.3 dS m⁻¹, SAR 0.4 (meq l⁻¹)⁻¹/²

TW – pH 9.5, EC 3.0 dS m⁻¹, SAR 38.5 (meq l⁻¹)⁻¹/²

Brackish water and agro-based industrial effluents are marginal quality water for agricultural use. At present, approximately 20 million hectares of arable land worldwide are reported to be irrigated with effluent. It is particularly common in urban and peri-urban areas of the developing world. Effluent use in agriculture has certain benefits by providing water and nutrients for the cultivation of crops, ensuring food supply to cities and reducing the pressure on available fresh water resources. Studies on farmer's field have clearly indicated that alternate irrigation with lime treated coffee pulp effluent and fresh water with microbial culture provided higher bunch yield in banana plant crop (details available in separate chapter). Besides banana other short duration crops like baby corn, fodder maize and any perennial fodder grass can be raised and irrigated with treated coffee effluent.

Molasses based distilleries are considered to be one of the most polluting agro-based industries due to generation of large amount of foul smelling, brown coloured waste water also called as spent wash with very high BOD and COD levels. However, it is very rich in plant nutrients containing not only major ones but also all the micronutrients in appreciable amounts. These nutrients taken up by sugarcane and in organic form if recycled could save a lot on expenditure of fertilizers. Approximately 40 million m³ of distillery spentwash are discharged annually from 285 alcohol distilleries in India which can be productively used as a nutrient source for crop production. 1 m³ of spentwash contains - 1.0 kg N, 0.02 kg P₂O₅ and 10.0 kg K₂O. In addition, it supplies around 2 kg Ca, 1.5 kg Mg and 1.5 kg S. Hence, it been regarded as liquid fertilizer by sugar and alcohol industries. Spent wash is nothing but waste water generated by distilleries during the distillation of fermented molasses to ethyl alcohol using specific strains of yeast. It is a dark brown coloured liquid containing residual nutrients from sugarcane and yeast cells, it does not contain any heavy metals or other toxic residues. Distillery spentwash is categorized into raw spentwash that comes out from the distillation unit which is reddish brown in colour characterized by low pH, high BOD and COD values. The spentwash that leaves bio-methanation plant after the anaerobic digestion is referred to as primary spentwash which is dark brown in colour with neutral to alkaline pH and has relatively lower BOD and COD values than the raw spentwash. The distillery spentwash is also a source of valuable plant nutrients such as N, K, Ca, Mg, S and micronutrients. Because of its non-toxic nature and containing plant nutrients, it can be considered as a liquid fertilizer.

Surface fertigation with water soluble fertilizers, and drip fertigation with distillery spentwash produced on par yields both in plant and in ratoon crop (refer chapter by Bhaskar and others). Thus, filtered spent wash can be safely used for fertigation through drip system hence it is possible to save exorbitant cost being incurred on fertilizers. One time land application of distillery effluent is beneficial in dry land where it is advocated to be applied once in 2 or 3 years and such application would bring down the cost on fertilizers without causing any decline in yields.

The brewing industry is one of the largest industrial users of water. It has been documented that approximately 3 to 10 litres of waste effluent is generated per litre of beer produced in breweries. Agricultural use of treated wastewater, therefore, might represent a unique opportunity to solve both the problems of water supply for irrigation and the disposal of treated water at the same time. Application of 150 per cent recommended N through TBWW (50 per cent as basal and 50 per cent in three irrigations) is beneficial in giving higher grain yields of maize (39.6 q ha^{-1}) compared to lower grain yield (30.5 q ha^{-1}) in plots receiving fresh water + RDF (find details in following chapter). Results have amply showed that there was no deleterious effect of either treated or untreated effluent on yielding potentials of both grain and stover. Further, because of N content effluent, the crude protein content of grain also improves.

Besides improving the productivity, the effluents are also beneficial in preventing environmental pollution without affecting the soil health. Even the microbial analysis carried out in these studies have clearly shown that the beneficial micro flora was in fact either unaffected or in some cases improved their status. Economically the use of effluents would bring down cost of production to be incurred on costly input like fertilizers.

5. Drainage of Agricultural Fields

Irrigation and drainage go hand in hand otherwise production will hamper due to waterlogging and salinity (Chittapur, 2016). In UKP project experiments under Indo-Dutch project at Islampur, near Hunsagi, study revealed that open drains at 50 m distance after three years get chocked and become inefficient unless cleared and maintained every year (IDNP, 2003). While, sub surface drainage (SSD) at 30 m apart at 0.9 to 1.0 m depth with a slope of 0.1 to 0.2 per cent and maximum length of laterals of 295m using 8-10 cm dia perporated PVC pipes covered with 60 mesh nylon net and let open into a natural nala was effective in reducing salinity and water table front and raised cropping intensity form 72 to 157 per cent. Later the work was extended on farmers' fields under Water productivity project of RKVY, GoK. In TBP, SSD drains (perforated PVC pipes of 0.10 m dia with filters) spaced at

10 m intervals laid at a depth of 0.75 m from the surface revealed that soil salinity decreased from 8.4 ds/m to 2.51 dS/m and water table receded from 50 cm to 87 cm after 2 years of installation (Vishwanath, 2016). Further, rice yields increased from 2.18 to 7.0 t/ha which was about 69 per cent more over its initial value with a cropping intensity changing from 143 to 191 per cent. Where water table is within the threshold limits and inflicted by salinity, intervention through land drainage is essential to improve the productivity in canal command.

However, the soil salinity showed an increasing trend after *kharif* 2000. This was due to tendency of the farmers to block the drain outlets in order to save water and possibly nutrients. Surprisingly this small change in the SSD did not affect crop performance even after two years of such practice. However such blockages of drain may become detrimental once the soil salinity build-up reaches beyond a threshold value of the crop. The nutrient loss (mainly N) was monitored and estimated to be about 7.0 kg/ha/season for the design drainage coefficient of 1.0 mm/day in the Vertisols of TBP, which is only around 5 per cent of the N applied for the rice crop. This can further be reduced by adopting selective blocking (controlled drainage), especially during fertilizer application.

Conclusion

Agricultural water productivity is still very low (ranges from 0.28 to 1.60 kg m^{-3} in different crop species) in India due to low grain yields as a result of poor irrigation and cultural management practices. This provides ample scope for raising the levels of water productivity by increasing crop yields or by irrigation water saving. In the former fertilizer management and use of efficient cultivars and cropping systems based on available water supplies play a major role. Economic use of irrigation water intends at water supply to the crop need and minimising evaporation and percolation losses while achieving a high yield and inducing water deficit at non critical stages of crop growth while providing supplemental irrigation at critical stages. Practices that save water indirectly by increasing the crop production without affecting ET and those that decrease ET as well as boost production will increase water productivity. Besides, socioeconomic factors such as irrigation institutional reforms, privatization of wells, government policy (*e.g.* electricity supply) and response of farmers to water crisis and incentives also influence water productivity.

Technologies, therefore, should be revitalized for efficient use of irrigation water, minimum groundwater contamination, recharging of groundwater through storage structures *etc.* It is necessary to prevent water crisis by making best use of the available technologies and resources to conserve the existing water resources, convert them into utilisable form and make efficient use of them for agriculture. In agriculture, among various management strategies water conservation techniques, protective irrigation concept, crop need based supply, alternate water efficient cropping system techniques *etc.* can play vital role in addressing water crisis. Although, considerable progress has been made in the improvement of high water efficient and productive crops; the achievements are not satisfactory to this crisis. At large policy reforms should impose water use as a costly commodity rather than as a cheap resource.

References

Ahmadi, S.H., Andersen, M.N., Plauborg, F., Poulsen, R.T., Jensen, C.R., Sepaskhah, A.R., Hansen, S., 2010. Effects of irrigation strategies and soils on field grown potatoes: gas exchange and xylem [ABA]. Agric. Water Manage. 97, 1486–1494.

Ahuja, I., de Vos, R.C.H., Bones, A.M., Hall, R.D., 2010. Plant molecular stress responses face climate change. Trends Plant Sci. 15, 664–674.

Ajay kumar, Singh, S. K., Kaushal, K. K., Purushottam, 2015. Effect of micro irrigation on water productivity in system od rice (*Oryza sativa*) and Wheat (*Triticum aestivum*) intensification. Indian J. Agril. Sci., 85 (10):1342-8.

Ali, H., Ahmad, S., Mehmood, M.J., Majeed, S., Zinabou, G., 2012. Irrigation and Water Use Efficiency in South Asia. Global Development Network, New Delhi. Alonso, J.M., Ecker, J.R., 2006. Moving forward in reverse: genetic technologies to enable genome-wide phenomic screen in Arabidopsis. Nat. Rev. Genet. 7, 524–536.

Ali, M.H., Talukder, M.S.U., 2008. Increasing water productivity in crop production—a synthesis. Agric. Water Manage. 95, 1201–1213.

Amritha Patel, 2013.

Bhattacharya, A. K., 1998, Drainage of agricultural lands. In. *50 Years of Natural Resoure Management Research*, Division of Natural Resoure Management, ICAR, New Delhi:347-62.

Bouman, B.A.M., Tuong, T.P., 2001. Field water management to save water and increase its productivity in irrigated lowland rice. Agric. Water Manage. 49, 11–30.

Camp, C.R., 1998. Subsurface drip irrigation: a review. Trans. Am. Soc. Agric. Eng. 41, 1353– 1367.

Chandrashekhar, C. P., 2009, Resource management in sugarcane (*Saccharum officinarum* L.) through drip irrigation, fertigation, planting pattern, and LCC based N application and area - production estimation through remote sensing. Ph. D. Thesis. Submitted Univ. Agril., Sci., Dharwad, Karnataka.

Chittapur, B. M., 2016, Climate smart agriculture: lessons learnt, technological advances made and research priorities in SAT.In. *Climate Smart Agriculture: Status and Strategies*, Eds. B. M. Chittapur, A. S. Halepyati, M. R. Umesh and B. K. Desai, University of Agricultural Sciences, Raichur, Karnataka, India:30-40.

Collins, M.J., Fuentes, S., Barlow, E.W.R., 2010. Partial root zone drying and deficit irrigation increase stomatal sensitivity to vapour pressure deficit in anisohydric grapevines. Funct. Plant Biol. 37, 128–138.

Dhawan, B.D., 2002. Technological Change in Indian Irrigated Agriculture: A Study of Water Saving Methods. Commonwealth Publishers, New Delhi, India.

Du, T., Kang, S., Zhang, J., Li, F., 2008. Water use and yield responses of cotton to alternate partial root-zone drip irrigation in the arid area of north-west China. Irrig. Sci. 26, 147–159.

Fekadu, Y., Teshome, T., 1998. Effect of drip and furrow irrigation and plant spacing on yield of tomato at Dire Dawa, Ethiopia. Agric. Water Manage. 35, 201–207.

Fereres, E., Soriano, M.A., 2007. Deficit irrigation for reducing agricultural water use: integrated approaches to sustain and improve plant production under drought stress special issue. J. Exp. Bot. 58, 147–159.

Fischer, E.M., Seneviratne, S., Schr, C., 2007. Contribution of land-atmosphere coupling to recent European summer heat waves. Geophys. Res. Lett. 34, 606–707.

Geerts, S., Raes, D., 2009. Deficit irrigation as an on-farm strategy to maximize crop water productivity in dry areas. Agric. Water Manage. 96, 1275–1284.

Grimes, D.W., Walhood, V.T., Dickens, W.L., 1968. Alternate-furrow irrigation for San Joaquin Valley cotton. Calif. Agric. 22, 4–6.

Gulati, Ashok, Ruth Meinzen-Dick and Raju, K.V., 2005. Institutional reforms in Indian Irrigation, Sage Publications, New Delhi.

Hanson, B.R., Schwankl, L.J., Schulbach, K.F., Pettygrove, G.S., 1997. A comparison of furrow, surface drip, and subsurface drip irrigation on lettuce yield and applied water. Agric. Water Manage. 33, 139–157.

Home, P.G., Panda, R.K., Kar, S., 2002. Effect of method and scheduling of irrigation on water and nitrogen use efficiencies of okra (*Abelmoschus esculentus*). Agric. Water Manage. 55, 159–170.

Howell, T.A., 2001. Enhancing water use efficiency in irrigated agriculture. Agron. J. 93, 281–289.

INCID, 1998. Sprinkler Irrigation in India. Indian National Committee on Irrigation and Drainage (INCID), New Delhi.

Jayakumar, M., Surendran, U., Manickasundaram.,2014. Drip fertigation effects on yield, nutrient uptake and soil fertility of Bt Cotton in semiarid tropics. Int. J. Pl. Prod. 8(3): 1735-8043.

Jensen, M.E., 1984. Improving irrigation systems. In: Engelbert, E.A., Scheuring, A.F. (Eds.), Water Scarcity: Impacts on Western Agriculture. Univ. California Press, Berkeley, pp. 218–236.

Kang, S.Z., Hu, X.T., Cai, H.J., Feng, S.Y., 2004. New ideas and development tendency of theory for water saving in modern agriculture and ecology. J. Hydrol. Eng. 12, 1–7.

Kang, S.Z., Liang, Z.S., Pan, Y.H., Shi, P.Z., Zhang, J.H., 2000b. Alternate furrow irrigation for maize production in an arid area. Agric. Water Manage. 45, 267–274.

Kang, S.Z., Shi, W.J., Zhang, J.H., 2000a. An improved water-use efficiency for maize grown under regulated deficit irrigation. Field Crops Res. 67, 207–214.

Kang, S.Z., Zhang, L., Liang, Y.L., Hu, X.T., Cai, H.J., Gu, B.J., 2002. Effects of limited irrigation on yield and water use efficiency of winter wheat in the Loess Plateau of China. Agric. Water Manage. 55, 203–216.

Kirda, C., Cetin, M., Dasgan, Y., Topcu, S., Kaman, H., Ekici, B., Derici, M.R., Ozguven, A.I., 2004. Yield response of greenhouse grown tomato to partial root drying and conventional deficit irrigation. Agric. Water Manage. 69, 191–201.

Kirda, C., Topcu, S., Cetin, M., Dasgan, H.Y., Kaman, H., 2007. Prospects of partial root zone irrigation for increasing irrigation water use efficiency of major crops in the Mediterranean region. Ann. Appl. Biol. 150, 281–291.

Kukal, S. S., Yadvinder Singh, Jat, M. L., Sidhu, H.S., 2014, Improving water productivity of wheat – based cropping systems in South Asia for sustained productivity. Adv. Agron. 127; 157-

Kukal, S.S., Aggarwal, G.C., 2002. Percolation losses of water in relation to puddling intensity and depth in a sandy loam rice (Oryza sativa) field. Agric. Water Manage. 57, 49–59.

Kulkarni, S.A., 2005. Looking beyond eight sprinklers. In: Paper Presented at the National Conference on Micro-irrigation. G.B. Pant University of Agriculture and Technology, Pantnagar, India: 3–5 June 2005.

Kumar, A., Singh, A.K., 2002. Improving nutrient and water use efficiency through fertigation. J. Water Manage. 10, 42–48.

Li, F.S., Wei, C.H., Zhang, F.C., Zhang, J.H., Nong, M.L., Kang, S.Z., 2010. Water use efficiency and physiological responses of maize under partial root-zone irrigation. Agric. Water Manage. 97, 1156–1164.

Liu, F., Shahnazari, A., Andersen, M.N., Jacobsen, E.E., Jensen, C.R., 2006. Effects of deficit irrigation (DI) and partial root drying (PRD) on gas exchange, biomass partitioning, and water use efficiency in potato. Sci. Hortic. 109, 113–117.

Mahender Kumar, R., Ravindra Babu, V., 2016, Advances in rice production technologies in view of changing scenarios of climate. In. *Climate Smart Agriculture: Status and Strategies*, Eds. B. M. Chittapur, A. S. Halepyati, M. R. Umesh and B. K. Desai, University of Agricultural Sciences, Raichur, Karnataka, India:30-40.

Minhas, P.S., 1998, Use of poor quality waters. In. *50 Years of Natural Resoure Management Research*, Division of Natural Resoure Management, ICAR, New Delhi:327-46.

Morison, J.I.L., Baker, N.R., Mullineaux, P.M., Davies, W.J., 2008. Improving water use in crop production. Philos. Trans. R. Soc. B. 363, 639–658.

Narayanamoorthy, A., 2006. Potential of Drip and Sprinkler Irrigation in India. Gokhale Institute of Politics and Economics India.

Naresh, R.K., Singh, S.P., Chauhan, P., 2012. Influence of conservation agriculture, permanent raised bed planting and residue management on soil quality and productivity in maize–wheat system in western Uttar Pradesh. Int. J. Life Sci. Biotechnol. Bioresour. 1, 26–34.

Oweis, T., Hachum, A., 2006. Water harvesting and supplemental irrigation for improved water productivity of dry farming systems in West Asia and Africa. Agric. Water Manage. 80, 57–73.

Pawar, D.D., Bhoi, P.G., Shinde, S., 2002. Effect of irrigation methods and fertilizer levels on yield of potato. Indian J. Agric. Sci. 72, 80–83.

Pereira, L.S., Oweis, T., Zairi, A., 2002. Irrigation management under water scarcity. Agric. Water Manage. 57, 175–206.

Postel, S., Paolak, P., Gonzales, F., Keller, J., 2001. Drip irrigation for small farmers. A new initiative to alleviate hunger and poverty. International Water Resources Association Water Int. 26, 3–13.

Prihar, S. S., Khera, K. L., Gajri, P. R., 1976, Effect of puddling with different implements on yield and water expense in paddy. J.Res. Punjab Agril.Univ., Ludhiana 13:240-45.

Prihar, S.S., Khera, K.L., Sandhu, K.S., Sandhu, B.S., 1976. Comparison of irrigation schedules based on pan evaporation and growth stages in winter wheat. Agron. J. 68, 650–653.

Puspa, K., 2006, Irrigation management for improving productivity nutrient uptake and water use efficiency in aerobic rice. *Ph.D thesis*, UAS, Bengaluru, Karnataka.

Rajput, T. B. S., Patel, N., 2002, Water soluble fertilizers-oppurtunities and challenges. FAI Annual Seminar, December 2002 pp SII-3/1-9.

Sadras, V.O., 2009. Does partial root-zone drying improve irrigation water productivity in the field? A meta analysis. Irrig. Sci. 27, 183–190.

Sadras, V.O., Angus, J.F., 2006. Benchmarking water-use efficiency of rainfed wheat in dry environments. Aust. J. Agric. Res. 57, 847–856.

Sandhu, B. S., Khera, K. L. and Singh Baldev, 1982, A note on the use of irrigation water and yield of transplanted rice in relation to timing of last irrigation. Indian J. Agril. Sci. 52: 870-71.

Sabreen Kh. Pibars,H. A., Mansour, Imam, H. M., 2015. Effect of organic manure fertigation on sesame yield productivity under drip irrigation system. Global Advanced Res. J. Agric. Sci.,4(8): 378-386.

Sepaskhah, A.R., Ahmadi, S.H., 2010. A review on partial root-zone drying irrigation. Int. J. Plant Prod. 4, 1735–6814.

Shani-Dashtgol, A., Jaafari, S., Abbasi, N., Malaki, A., 2006. Effects of alternate furrow irrigation (PRD) on yield quantity and quality of sugarcane in southern farm in Ahvaz. In: Proceeding of National Conference on Irrigation and Drainage Networks Management. Shahid Chamran University of Ahvaz. 2–4 May, pp. 565–572.

Sharma, B.R., Amarasinghe, U., Xueliang, C., 2009. Assessing and improving water productivity in conservation agriculture systems in the Indus-Gangetic Basin. In: Lead paper for the 4th World Congress on Conservation Agriculture-innovations for Improving Efficiency, Equity and Environment, Irrigated Systems, National (Indian) Academy of Agricultural Sciences, NASC Complex, Pusa, New Delhi, India; 4–7 February 2009.

Sitaram Singh, Chaudhary, T. N., Batt, R. K., Bhatnagar, P. R., Saha, B., Patil, N. G., 1998, Irrigation water management. In. *50 Years of Natural Resoure Management Research*, Division of Natural Resource Management, ICAR, New Delhi:265- 310.

Soundharajan, B., Sudheer, K.P., 2009. Deficit irrigation management for rice using crop growth simulation model in an optimization framework. Paddy Water Environ. 7, 135–149.

Srinivas, K., 2004, Fertigation in horticultural crops. In. Crop Improvement and Production Technology of Horticultural Crops. I :506-25. Chadda, K. L., Ahloowalia, B. S. Prasad, K. V. and Singh, S. K. (Eds.). Proc. First Indian Hort. Cong. 2004 held on 6-9 November 2004, New Delshi, India.

Thind, H.S., Aujla, M.S., Buttar, G.S., 2008. Response of cotton to various levels of nitrogen and water applied to normal and paired sown cotton under drip irrigation in relation to check-basin. Agric. Water Manage. 95, 25–34.

Thomson, T. L., White, S. A., Walsworth, J., Sower, G. J., 2003, Fertigation frequency for subsurface drip-irrigated broccoli. *Soil Sci. Soc. Am. J.*, 67: 910-18.

Tiwari, K.N., Singh, A., Mal, P.K., 2003. Effect of drip irrigation on yield of cabbage (*Brassica oleracea* L., va.*capitata*) under mulch and non-mulch conditions. Agric. Water Manage. 58, 19–28.

Vanishree, H., 2016, Yield maximation in transplanted pigeonpea (*Cajanus cajan* L.) through drip fertigation. M. Sc. (Agri.) Thesis. Submitted to University of Agril. Sciences, Raichur, Karnataka.

Verma, S.K., Shrivastava, N.C., 1992. Sprinkler irrigation is advantageous in alkali soils. Indian Fmg., 42, 37–38.

Viets, F.G., 1962. Fertilizers and the efficient use of water. Adv. Agron. 14, 233–264.

Vishwanath, 2015, AICRP on Management of Salt Affected Soils and Use of Saline Water in Agriculture, Agriculture Research Station, Gangavati, Karnataka. Pp 1-4 (personal communication).

Von, W.,S., Chieng, S., Schreier, H., 2004. A comparison between low-cost drip irrigation, conventional drip irrigation, and hand watering in Nepal. Agric. Water Manage. 64, 143–160.

Wang, Y., Liu, F., Andersen, M.N., Jensen, C.R., 2010. Improved plant nitrogen nutrition contributes to higher water use efficiency in tomatoes under alternate partial root-zone irrigation. Funct. Plant Biol. 37, 175–182.

Yuan, L.P., 2003. Recent progress in breeding super hybrid rice in China. In: Virmani, S.S., Mao, C.X., Harby, B. (Eds.), Hybrid Rice for Food Security, Poverty Alleviation and Environmental Protection. International Rice Research Institute, Los Baños, Philippines, pp. 3–6.

Zhang, X., Chen, S., Sun, H., Pei, D., Wang, Y., 2008. Dry matter, harvest index, grain yield and water use efficiency as affected by water supply in winter wheat. Irrig. Sci. 27, 1–10.

Zhang, H., Oweis, T., 1999. Water yield relations and optimal irrigation scheduling of wheat in the Mediterranean region. Agric. Water Manage 38, 195–211.

2

Concepts of Soil-Water-Plant-Atmosphere Relationship in Water Management

M.R. Umesh and Jagadish

The knowledge of relationship between soil-water and plant growth is important for proper water management. Water is considered to be both boon and bane resource for crop plants. Excessive water can be just as detrimental as lack of water. Soil moisture available for crop production varies substantially across geographic regions and from year to year. Not only are there large differences in annual precipitation, but the efficiency of water use also varies due to atmospheric demand of water, dry regions with low humidity and high winds have very high evaporative demand. In such areas, water loss rates are so high that crops may not compete well with the atmosphere for water.

Properties of Water

Each molecule of water consists of two atoms of hydrogen and one atom of oxygen. Water is a bipolar liquid. Oxygen atom is much bigger in size than hydrogen atom. Chemical bonds are formed by pairs of electrons. Two pairs of bonds in H-O-H are formed by sharing of electrons between hydrogen and oxygen. The electron of each hydrogen atom is shared with each of two unpaired electrons form the oxygen atom. The two hydrogen atoms link the oxygen atom at an angle of 105°. The properties of water arise from the hydrogen bonding and tetrahedral arrangement of electron pairs around oxygen atom. The size of one water molecule is 3×10^{-10} m. one cubic centimeter of water in its liquid state contains 3.4×10^{22}

molecules. The weight of one cubic centimeter of water is 980 dynes. Its density is one. Its surface tension with glass surface is 72.75 dynes at 20°C.

Soil as Water Reservoir

Soil acts like sponge in the way that takes in, transmits and stores water. The key understanding of water storage and movement is to consider the distribution of large and small pores in the soil. Small pores hold water against the forces of gravity while larger pores tend to drain freely and provide pathways for roots and air. Silt loam with good tilth is about 50 per cent pore space and 50 per cent mineral matter. Ideal growing conditions occur when the pore space is about half filled with water and half with air. Large pores are essential for providing natural internal drainage as well as serving as pathways for root exploration.

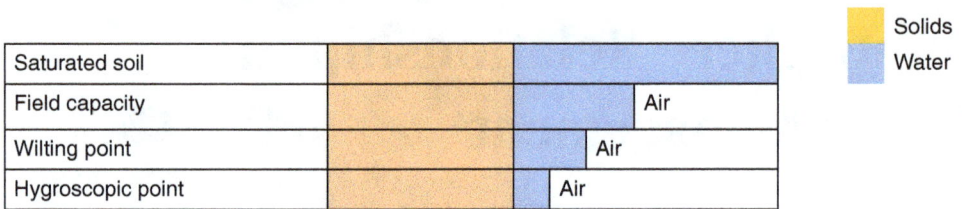

Figure 2.1: Relative Proportion of Solid, Water and Air in Soil at various Phases.

Saturation occurs when virtually all the pores in a soil are filled with water, this condition occurs following long duration of heavy rains or flooding. If saturation persists, roots deplete oxygen and carbon dioxide which roots cannot use, builds up. Oxygen does not readily diffuse from the atmosphere through soil water and plants will die if their roots remain too long without oxygen.

Field capacity is a condition that exists after a soil has been thoroughly wet and then allowed to drain freely for 48 hours. This point defines the water content of the soil when it is holding the maximum amount of plant available water. Wilting point is the water content at which plants can no longer extract water from the soil. Fortunately the wilting point is rarely occur in agriculture soils in humid regions. It will be more pronounced in tropical and sub-tropical regions. Plants that extract enough water to reach the wilting point soon die if soil water content stays this low for any length of time.

Water Retention in Soil

Soon after irrigation or heavy rainfall soil becomes saturated with water held in between soil particles. Both macro (>0.08 mm size) and micro pores (<0.08 mm) are filled with water. However soil will not hold entire quantity from heavy rainfall, excess water entered into soil will move away from gravity known as gravitational water. Upon time, soil water present in macro pores will start empting due to evaporative demand in the form of water vapours. Soil water retained only in micro pores is known as field capacity. This stage will arise after 48 hours of irrigation or rainfall depending upon initial soil moisture, texture and water holding capacity of soil. As time passes more soil water losses due to evaporation and plant transpiration

Plant unavailable ⇄ Plant Available ⇄

Saturation Field   Field Capacity   Wilting Point

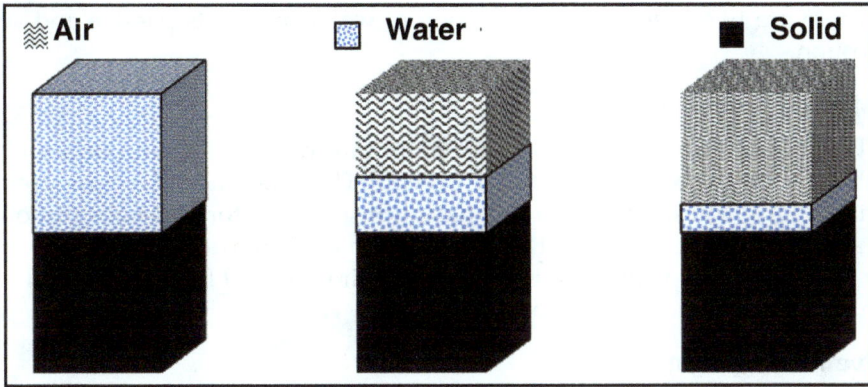

Figure 2.2: Share of Solid, Water and Air at different Situations.

demand, water retain only in micro pores. Further, plant consumes more and more water from soil subsequently quantity of water held by soil becomes much lower. It results in wilting of plants.

Soil water is get adsorbed on clay surface by hydrogen bonding and Van-der-Waal forces and some get adsorbed by exchangeable ions getting hydrated. Number of layers attracted each other depends on type of clay minerals and cation content. The water molecules in contact with lattice layer of clay mineral form covalent hydrogen bonds with oxygen in tetrahedral layer. The covalently bonded water molecules are strongly polarized so that a second layer of water molecules can form similar to first layer of water molecules. In this manner thick water films are formed however, thickness of layers varies with clay minerals *viz.,* least in Kaolinite and higher in Montmorillonite clay. The energy in which first layer of water molecules held so tightly and it decrease with subsequent layers. The attraction of soil particles and water molecules within soil is due to processes of cohesion and adhesion.

Adhesive Forces

Attraction of water molecule and soil particles is called adhesion. It occurs only at moisture contents above maximum cohesion. As moisture content increases the attraction between the water particles decreases, water is held less tightly by the particles and it is attracted in the surface of the object to form connecting films between it and the soil. The adhesion of soil to water is due to formation of films. The adhesion becomes maximum when the moisture content is such that it satisfies the films between individual particles. For ideal soil the moisture content at maximum adhesion is always more than the moisture content at maximum cohesion.

Cohesive Forces

These forces act on water molecules alone. At high soil moisture cohesive forces are higher, it gradually decreases as adhesive force increases. In extreme dry soil cohesive forces tend to become zero. Both forces are similar at field capacity wherein no much soil moisture tension occurs. Extent of available water depends on adhesive forces.

There are two theories on absorption of water by plants passive and active absorption.

Passive Absorption

Translocation of water from root hairs into different plant parts without metabolic energy is known as passive process. The difference in diffuse pressure deficit (DPD) of soil solution and root hairs is responsible for water absorption. As long as DPD of plant parts are at higher level water absorption is more. The major force responsible for water absorption is transpiration pull from leaves and some extent by green portion of stem and fruits.

Active Absorption

The metabolic energy is required to absorb water from soil. It takes place due to root activities. In contrast to passive absorption the solute concentration between roots and soil solution. The active absorption is mainly by osmotic and non-osmotic mechanisms at various levels.

Water Movement in Soil

The soil is saturated with water after an irrigation or rainfall, and it flows from this saturated soil. The movement is mainly due to gravitational force. This follows the Darcy's law defined as Velocity of a fluid in permeable media is directly proportional to the hydraulic gradient. Rate of flow increases with an increased depth of water above the bottom of the soil and decreases with an increased depth of soil through which water flows.

$$Q = KA \frac{H_1 - H_2}{L} = Kai$$

where Q is quantity of water flow/time

$$i = \frac{H1 - H2}{L} = H/L$$

A = Cross sectional area of flow

L = Length of soil column

K = Hydraulic conductivity

H = Length of water head or difference in length of water head= $(H_1 - H_2)$

When excess water in a field after rainfall or irrigation water has gone away from it after one or two days, the field comes to field capacity. Volume of water starts lowering in soil at this stage and water rarely flows in liquid form. The forces which act at this stage are mechanical pressure and gravitational forces and some extent osmotic and absorptive forces. Water movement is from high matric potential to low matric potential. Higher the water content in the soil greater the matric potential gradient. Vapour movement is due to differences in vapour pressure from one area to another in the soil. Water in the form of vapour in soil is very small.

Available Water

Plant available water is the amount of water held between field capacity and the wilting point. The inherent plant available water holding capacity of a soil is an important property when assessing productivity potential. Plant available water holding capacity is closely related to soil texture. For instance, sandy soils hold less than 25 per cent as much plant available water as a silt loam with an equivalent rooting depth. The soils that hold the most water available to plants are silt loams, silty clay loams and some clay loams. Figure 2.3 show the relationship between field capacity, wilting point and plant available water for these different soil textures.

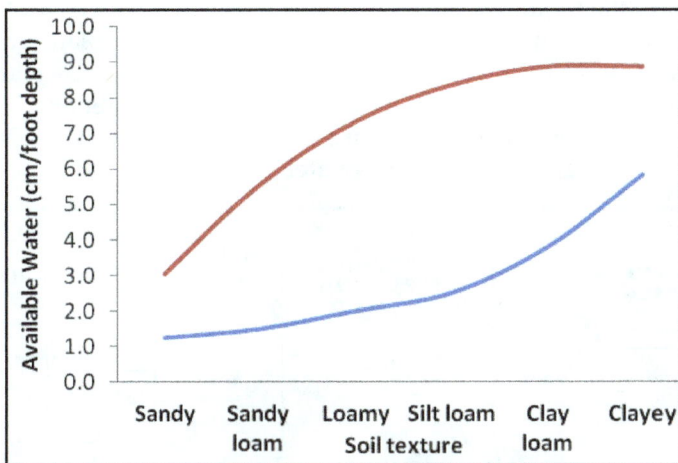

Figure 2.3: The Relationship between Field Capacity, Wilting Point and Plant available Water for different Soil Texture.

Mechanism of Water Absorption

In different crop plants the root system varies in terms of root length, root depth ad root density which is further influenced by plant genetic constitution and edaphic factors. The roots of cereal crops occupy more surface area of soil than the roots of other crop plants. Cereal roots extend up to 200-4000 cm/cm^2 of soil surface area against 15-200 cm/cm^2 for non-cereals. Water absorption takes place due to differences in water potential between soil to outer surface of root, then outer surface to inner side, flow from inner roots to leaf then air. Quantity of water absorption by roots is dependent on the supply of water at the root surface. Primary site of water absorption is root hairs. For sufficient water availability either root has to reach site of soil water or else water has to move to root surface. Under unsaturated condition, resistance to water movement in the soil results into decline of rate of water movement in the soil. Under water stress condition roots develop and reach deeper layers in search of water to fulfill transpirational losses of plant systems.

The movement of water from root hairs to functional organs of the plant includes stem, leaves and reproductive organs through xylem vessels by transpiration pull. The water potential gradient will arise from the soil through the plant and to the atmosphere. The water in response to gradient moves from the xylem strands of the leaf across the mesophyll tissue and through the cell walls bordering the sub-stomatal cavities where liquid water vaporizes and diffuse out of leaves. Within plant system water moves either by symplast and apoplast processes. These are characterized by movement of water between cells and within cells.

Available Water in Soil

After two to three days of rainfall downward movement of soil water ceases or completely stopped due to gravitational forces. Macro pores are filled with air and micro pores are filled with water. At this stage soil water held by 1/3 atmosphere soil moisture tension and maximum water is available for plant growth. Upon plant absorption and evaporation water film around soil particle start shrinking. It reaches a stage where plants fails to take up any more water and start showing wilting symptom known as wilting point. The soil water between field capacity and wilting point is known as available water.

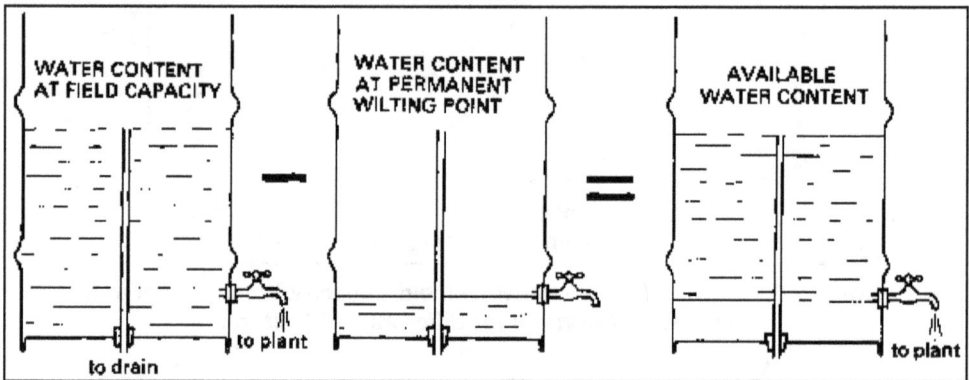

Figure 2.4: Comparison of Water Content at Field Capacity and Permanent Wilting Point and available Water Content.

Soil Water Potential

It is an important characteristic of soil water supply capacity. It is defined as difference between the free energy of soil water and that of pure water in standard reference. The total soil water potential is governed by gravitational, matric, osmotic and pressure potential.

$$\Psi_{sw} = \Psi_g - \Psi_m + \Psi_p - \Psi_o$$

Gravitation potential (Ψg) is force of gravity acting on soil water and removes excess water especially water in the macro pores and is expressed as gravitational potential. It is +ve.

$\psi g = \psi h$

where,

h = Height if soil water above the reference elevation

G = Acceleration due to gravity

Matric potential (ψm) it is due to adsorption and capillarity. Capillarity is due to the attractive force of water for the solids on the walls of the pores through which it moves and surface tension of water which is largely due to attraction of water molecules for each other. Matric potential is –ve. Osmotic potential (ψo) is due to presence of solutes in the soil water which reduce the free energy of water as solute ions attract water molecules. Osmotic potential has no effect on mass movement of water in soils. It affect the uptake of water by plants. It is always –ve. Pressure potential (ψp)is result of hydrostatic pressure acting on soil water. Increase pressure increases the water potential. Pressure potential is +ve. Matric and osmotic forces hold water molecules strongly to soil particles thus lowering the potential energy of water in a bulk in free state thus free energy of soil water is decreased. The retention and movement of water are thus consequences of energy effects. This energy concept indicates the difficulty or ease with which water can be extracted from the soil.

Plant Water Potential

Energy status of water in plants is the ease with which water is available to plant cells. Maximum value of water potential is zero. So water potential on plant tissue is always less than zero. Hence negative number. Higher the water potential of plant easier the availability of water to plant cells. It is measure in bars or Pascals. Leaf water potential at field capacity ranges from -0.2 to -0.8 MPa. Water potential is directly measured with the pressure chamber apparatus.

Magnitude of Water Potentials in SPAC

Component	Water Potential (bars)
Soil	-0.1 to -20
Leaf	-5.0 to -50
Atmosphere	-1000 to -2000

The water absorption needs to be considered in the system of SPAC rather than of the roots alone. The flow rate of water in this system can be explained by the following equations.

Flow from soil to outer of root= (ψ soil- ψ root surface)/r soil

Flow from outer to inner root= (ψ root surface – ψ xylem)/r root

Flow from root xylem to leaf = (xylem – ψ leaf)/r plant xylem + r leaf

Flow from leaf to air = (ψ leaf- p air)/r leaf + r air

Where, ψ is the water potential at various sites of the system and r is resistance offered at different sites of system.

Canopy Temperature

During day time plants absorb short wave radiation which tend to build up heat in canopy, transpiration rate will cool up by way of exchange of water molecules. In water stressed plants transpiration rate is much lower than well-watered plants. It may result in high canopy temperature. The canopy temperature has been advocated as index of water stress in plants. It is measured by infrared gun thermometer. The increased stress in plants will then result in more and more increase in canopy temperature. The difference between canopy and air temperatures can be related to the vapour pressure deficit of the air the net radiation and the aerodynamic and crop resistance.

Infiltration Rate

Infiltration is the process of entry of water into the soil. The water flux into the soil is called infiltration rate. It is expressed in millimeter per hour. Infiltration rate is the maximum rate at which water penetrates into the soil at a given moment, the surface being in contact with water at atmospheric pressure. Infiltration rate is dependent on the quantity and quality of soil pores at the soil surface. It is naturally higher on well-structured or sandy soils. It can be lowered by the formation of soil crusts. Infiltration is best managed by protecting the soil surface with organic matter and by avoiding practices that destroy soil structure. It is estimated by double ring infiltrometer. Cumulative infiltration plotted against time shows a curvilinear relation. After a long time, the infiltration rate tends to approach a constant value.

Water applied to the soil at rates higher than infiltration capacity is lost as surface runoff or accumulate on the soil. Soil texture and vegetative cover influence infiltration rate of given soil. Other factors include initial soil moisture content, hydraulic conductivity, soil structure, presence of impeding soil layers, exchangeable cations, organic matter, kind of clay and tendency to crust even in the same textural class. Sandy soils IR is 7.0 to 14.4 mm/h, sand loamy soils 3.8 to 7.6 mm/h and clayey soils 1.3 to 3.8 mm/h depending on initial soil moisture content. Factors influence infiltration rate is soil texture and structure, vegetation, soil roughness, hydraulic conductivity of soil, soil profile arrangement, exchangeable cations, organic matter content, clay content, tendency to crust *etc.* However, role of each factor varies from region to region.

Soil Permeability

When saturated soil conditions occur water moves by a different flow path. Under saturated conditions positive pressure is the driving force for water movement rather than the suction pull that occurs in unsaturated soils. Under pressure the paths of least resistance the larger pores (macro pores) are responsible for faster movement. Saturated flow occurs faster in sandy soils and well-structured soils. Soil permeability is an important characteristic of soil to determine the suitable drainage methods. For arable crops high soil permeability is important for well aeration. Soil management techniques that encourage good soil permeability are compaction avoidance and soil aggregate protection in fine textured soils.

Soil Moisture Measurement

Quantification of soil moisture is important for scheduling of irrigation *i.e.*, when to irrigate and how much to irrigate. Many of the methods are having some limitations and a few of them are widely accepted at scientific community. These are categorized into direct (measuring amount of water) and indirect (measuring energy status of soil moisture).

Direct methods- oven dry weight method, volumetric method, alcohol burning method.

Indirect methods- Neutron probe, gamma ray attenuation method, tensiometric, pressure plate apparatus, gypsum block or electrical resistance blocks.

3

Water Efficient Crops and Cropping Systems

M.R. Umesh and Jagadish

In Hyderabad- Karnataka, predominantly an agrarian region known in the rest of the state as well as in the country as 'pulse-bowl' and 'rice-bowl', climate is a real challenge for profitable and sustainable agriculture as in the entire north Karnataka. Except for the prestigious Upper Krishna Project and the Tunga Bhadra Project, the projects conceived to cater water needs of this traditionally sun parched and drought prone area, life has been an eternal struggle for food and livelihood security. Even the little industry that is slowly raising its head is agro-based and hence their fate is also monsoon dependent and struggle for respectable life for industrial labour is also not an easy one. Therefore, the State Government started Farm University in the region in 2009 as an ultimate institute for greening the life of farmers through sustainable and productive agriculture. However, the challenges before the scientific fraternity are diverse yet unique, and need concerted efforts addressing the issues through both short term as well as long term research strategies. Following is the prelude to the lessons Learnt, technological advances made and research priorities for climate smart agriculture.

Predominant Monocropping

In the TBP command where groundnut was leading after development of the first ever interspecific cotton hybrid, *viz.*, Varalakshmi, in the world at Agricultural Research Station, Dharwad, Karnataka during eighties and Jayalakshmi (DCH 32) which followed it later the entire TBP command turned into cotton bowl with world record cotton productivity. However, the success did not last long, as the greed for money led to monocropping of cotton and consequent pest build up eliminated the crop from the region paving way for paddy with large acreage being taken over by immigrant Andhra farmers. Even today rice occupies premier position among crops in the region. As problems with paddy increased as an alternate, sizable area also came under chilli, and of late once again under cotton with the advent of BG I and BG II hybrids.

In drylands the typical climate and soil, and ever increasing prices led way to monopoly of pigeonpea, and today pigeonpea *dal* from this region goes to other parts of the state as well as country. Even the multinational companies are directly trading with farmers and sending the packed *dal* to other parts of the world. In the process crop diversity was the casulty and groundnut, pundi, *rabi* sorghum, millets, *kharif* pulses, *etc.*, almost became extinct.

Growing Problems of Waterlogging, Salinization and Alkalization

Water is gold; and it is not a luxury. But who cares. When it flows in the canal, similar to paddy every arable crop here is inundated consequently not only the problem of waterlogging and salinization increased but it continues at alarming rates and now waterlogging, salinization and alkalization are the highest in the country.

Undependable Monsoon

The ever undependable Indian monsoon with changing climate has its own owes on agriculture and humanity in the region. No doubt, delayed monsoon is an accepted fact but it not only affects selection of current season crops and/ varieties because of reduced growing period but also reduces the choice and success of succeeding crop even in commands. That apart, during 2009, monsoon not only delayed in the entire northern Karnataka, but subsequent floods due to heavy showers led to relocation of many river bank settlements and villages. In the sugarcane belt the crop was lost due to inundation, while in remaining areas cropping was delayed and the questions from farmers were whether the traditional crop like pigeonpea can still be planted, what should be the suitable management practices, or else, whether they have to wait till the onset of *rabi*?

Delayed rains in the catchments delay inflow into reservoirs and consequently letting of water in the canals during the season gets delayed inordinately. Then the questions are whether same paddy cultivar could be still cultivated, if so what are the management practices and what about the tail end farmers in the commands? Further, when the first crop is delayed whether second crop of paddy could be still prevailed with, and if so which cultivars, and if not, what else is more profitable, and what are the management options? In fact, the canals which once flooded the fields for nine months are now not flowing even for six months (July end to January) these days.

Water Efficient Crop Varieties for Vagaries of Climate

In north-eastern transitional zone of Karnataka, *kharif* pulses like blackgram, green gram, yellow gram, sesame *etc.*, either as sole crop in a double cropping system or as intercrops with long duration pigeonpea (cv. BSMR 736) were once common.

But the failing rains in the beginning of the season and havoc created by powdery mildew reduced their intensity and blackgram has almost become extinct from this area once known as its fortress in the state. Today, soybean is emerging as a most potential climate smart crop both as sole and as an intercrop during rainy season with highest acreage (1 25 000 ac). A single cultivar, JS 335, is responsible for this revolutionary change. The area has also the advantage of absence of rust disease, which is affecting production in traditional soybean area. Fortunately, today rust resistant cultivars are available and soybean cultivation will go unhindered in the region at least for some period. And, fascinated farmers are already toying with the idea of summer soybean.

In rainfed areas, cultivation of long duration pigeonpea varieties which was *in vogue* slowly paved way for medium and early types due to the menace of *Helicoverpa* later in the season and also due to high frequency of end season drought. Starting from GS 1 and BDN 2, to Maruti (ICPL 8863), BSMR 76, now to Gulyal local, TS 3R and the latest in the list is GRG 811 which is in between Maruti and TS 3R in duration but resistant both to wilt and sterility mosaic virus are ruling the day. These are climate smart and disease smart cultivars.

Similarly in paddy, during rainy season BPT 5204 *i.e.* Sona Masuri, a long duration (165-170 days) cultivar is common. But delayed water release and occurrence of cold at flowering and consequent spikelet sterility in BPT 5204 led to Gangavati sona (ICGV 05-01), a high yielding, tolerant to sheath blight, neck blast and BLB during *Kharif* and sheath blight and brown spot during summer, resistant to shattering and lodging, can withstand salinity and importantly it is about 10 days earlier to BPT 5204. The cultivar originally found favour for summer is becoming popular even during rainy season. Similarly for summer this year IET 19251 was released which is a dual season cultivar, matures in 115-120 days, has a potential of yielding 6 t ha⁻¹, resistant to brown spot and sheath rot, tolerant to false smut and moderately resistant to leaf blast, and importantly is still earlier, and hence can even be planted during rainy season under extreme delayed planting conditions of mid September (Anon., 2015a).

Nobody can deny the climate smartness of cotton. In this TBP and UKP commands, in spite of scarcity and delayed water release farmers could harvest 20-25 q ha⁻¹ cotton with single irrigation. This year in spite *kharif* being dry, farmers replaced paddy in Kasabe camp of Raichur, and with single irrigation they are excepting 20 q ha⁻¹ kapas yield in first picking and additional 3-4 q ha⁻¹ in subsequent pickings (Pyati, 2015). Other crops like maize, pearlmillet and sunflower being photoinsensitive have shown promise. Pearlmillet and sunflower are predominant in rainfed region.

Labour being a problem, BGM 2, a chickpea cultivar, growing to a height of 50 cm with pods borne terminally is found suitable for machine harvesting and hence released for general cultivation by the UAS, Raichur during 2014.

Laser Levelling Technology

Under conservation agriculture, farmers were introduced to laser leveller and today it is the much sought after machinery both by rainfed farmers and of command areas because of its ability to increase water use efficiency which is of paramount importance. Uniform water distribution, reduced runoff and erosion control are attracting farmers. High potential of cotton mentioned previously in paddy fallows in Kasabe camp is due to laser levelling (Pyati, 2015). This also enabled large plots in case of paddy and thus eliminated mid bunds resulting in overall increased yield and saving of water compared to traditional system where due to uneven land more water was required.

Drainage, Bio-drains and Conjunctive Use of Saline Water

Irrigation and drainage go hand in hand otherwise production declines due to waterlogging and salinity. In UKP project experiments under Indo-Dutch project at Islampur, near Hunsagi, revealed that open drains at 50 m distance after three years get chocked and become inefficient warranting clearing and yearly maintenance (IDNP, 2003). While, sub surface drainage (SSD) at 30 m apart at 0.9 to 1.0 m depth with a slope of 0.1 to 0.2 per cent and maximum length of laterals of 295 m using 8-10 cm dia perforated PVC pipes covered with filter of 60 nylon mesh let into a natural *nala* was effective in reducing salinity and water table front and raised cropping intensity form 72 to 157 per cent. These results are now taken to farmers' fields under Water productivity project of RKVY, GoK, by UAS, Raichur.

In TBP SSD drains (perforated PVC pipes with filters of 0.10 m dia) spaced at 150 m intervals laid at a depth of 0.75 m from the surface decreased soil salinity from 8.4 ds/m to 2.51 dS/m and water table receded from 50 cm to 87 cm after 2 years of installation, and increased rice from 2.18 to 7.0 t/ha, about 69 per cent over the initial value. Further, cropping intensity also increased from 143 to 191 per cent.

Use of saline water up to 4 dS/m in direct mode had no adverse effect on cotton yield. Conjunctive Use of 4 irrigations with saline water (4-6 dS/m) during canal lean period (June) and then switching over to good quality canal water later (August) produced highest kapas yield (22.1q/ha) compared to a crop receiving good water but sown during August (12.6 q/ha). The salt balance remained favourable and did not cause any concern.

Further, evaluation of tree/grass species for the control of seepage, rising water table and soil salinity in commands (bio-drainage) revealed that *A. nilotica* was the most promising at all salinity levels ranging from < 5 to > 15 dS/m whereas *C. equisetifolia* promising initially registered high mortality and cease of growth after 6-8 years. In contrast, *H. binata* less promising initially became promising after 6-8 years. In terms of seepage control, *A. nilotica* and *C. equisetifolia* were effective in arresting emerging seepage flows from the canals. *A. nilotica* and *C. equisetifolia* intercepted seepage over 80 per cent and remained most promising over other species. The grasses in between complimented the effects. The water table receded significantly underneath the plantation while increased at the rate of 10 cm rise outside the plantation area. *A. nilotica* followed by *C. equisetifolia* also improved soil organic carbon and porosity, while bulk density decreased. Trees improved hydraulic conductivity and infiltration rate and brought about a significant change in soil stability by improving aggregates, and decreased soil and water erosion.

Under extreme conditions of high water table (WT) and salinity (WT: 0.75 to 1.0 m and saline 10-12 dS/m), *Acacia ferugenia, Albizzia lebbeck, Glyricidia maculata* and *Casuarina equisetifolia* were most tolerant, while *Dalbergia sissoo, Inga dulse, Eucalyptus hybrid* and *Pongamea pinnata* were moderately tolerant to salinity level up to 10-12 dS/m and WT up to 0.75 m. That apart, all the tree species also enriched the soil nutrient pool (NPK) and organic carbon. Among fruit species mango, custard apple, guava and pummel were found unsuitable for soil salinity in the range of 8-15 dS/m and WT 0.40-0.70 m. Jamun and sapota survived and grew better under relatively lower salinity and shallow water table conditions, whereas wood apple was promising under relatively high salinity but deeper water table conditions. Pomegranate and ber maintained a moderate survival and steady growth rate in low salinity and shallow water table conditions.

Transplanting Technique

Transplanting vegetables like chilli, tomato *etc.* under assured moisture or paddy under irrigation is a common practice. Transplanting of fingermillet under late sowing conditions or in case of pearlmillet to fill the gaps as a contingency measure has been suggested. Recently, in the north-eastern transitional and dry zones of Karnataka, transplanting of 25-30 days old poly bag (of 5-6" dia/height, for successful establishment the boll of earth with roots should be intact) raised pigeonpea produced more yield than the conventional drilling/dibbling. Nipping of 5-6cm top at 20-25 days after transplanting may be followed in case of excessive vegetative growth which improves branching and flowering leading to higher harvest index. Particularly under drip method of irrigation, the practice gave nearly three times higher yields (16 – 18 q ac^{-1}) over normal drill sowing (5-6 q ha^{-1}). The width of the row could be increased to 6- 8 feet under drip and the inter row space raising an intercrop in the beginning is advantageous. Presently, the practice is attracting the attention of neighbouring state farmers.

Rajkumar and Gurumuthy (2008) revealed the scope of transplanting in cotton. Subsequent studies in TBP and UKP confirmed high seed cotton yield (32 per cent) with transplanting of cotton at 90 cm X 90 cm space over farmers' practice of dibbling due increased sympodials, bolls and seed cotton yield per plant. The cost of transplanting was covered by the increased income (39 per cent) realized in the technique (Honnali and Chittapur, 2013). Importantly, transplanting ensures efficient use of water and growing season and is more advantageous in UKP region, where release of water is always delayed resulting in low productivity.

Direct Seeded Rice (DSR)

DSR though not new to farmers in many areas of Karnataka *e.g.* Western Ghats (Belgaum, Dharwad and Karwar districts) of late seems inevitable due to severe water and labour shortages and high cost of production constraints in many areas of Hyderabad-Karnataka. Efforts of validation of DSR technology of CIMMYT and University of Agricultural Sciences, Raichur, had shown promise for its out-scaling through innovative strategies in the areas where water supplies are limited and farmers do not get sufficient water at right time and constrained with ON-OFF canal water supply (GoK-CGIAR Project Progress Report -2013 CIMMYT). Response to early dry seeding to take advantage of early rains received before canal supplies was met with success. Success of Kasabe camp, Raichur today has spread over 90,000 acres in TBP and UKP commands. In addition to increase in net income, timely sowing, reduced seed rate by half, reduced fuel consumption by 40-50 l/ha, reduced water use by 25-35 per cent, reduced GHGs, increased NUE are the other benefits.

Cropping System Diversification and Intensification in Commands

Paddy is banned for its heavy irrigation requirement in UKP command but it covers large area. Over rice-rice as an alternate transplanted *Bt* cotton recorded maximum seed cotton equivalent yield, while *Bt* cotton-seame/greengram (summer), maize – chickpea and chilli + onion were at par with rice-rice (Honnali and Chittapur, 2014). However, Protein yields were higher with maize-chickpea while carbohydrate yield was higher with rice-rice but the highest land utilization index was observed with *Bt* cotton – sesame cropping system. Thus, arable cropping is more sustainable and productive over rice-rice.

Identification of crops does not end with crop cultivars. Fig, pomegranate, custard apple even date palm are catching up with large plantations. Farmers from Ballari, Karnataka first ventured with pomegranate and ber by bringing planting material from Jaisalmer during 80's. Today, the region has grown as a major exporter for pomegranate. While, University is yet to make a head way in date palm, a grower near Ballari has 15 acres of seven year old date palm plantation which he established as an alternative to increased salinity and alkalinity as well as for water scarcity that he thought to happen in near future. Similarly, one cannot get such a quality acid lime as you get in northern dry zone. In this area where ber prevailed earlier and Anab-e-shahi made a presence on the degraded lands with no soil except the filled in pit/trench, Thomson seedless is ruling the day. Today farmers follow Australian or French production practices for grape and hundreds of acres of wine yards alongwith winery are coming up in this region. Are these new crops not climate smart?

Mango, amla, tamarind, custard apple, sweet lemon, biofuels plants, mentha, safed masli, coleus, beet root *etc.*, are other new climate smart crops promising bounty to farmers (Table 3.1).

Table 3.1: Promising Unconventional Crops in Northern Karnataka

	Common Name	*Reference*
Safed masli	Chlorophytum borivlianum	Somanath (2008)
Stevia	*Stevia rebaudiana* Bertoni.	Aladkatti (2011)
Aswagandha	*Withania somnifera* Dunal	Chandranath (2006) and Kubsad (2008)
Coleus	*Coleus forfkohlii* briq.	Mastiholi (2008)
Senna	Cassia angustifolia	Rathod (2009)

Promising Crops and Cropping Systems for Paddy Fallows

Relay cropping is such an innovative strategy to replace traditional monocrop of cotton, transplanted double cropping of rice and rice-fallow with direct seeded rice-mustard, direct seeded rice-maize, direct seeded rice-chickpea, and direct seeded rice-greengram. In addition, maize-fallow system is being intensified with maize-zero tillage chickpea. In recent years, traditional rice-rice system in South India is challenged by non-availability of water to grow 2nd rice crop. Maize has emerged as an obvious choice, as it can be grown with less than 1/3 amount of water and has potential to maintain farm profitability at par or better. However, the *spring* maize is prone to heat stress during reproductive phase, as temperature peaks during month of March/April during which availability of water is also invariably a challenge. Maize production linearly decreases with every accumulated degree day above 30°C (Lobell *et al.*, 2011). However, the efforts made under Conservation agriculture and Heat resilient maize projects of CIMMYT, Mexico and UAS, Raichur attracting farmers towards maize after rice. Half a dozen of maize hybrids are already identified for release (Kuchunur, 2015).

Sesame exports are on the rise, however *Kharif* produce being poor as the crop is caught in the rains at maturity, while summer sesame is becoming popular as it is disease-free and of good quality. Paddy fallows are best suited for sesame both in irrigation commands and in hilly region. Experiments revealed high potential of the crop; nevertheless cultivars performance and response to agronomic practices vary (Prasannakumar, 2011). At Agricultural Research Station, Mugad, Karnataka, of the three cultivars, DS 1 and E 8 were comparable and were superior to DSS 9 both in terms of seed yield and income, while increasing dose of fertilizer had no significant influence (Prasanna kumar *et al.*, 2014). Higher population (with 30/40 cm X 10 cm) recorded higher seed yield and net income (over 30/40 cm X 20 cm). Among all, DS 1 recorded significantly higher seed yield (1393 kg ha^{-1}) with recommended NPK (40:25:25 kg ha^{-1}, respectively) and spacing (30 cm X 10 cm). Combined application of ethepan and Planofix hastened flowering (37.13 days) while chemical fertilizer alone + nipping of plants at 25 days delayed flower initiation (39.3 days), INS (RDF +FYM) + borax (37.63 days) and fertilizer alone + FeSO$_4$ further delayed time of peak flowering (41.9 days) (PrasannaKumar, 2011). Seed yield with INS was higher (1300 kg ha^{-1}) than with fertilizers alone (1040 kg ha^{-1}) and top dressing with DAP at 25 kg ha^{-1} after 25 DAS proved advantageous.

In a trial on farmers' fields wherein four mustard varieties were evaluated for their production potential in 100 acres in the paddy fallows during 2014-15 at Vijaynagar and Maramma camps of Raichur district under RKVY project (Krishnamurthy, 2015). The yield ranged from 460-494 kg/ha in Pusa mustard-30, 361-615 kg/ha in Pusa mustard -25, and 250-360 kg/ha in NRCBH-101 against 150-200 kg/ha in local variety. While, in a trial at Agricultural Research Station, Siruguppa Pusa mustard 25 (688 kg ha^{-1}), Pusa mustard 26 (681 kg ha^{-1}), Pusa Agrani (642 kg ha^{-1}), among the eight cultivars evaluated (yield ranged from 688 to 357 kg ha^{-1}) found promising (Basavanneppa, 2015).

In situ Green Manuring

Efforts are required to increase soil organic carbon through in situ green manuring as traditional source of FYM is becoming scarce. Maize – safflower, sequence is one of the predominant cropping sequences under rainfed conditions of northern transitional zone of Karnataka and both cops being exhaustive, organic recycling through green manuring of intercropped sunnhemp (1 maize:2 green manure) 50 DAS without adversely affecting maize during normal years greatly benefitted safflower (Nooli, *et al.*, 2002, Biradar, 2008). Sunnhemp recorded significantly higher phytomass (11.38 t/ha), biomass production (2.02 t/ha) and, N accumulation (60.08 kg/ha) than *dhaincha* (37.87 kg/ha) and cowpea (37.57 kg/ha). There was also positive residual effect sunnhemp on succeeding chickpea besides increasing soil organic carbon.

Among the intercropped multicut green manures in maize, *Medicago sativa* (Lucerne) recorded significantly higher plant height (40.2 cm), greatest accumulation of total green matter (6.16 t/ha) and total dry matter yield (1.08 t/ha) followed by *Stylosanthes hamata* and *S. scabra*. Similarly, total N and P accumulation (34.26 and 5.01, kg/ha, respectively) were significantly higher in *Medicago sativa*. Consequently, maize yields with *Medicago sativa*, *Stylosanthes hamata* or *Stylosanthes scabra* intercrops and sole maize were at par in normal years with beneficial effects on succeeding crops but maize was significantly affected in the years of stress and inadequate rainfall. Similarly, experiments on *in situ* green manuring revealed its possibility in chilli+cotton (Hongal, 2001), hybrid cotton (Biradar, 2000) and paddy (Halepyati, 1989, Matiwade 1992) in the transitional and hilly zone of Karnataka.

Selection of Efficient System

At Bijapur, chilli – onion under rainfed condition is catching up. And, onion in wide rows during *kharif* and relay planting of *rabi* sorghum is also becoming popular (Surkod, 2015). That apart, area under pigeonpea is increasing due to its commercial value and because of availability of climate smart cultivars. Thus, the overall crop diversity is increasing which is a welcome sign in comparison to irrigated ecosystem. Ultimately, the major factor that determines farmers' choice is productivity/and economic viability. But, where different systems are made up of different crops, variety of options available for comparison. For instance, yield comparisons could be made in a cereal – pulse – legume combination such as sorghum + pigeonpea, mung – sorghum, and maize – chickpea systems. However, this may not be meaningful in case of sorghum – safflower system involving a cereal – oilseed crop. Other possible comparisons could be nutritional output, biological efficiency (*i.e.* the efficiency with which systems utilize environmental resources), net total energy accumulated in the system, or economic assessment (Honnali and Chittapur, 2013). The later approach has relevance in cash crops and in subsistence situation. Willey (1987) found that among the sorghum + pigeonpea, mung – sorghum, maize – chickpea and sorghum – safflower, sorghum + pigeonpea has slightly lower net returns than maize – chickpea but because of lower input costs (mainly because of not to establish a second crop), it has greater rate of returns to inputs hence could be a more attractive proposition. In contrast, sorghum – safflower has high costs (mainly because both crops have a high fertilizer requirement) and, therefore, a lower net return and lower rate of returns than the other two double cropping systems.

Another important aspect that needs consideration in selection of cropping system for a dry farming situation is the risk of failure. One of the problems in trying to ensure full use of potential growing period is that as the required growing period for a given system increases, the potential of end season water stress also increases. For instance, on Vertisols with 990 mm rainfall, after 91 days rainy season sorghum crop, there is still quite a high probability (73 per cent) of having sufficient water stored in the soil profile for a sequential second crop (assumed to be 100 days).

After taking in to account those years when there is sufficient water, but surface soil conditions are too dry to establish the crop, this probability drops to 60 per cent. If the rainy season sorghum has 105 day growing period, the probability of getting good growing condition or sufficient stored soil water for second crop; the overall probability of success, therefore, drops substantially to only 27 per cent. On the other hand, a relay system of planting based on the 105 day first crop and 14 day overlap does of course give the same probability of success as a sequential system based on 91 day crop (provided there is no injury to second crop at harvest of first crop). The advantage of sorghum + pigeonpea system in which pigeonpea acts as an already established second crop is very striking; there is extremely high probability of success and the 105-day first crop can be accommodated as easily as the 91-day crop.

Integrated Farming Systems

The integrated farming system (IFS) 'an agriculture that is sustainable and efficiently productive and allows the welfare of man, animal and plant' is highly climate smart and a boon to small and resource poor farmer for its high productivity with substantial fertilizer economy. It relies on organic recycling for maintaining soil productivity and livestock plays a key role in the system wholeness. The dairy and small ruminants (goat/sheep) are prominent. Because 85 per cent of Indian farmers and 98.4 per cent Raichur farmers practicing crop based cropping system research on IFS initiated at Siruguppa, Ballari during nineties was renewed under UAS, Raichur, in both for rainfed and irrigated situations on the farms under RKVY, GoK project and under ICAR project, and on SC and ST farmers under SCP/TSP project and on 100 acres blocks under each RSK, through RKVY, IFS funding. Crossbred cows, shirohi, Jmnapari and barberi goats, Giriraj poultry, horticulture crops involving flowers, fruits, and vegetables, agricultural crops comprising commercial and food species, timber species on bunds, bio-digester, and vermicompost units formed predominant components of IFS unit.

In a study farming system revealed higher system productivity, net returns and gross returns with cropping + goat, + poultry + fishery system with 206 man days ha^{-1} year^{-1}, while IFS comprising cotton + onion, maize + fodder cowpea- chickpea, drumstick, curry leaf, banana registered significantly higher cotton equivalent yield over conventional cropping of cotton alone. Integration of cow component increased milk yield from the system. Poultry also found to go well with certain other system combination (Naik, 2014). Further, it was found that the overall food security was improved with integration of allied enterprises *viz.*, goat, poultry, fishery and horticultural crops. Nevertheless, system being dynamic and interactive with availability of resources, family requirement, policies, customs and market price need continuous upscaling and moderation for a situation and no model can be recommended as a sustainable eternal module. UAS, Raichur among farm universities, however, takes pride to be leader in developing sustainable models.

References

Anonymous., 2015a, Proceeding of ZREAC and ZREFC meeting, UAS, Raichur.

Arunachalam, N., 1996. Input management in dryland agriculture. In: Recent Advances in Weather Forecasting and Dryland Management(*Ed.* N. Balasubramaniam). Short training course, 10-23 April, 1996.Centre for advanced studies in agronomy. TNAU, Coimbatore, pp. 25-28.

Biradar, S. A., 2008, *In situ* green manuring of intercropped legumes on the performance of maize-chickpea/safflower cropping system under rainfed condition. Ph.D Thesis, submitted to University of Agricultural Sciences, Dharwad, pp. 1-338.

Halepyati, A.S., 1989, Studies on *Sesbenia rostrata* (Brem and Oberm) and nitrogen substitution in rice (*Oryza sativa* L.) production. Ph.D Thesis, submitted to UAS, Dharwad.

Hongal, M., 2001, Effect of green manuring and levels of N on performance of chilli + cotton intercropping system. M.Sc. Thesis, submitted to University of Agricultural Sciences, Dharwad.

Honnali, S. N., Chittapur, B. M., 2013, Enhancing Bt Cotton (*Gossypium* spp.) productivity through transplanting in Upper Krishna Project (UKP) command area of Karnataka. *Indian J. Agron.*, 58 (1):105-108.

IDNP, 2003, Final Report. Indo-Dutch Network Operational Research Project (IDNP) on Drainage and Water management for Salinity Control in Commands, pp. 1-187.

Kuchnur, P. 2015, Performance of heat resilient CIMMYT maize hybrids in UKP and TBP commands (Personal communication).

Lobell, D.B., Bänziger, M., Magorokosho, C., Vivek, B., 2011, Nonlinear heat effects on African maize as evidenced by historical yield trials. Nature Climate Change 1:42-45.

Matiwade, P, S., 1992, Green manuring of rice (*Oryza sativa* L.) with *Sesbania rostrata* substitution and sustained production. Ph.D Thesis, submitted to University of Agricultural Sciences, Dharwad.

Naik, V. S., 2014, Development of cotton based integrated farming system models for irrigated ecosystem of north-east Karnataka. Ph. D. Thesis, submitted to University of Agricultural Sciences, Raichur, Karnataka, pp. 1-213.

Nooli, S. S., Chittapur, B.M., Hiremath, S. M., Chimmad, V.P., 2002, Effect of in situ green manuring of intercropped legumes in maize on the performance of succeeding safflower crop. J Oilseeds Res., 19(2):254.

Prasanna kumar, B. H., 2011, Agrotechniques to enhance productivity of sesame (*Sesamum indicum* L.) during Kharif and summer.Ph.D. Thesis, submitted to University of Agricultural Sciences, Dharwad, pp. 1-444.

Prasannamkumar, B. H., Chittapur, B.M., Hiremath, S.M., Malligawad, L. H., 2014, Enhancing productivity of summer sesame (*Sesamum indicum*) on paddy fallows through optimization of fertilizer, planting geometry and genotypes. Indian J. Agric. Sciences, 84 (6): 720-4.

Pyati, P., 2015, Performance of Bt cotton under nutrient omission situations in paddy lands of Kasabe camp of Raichur in TBP command (Personal communication).

Rajkumar, D., Gurumurthy, S., 2008, Effect of plant density nutrient spray on the yield attributes and yield of direct sown and polybag seedling planted hybrid cotton. Agriculture Science Digest, 28 : 174-177.

Surkod, V. S., 2015, Experience with sand mulching on crop performance and soil moisture dynamics on Vertic inseptisols, Regional Research Station, Bijapur, Karnataka (Personal communication).

Willey, R. W., 1987. Cropping systems for Drylands.*In*: Challenges in Dryland Agriculture - As global Perspective. Proc. of Int. conf. on dry land agric, Amarillo/Bushland. Texas. USA, August 15-19. pp. 77-774.

4

Real Water Saving in Rice Based Cropping Systems

B.K. Desai, S.R. Rajesh and U.N. Santhosh

Scarcity of water for agricultural production is becoming a major problem in many countries, particularly the world's leading rice-producing countries, China and India, where competing and growing demands for freshwater are coming from other sectors. Also, climate change in some areas may be reducing rainfall, thereby creating short-term problems even if only a cyclical rather than a permanent change. In many areas, rainfall patterns are becoming more and more unreliable, with extremes of drought and flooding occurring at unexpected times.

Government fiscal problems have put greater budgetary pressure on irrigation departments so that their operation and maintenance expenditures are curtailed, and new investments to create more irrigation capacity have declined. To alleviate fiscal burdens, efforts are being made to turn the ownership and operation of irrigation facilities over to the private sector, but this move, although having some policy logic, is likely to leave smaller and poorer producers less able to ensure themselves of water for their crops.

All of these trends make water-saving a high priority for the agricultural sector in the years ahead. Because irrigated rice production is the leading consumer of water in the agricultural sector, and rice is the world's most widely consumed staple crop, finding ways to reduce the demand for water to grow irrigated rice should benefit both producers and consumers. If rice production can be increased while water consumption is reduced, this will make changes in agricultural practice more attractive.

Getting farmers to adopt water-saving methods in rice production is impeded by the fact that, so far, there has been little or no associated increase in yield and profitability that would compensate farmers for their greater labor and management effort. Small increases in the range of 5-10 per cent may not suffice to justify the added cost and inconvenience. As long as water has been freely available, farmers have used water in ways that reduce their labor requirements.

In Tungabhadra Project (TBP) area, rice is the staple food stuff and also most important crop economically. The most common rice-growing system in this area is lowland and irrigated. The command area on the left bank of the river is entirely in Karnataka, whereas right bank command area continues in Andhra Pradesh. The total command area in Karnataka for which the potential has been created is about 3.62 lakh ha., 2.44 lakh ha of which is on the left bank and 1.18 lakh ha on the right bank.

The project was originally designed for protective irrigation, which means that a very limited amount of water has to be thinly spread over a larger area in order to benefit many farmers. Therefore land was 'localized' for certain crops. As per localization pattern, about 36 per cent is in *kharif* light and *rabi* light crops followed by cotton (12.3 per cent), paddy (8.7 per cent), sugarcane (3.5 per cent) and garden (2.6 per cent). However, the areas actually irrigated under the crops deviate from the localization pattern.

In the tail-end regions of Raichur taluka, taking rice-rice is very risky owing to canal closure from March 31. This opens the possibility of growing another second crop with low water requirement. Even then the farmers invariably go for second rice crop or keep the land fallow from December since, planting second crop will be delayed by about 2-3 weeks due to number of factors like canal opening during *kharif* is only after July II fortnight, first crop of rice (Var. Sona Mahusuri) is of long duration and harvested only in December 2nd week, presence of rice residues (stubbles), non-availability of labour *etc.*

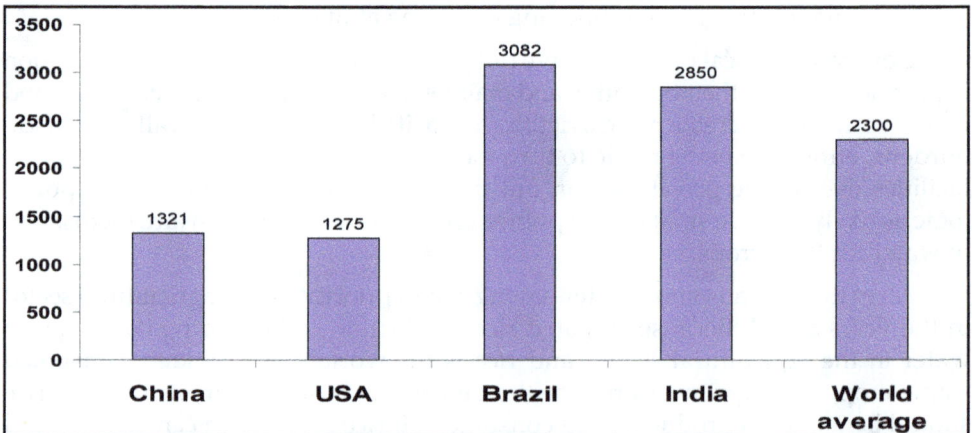

Rice: Average Virtual Water Content (l/kg)

Source: China water risk, Water foodprint network, Arjen Y Hoekstra and Ashok K Chapagam.

Thus, the farmers are facing difficulty in selection of a profitable sequence after paddy. Under such situations, there is ample scope for introduction of high valued arable crops and vegetables with the usage of conservation agricultural techniques or zero tillage technology which is popular in northern parts of our country.

The goal of sustainable intensification is to increase food production from existing farmland while minimising pressure on the environment. It is a response to the challenges of increasing demand for food from a growing global population, in a world where land, water, energy and other inputs are in short supply, overexploited and used unsustainably.

The Majority of the Rice is Transplanted ??

☆ Availability of labour for intensive cultivation

☆ Extensive irrigation infrastructure

☆ Mechanization

☆ Easy access to production inputs

☆ Better marketing and better weed management

☆ Best plant protection

But......?

☆ Now farmers are facing shortage of labour and water during the peak transplanting time and due to this transplanting is delayed sometime upto month of August which results in reduced yield.

☆ Sometime rice stands becomes patchy and optimum plant population can not be obtained.

☆ Due to intensive cultivation and long duration rice varieties wheat sowing is delayed sometimes upto January

Issues Related to Groundwater Depletion in Rice-Wheat (RW) Fields

☆ Water table going down in the RW belt of NW India

☆ Amount of water we need to save to arrest the declining water table

☆ Methods for reducing ground water depletion

    a) Methods for increasing recharge of ground water - Construction of recharge wells, adoption of soil conservation practices, rainwater harvesting, renovation of village ponds

    b) Methods for reducing withdrawal of ground water - Delaying transplanting of rice, diversification from rice to other crops, technologies like AWD, laser land levelling, raised beds and mulching

What do we mean by "Water Saving" in RW System

☆ Water saving – To cease unsuitable over exploitation of surface and groundwater resources and increase the amount of water available for non-agricultural purposes (*e.g.* Urban, environmental, recreational).

☆ Thus water saving in RW system has the dual goals of using less water than is currently being used, while increasing production.

☆ From farmer point of view- "Water saving" using less irrigation water to grow a crop ideally with the same or higher yield, thus increasing irrigation water productivity (g grain/kg irrigation water).

☆ Real water saving occurs when losses that cannot be recaptured are reduced or eliminated, however the magnitude of any water saving can vary considerably depending on the spatial and temporal scales of interest. (Loeve *et al.*, 2002)

☆ Real water saving is that which is achieved by reducing the unproductive water losses by soil water evaporation (Jalota *et al.*, 2009) or evapotranspiration (Seckler, 1996).

Saving Water in RW Fields

☆ **Generic approaches to saving water (Ws) or increasing water productivity (Wp) that can benefit both rice and wheat at the field scale**

- Laser land levelling, drainage recycling systems, the ability to forecast rainfall and irrigation water availability, and improved reliability of water supply system.

☆ **Reducing seepage and percolation losses in RW fields**

- Reducing percolation losses itself a real water saving

- Confining rice to less permeable soils, reducing ponding depth and AWD irrigation, laser levelling and raised beds.

☆ **Puddling for rice**

- There are some reports of similar yields for transplanted or direct seeded rice with and without puddling (Aggarwal *et al.*, 1995; Humphreys *et al.*, 1996; Kukal and Aggarwal 2003).

☆ **Reducing evaporation losses**

- Reducing non-beneficial evaporation direct from the soil or free water lying on the field is a true water saving.

- Evaporation can be reduced by mulching, by changing time of crop establishment

☆ **Sowing planting date**

- In NW India the evapotranspiration of rice declines from around 800 to 550mm as the date of transplanting is delayed from 1 may to june 30

- Irrigation water savings (25-30 per cent or 720mm) can be achieved by delaying transplanting from mid-may to mid-june (Narang and Gulati., 1995)

- Recommended practice in NW India is around mid-june. But, many farmers plant earlier than this (57 per cent in punjab) because of external factors.

☆ Varietal duration

- Water can be saved by using varieties of shorter duration
- Earlier maturity allows earlier harvest
- Increasing the chance of timely establishment of a winter crop after rice and making more efficient use of stored soil water and winter rainfall instead of losing it as deep and surface drainage or transpiration by weeds

☆ Mulching

☆ A few reports on the effect of mulching of wheat saved 25-100mm water, to reduced number of irrigations by one or irrigation time by an average of 17 per cent (RWC-CIMMYT., 2003).

☆ Happy seeder, which combines the stubble mulching and seed drilling functions into the one machine (Blackwell *et al.*, 2004).

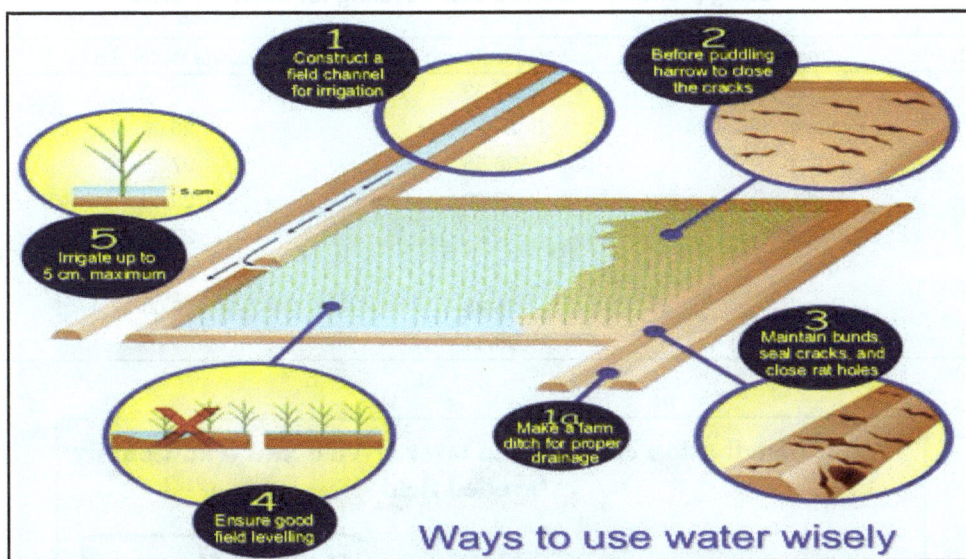

Ways to use water wisely

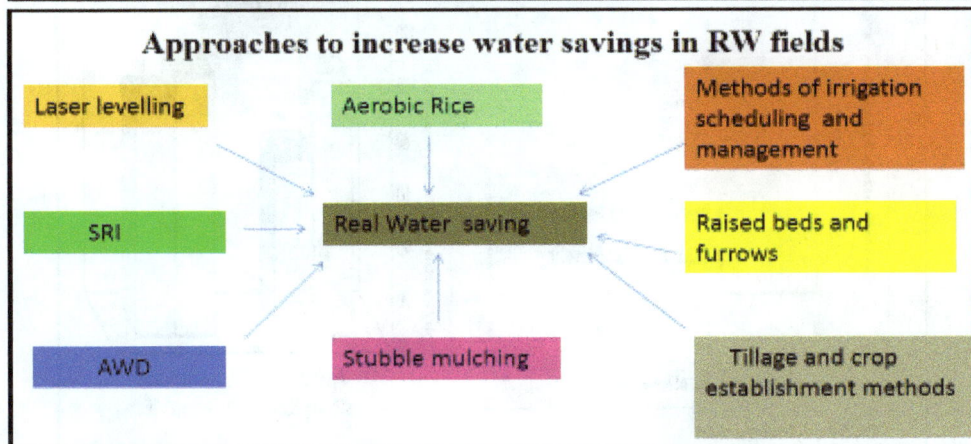

Approaches to increase water savings in RW fields

Advantages of Laser Levelling

☆ Water is distributed homogeneously to the field therefore, using efficiency of the present water will increase.

☆ It will be easier to control water and as a result surface and deep drainages will be lesser.

☆ Uniform vegetation cover is provided (Meral and Temizel, 2006).

☆ Water lost during the application of water is reduced by 25 per cent.

☆ Saving in irrigation water by 35-45 per cent (Singh *et al.*, 2008 and Chhatwal, 1999).

☆ This technology reduces weed problems and increases cultivable area by 3-6 per cent (Jat *et al.*, 2004).

Water Saving (Per cent) due to Laser Leveling for different Crops

Crops	Water Saving	Average Water Saving
Maize	22-33	27.10
Wheat	26-33	26.00
Cotton	26-43	27.25
Paddy	26-30	26.33
Berseem	27	27.00
Pea	25	25.00
Potato	25-28	26.00
Avg. Water saving	**26.64**	

Water distribution efficiency in laser leveled and traditionally levelled field

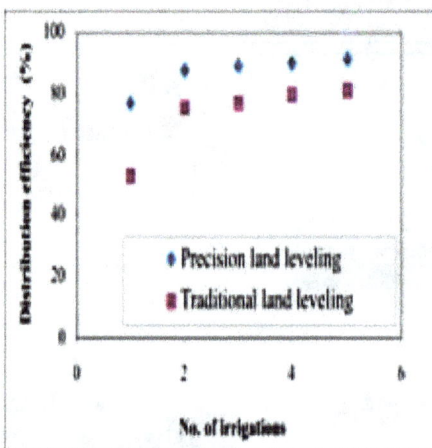

PDCSR., Modipuram

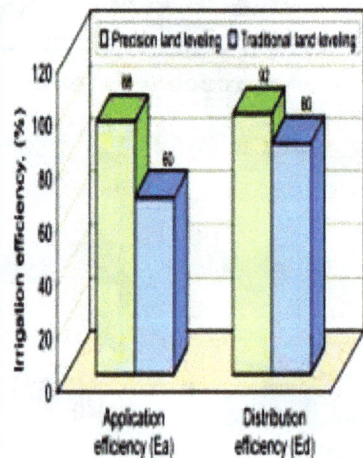

Jat *et al.*, 2006

Water Savings in different Crops at different Level of Adoption of Laser Levelling

Crop	Water Saved (cm)	Area (ha)	Water Saved (ha m) at different Level of Adoption (per cent)				
			10	*25*	*50*	*75*	*100*
Maize	9.5	1,52,560	1,447	3,618	7,235	10,853	14,470
Wheat	9.1	34,69,520	31,573	78,932	1,57,863	2,36,795	3,15,726
Cotton	9.5	5,41,060	2,949	7,372	14,744	22,116	29,488
Paddy	42.1	26,25,204	1,10,595	2,76,486	5,52,973	8,29,459	11,05,946
Total		67,88,344	1,46,563	3,66,408	7,32,816	10,99,224	14,65,631

Advantages of Bed Planting

☆ Management of irrigation water is improved.

☆ Bed planting facilitates irrigation before seeding and thus provides an opportunity for weed control prior to planting

☆ Plant stands are better.

☆ Weeds can be controlled mechanically, between the beds, early in the crop cycle

☆ Wheat seed rates are lower

☆ Less lodging occurs.

Wheat on Beds

The irrigation water savings are likely to be due to faster irrigation times and reduced deep drainage and therefore the magnitude of the irrigation savings is likely to depend on soil type and depth to the water table.

Yield Performance and Water Savings of different Crops grown in Raised Beds

Crops	Yield on Beds (t/ha)	Yield on Flat	Water Savings (Per cent over flat)	Yield Increase (Per cent over flat)
Maize	3.27	21.38	35.5	37.4
Urd	1.83	1.37	26.9	33.6
Mung	1.62	1.33	27.9	21.8
Green peas	11.91	10.4	32.4	14.5
Wheat	5.12	4.81	26.3	6.4
Rice	5.62	5.29	42	6.2
Okra	34.4	239.1	33.3	18.2
Carrot	36.3	28.6	31.8	16.9
Radish	34.7	26.7	29.4	30
Cabbage	33	27.8	26.8	18.7
Pigeonpea	2.2	1.5	30	46.7
Gram	1.85	1.58	27.3	17.1
Cauliflower	25.9	18.9	26.4	37
Average	-		31.2	24.2

PAU. Ludhiana (Connor *et al.,* 2012)

Potential of Water saving in rice through System of Rice Intensification

- SRI, originated in Madagascar during 1980s by Henri Laulanie
- Increased yields in SRI compared to conventional methods were reported by several researchers (Thiyagarajan *et al.*, 2005, Upoff, 2005 and Satyanarayana *et al.*, 2006)

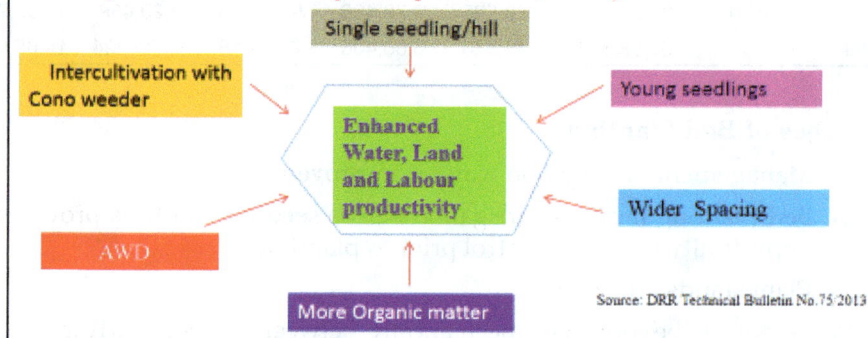

Source: DRR Technical Bulletin No.75/2013

Irrigation Water Savings by Bed Planting of Wheat Over Conventional Method

Tillage Option	Water Required at different Times of Irrigation (mm)					Water Saved over Conventional (per cent)
	Sowing	CRI	Max. Tillering	Grain Filling	Total	
2001-02						
70cm bed	57	49	41	23	170	46
80cm bed	55	49	40	21	165	48
90cm bed	55	48	39	21	163	48
Conventional	95	89	76	55	315	-
2002-03						
70cm bed	58	48	45	35	186	41
80cm bed	56	46	44	34	180	42
90cm bed	55	45	42	32	174	44
Conventional	94	85	79	60	318	-

Alternate Wetting and Drying – Smart Water Technique for Rice

By using a simple water level gauge and implementing smart but simple water management techniques, farmers can reduce water usage in paddy rice by 15-30 per cent without compromising yields.

About *Panipipe*

☆ A *Pani pipe* is a 40 cm length of 15 cm diameter plastic pipe or bamboo, with drilled holes, which is sunk into the rice field until 20 cm protrudes above soil level.

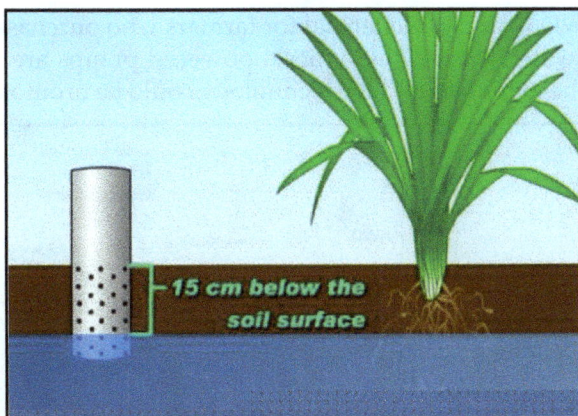

★ When the water level inside the *Panipipe* drops to 15 cm below ground level, the field is ready to be re-flooded. This threshold is called 'safe AWD' as it does not impact on yield.

★ By using *Panipipe* and implementing the smart but simple AWD technique, farmers save up to 30 per cent of the nearly 5,000 litres of water commonly used to produce 1kg of unmilled rice.

★ The savings in water has increased farmers' income by more than 30 per cent ; often from a net loss to a net gain.

★ In Vietnam and Bangladesh farmers reported yield increases of more than 10 per cent.

★ In the Philippines, more than 100,000 farmers have adopted AWD, which has also reduced conflicts over water in shared canal irrigation systems.

★ In Bangladesh, trials have shown reductions in water consumption of 15-30 per cent, translating into a reduction in pumping costs and fuel consumption and an increased income of US$67-97 per hectare.

AWD in Rice

★ Safe AWD maintains yield while giving very large irrigation water savings in transplanted rice on permeable soils with deep watertables, in comparison with continuously flooded rice.

★ The reduction in irrigation amount is likely to be due to reduced deep drainage, with little effect on ET and WPET.

★ In practical terms, adoption of safe AWD is generally not possible for farmers dependent on canal irrigation or electricity for groundwater pumping because of unreliable and limited supply of water or electricity. There is no incentive to adopt safe AWD because of the low prices of water and electricity.

☆ Safe AWD would be beneficial for farmers who purchase diesel to pump groundwater. Areas where diesel powered pumps are commonly used should be identified and the technology could be promoted immediately.

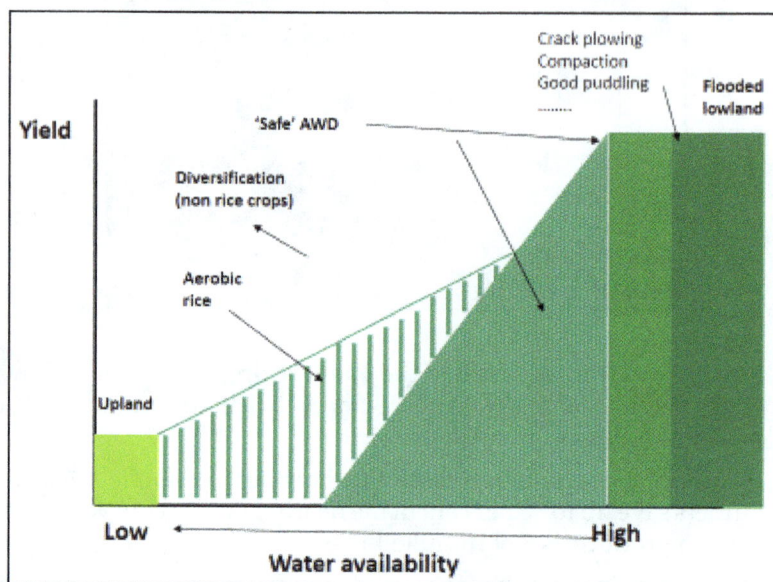

Various Response Options to Water Scarcity.

Zero till Wheat

☆ Adoption rates of zero till wheat are far higher than with any other improved technology for RW systems, with adoption on over 10 per cent of the RW area of India by 2003–2004.

☆ It increased profitability as a result of lower establishment costs.

☆ Irrigation water saving of at least 10 per cent (at least 20–30 mm) in comparison with conventional practice, while yields are generally slightly higher, leading to higher Water productivity.

☆ Experimental studies indicate that reductions in irrigation amounts of up to 30 per cent are possible.

Mulching

☆ Mulching offers the potential to reduce evaporation from wheat by 30–40 mm, which will reduce the number of irrigations needed by one in some years.

☆ Reduction in evaporation is offset by increased transpiration, and that under well-irrigated conditions transpiration efficiency is reduced, resulting in similar WPET with and without mulch.

☆ Under limited water conditions where mulching reduces water deficit stress and loss of yield, WPET is increased.

Replacement of Rice – Crop Diversification can Save Water in RW Cropping System?

☆ Replacing rice with another summer crop will greatly reduce the amount of irrigation water applied, with many benefits.

☆ In canal irrigated areas with saline ground waters, replacement of rice will reduce water depletion and the rate of water table rise and associated problems.

☆ But replacement of rice with other summer crops will not reduce the problem of groundwater depletion in areas with fresh groundwater, where groundwater is the main source of irrigation water and ET is the only source of water depletion, unless the alternative crop has lower ET than that of rice.

Indirect Approaches of water saving

☆ **Virtual Water Trade:** When a country imports a tonne of wheat or maize, it is in effect, also importing "virtual water", *i.e.* the water required to produce that crop. Trade in virtual water generates water savings for importing countries.

☆ Global water saving as a result of international trade of agricultural products has been estimated at about 350 billion m³/year. To maintain food security or food self-sufficiency, many countries in the arid and semi-arid regions have over-exploited their renewable water resources. Trade can help mitigate water scarcity if water-short countries can afford to import food from water-abundant countries.

☆ But political and economic factors are stronger drivers and barriers than water. Many countries view the development of water resources as a more secure option to achieving food security and livelihood of its population. Large water exporting countries may influence the policies of recipient countries. Therefore, there is a strong need to develop a set of principles/rules governing virtual water trade otherwise conflict may prevail over cooperation.

5

Physiological and Biochemical Adaptations to Mitigate Water Stress

Djanaguiraman Maduraimuthu and
Mukesh Kumar Meena

Modern agriculture is affected by several environmental factors such as water, temperature, salt, heavy metals and light. Plants grow in a complex environment constituting these environmental changes causes stress. Stress is an important phenomenon that limits crop productivity or destroys biomass (Grime, 1979). Environmental stress has certain characteristics such as severity, duration and number of exposure and combination of stress. Stress affects tissue/organ of plant, stages of development and genotype of specific characters. The response of plants may be either tolerance or susceptible which results in survival or death of plant. Stress was classified into two types *viz.*, abiotic and biotic stress. Abiotic stress was further classified into water, temperature, salt and chemicals. Water plays an important role in growth and development of all living organisms including plants. It constitutes 90-95 per cent of weight, act as solvent for many substances, involved in transpiration, photosynthesis, respiration and several other activities of plant. Thus, reduction or flooding during agriculture will alter the physiological and biochemical characters of the plant. In this chapter, physiological and biochemical adaptations of plants to mitigate water stress was discussed in detail.

Water Stress

Plant growth can be limited by water deficit and flooding. Water or drought in agriculture determines the insufficient water availability and is caused by intermittent to continuous periods without precipitation. Flooding or excess water was the result of soil compaction. The deleterious effect of excess water is a consequence of the displacement of oxygen from the soil.

Drought

Plants experience drought when water supply in the soil declines to meet the plant demand. It is induced when transpiration rate is higher than absorption rate. Transpiration will be higher with low air humidity, high temperature, high irradiance and strong wind. Low absorption occurs during low soil moisture, high salt concentration and low soil temperature. Plants transpire water far more than that used for metabolic activities. It directly affects the photosynthetic rate of the plant by stomatal conductance for water vapour and CO_2 entry. During drought, stomata closes to prevent water loss and also entry of CO_2 for photosynthesis.

Physiological Effects during Drought

Drought affects water potential induction, cellular dehydration and hydraulic resistance which are the primary effects. Secondary effects include reduced cell or leaf expansion, reduced cellular and metabolic activities, stomatal closure, photosynthetic inhibition, ROS production and ion toxication.

At cellular level it causes loss of turgor, hydrophilic amino acids fail to interact with water anymore (Campbell, 1991) with increase of intra cellular solutes, change in cell volume, denaturation of proteins (Bray, 1997) and disruption of thylakoid structure and PS II system results in reduction in photosynthesis. It disrupts the nucleic acids such as DNA and RNA by activation of DNAase and RNAase enzyme (Kessler, 1961). Enzymes such as nitrate reductase and RuBisCO are very sensitive to drought condition. During drought, uptake of nitrogen is reduced which results in reduction in the inducible enzyme nitrate reductase activity. RuBisCO activity reduced due to damage to thylakoid membrane and chloroplast. Nitrogenase is the enzyme involved in nodulation of leguminous crops and it gets affected during drought due to nodulation inhibition. It increases the activity of amylase enzyme involved in decrease of starch and increase in sugar content. At plant level, cellular growth is reduced with alteration in flowering time, increase in root-shoot ratio, stomatal closure, wilting, abscission, reduced transpiration and photosynthesis with accumulation of compatible solutes. Reduced cellular growth and reduced cell enlargement results in reduction of leaf area and shoot growth.

Accumulation of osmotic solutes would reduce the water potential of cell lower than that of soil. Solutes include proline, glycine-betaine, fructose, glycerol, sorbitol, mannitol, trehalose, raffinose and fructans.

Formation of reactive oxygen species like superoxide (O_2^-) molecule, singlet oxygen ($_1O^2$), hydrogen peroxide (H_2O_2) and hydroxyl radicals (OH^*) takes place in a number of sites as plasma membrane, mitochondrion and endoplasmic reticulum membranes (McKersie and Leshem, 1994), however, in vegetative tissues of plants, it is thought that the most prevalent cause of oxidative stress during periods of water limitation is the light-chlorophyll interactions which occur in the chloroplast (Farrant, 2000). During drought, stomatal closure limits the carbon dioxide availability for photosynthesis. In this situation, $NADP^+$ (electron acceptor in photosynthesis) becomes limited and ferredoxin selectively reduces oxygen and reactive radicals are produced owing to the electron transport by photosystem I (PS I) to O_2 (Mehler reaction) (Tambussi, *et al.*, 2000).

Flooding

Flooding or excess water causes poor aeration and results in anaerobic respiration and fermentation. This anaerobic respiration affects TCA cycle, oxidative phosphorylation and electron transport chain and results in degradation of mitochondria. Anaerobic respiration causes reduction in ATP and energy utilized water absorption by the plant. Production of alcohol, ethylene and lactic acid causes cytoplasmic acidosis and results in cell death and affects the protein synthesis. It also results in cell division, elongation and ion transport.

Physiological Mechanism of Flooding

During flooding, anoxia or hypoxia occurs leading to denitrification and ion toxicity. Anaerobic respiration products *viz.*, lactic acid and butyric acid alongwith H_2S causes unpleasant odour in waterlogged soil. During anoxia, ABA will be produced in root and transported to leaf. Accumulation of ABA in leaf causes stomatal closure.

Adaptation of Plants to Water Stress

Plants have various mechanisms that allow them to survive and live. Adaptation to the water stress is characterized by genetic changes in the entire population that have been fixed by natural selections over many generations. In contrast, individuals adopt themselves to water stress by altering their physiology and morphology of plants. This can be made possible with repeated exposure of the plant to the water stress. This method of adaptation is called acclimation or phenotypic plasticity and represents non-permanent changes. The evolution of adaptive mechanism in plants to water stress area as follows:

Morphological Adaptation

Drought Escapers

Many desert plants like ephemerals escape drought by completing its life cycle. Their seeds dormant during dry season and germinate, grow and flower during a period of adequate moisture.

Water Spenders

Plants like alfalfa, palms extend their roots deep down to the water table, aggressively consume water and avoid drought. A potential deep rooting system is advantageous for water spenders to avoid drought.

Water Collectors

Succulents such as cacti, agave and other CAM plants resist drought by storing water in their succulent tissues. They have the property of thick cuticle and stems. Leaves are modified into needles and scales. Stomata closes during day and opens during night for CO_2 uptake. Thick cuticles helps in improving drought tolerance in tobacco plants (Cameron *et al.*, 2006) as observed in Figure 5.1.

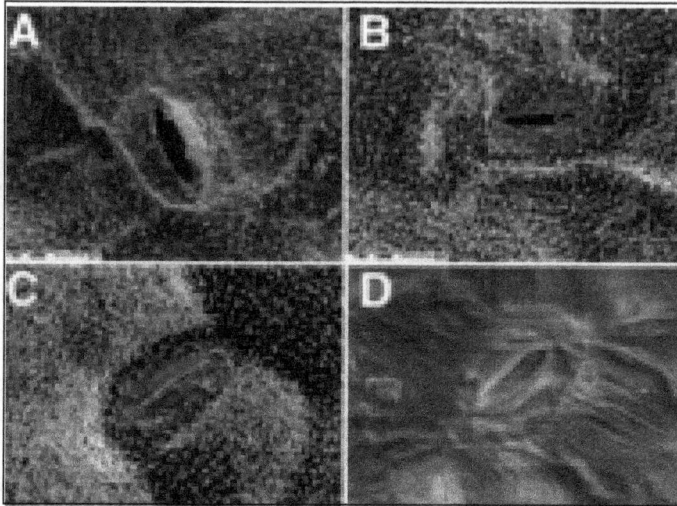

Figure 5.1: SEM of the Adaxial Surfaces of Leaves from Three Tree Tobacco Plants after Exposure to One Drying Event (A), Two Drying Events (B), Three Drying Events (C), and after Removal of the CH2Cl-Soluble Wax Fraction (D). Magnication 32,000. Bar 5 5 mm.

Water Savers

Non- succulent desert plants show many adaptations to reduce water loss through transpiration such as smaller laminar area, sunken stomata, thick hairy covering on surface of leaves and shedding of leaves.

Other Adaptations

Waxing, rolling, anatomy architecture

Physiological and Biochemical Mechanism

Osmotic Adjustment (OA)

It is a physiological trait that makes plant tolerant to drought. It is defined as the active accumulation of solutes within the plant tissue in response to lowering water potential in soil. As the water potential decreases organic compounds that do not interfere with enzyme functions get accumulated. They are proline, glycine betaine, sugars, sorbitol, mannitol, *etc.* These will maintain the water potential equilibrium within cell and doesn't affect other metabolic activities of the tissue and non-toxic. OA lowers the osmotic potential (OP) and helps to maintain dehydration tolerance.

They offer protection against dessication in two ways:

During the course of cellular water loss, solutes are actively accumulated and cause reduction in OP. This reduces the outflow of water from cell thereby reducing turgor loss and allows stomatal opening and expansion growth to continue at lower water potential. They protect protein complexes in cell organelles and cytosol against dehydration damage by keeping them hydrated.

Proline

The accumulation of proline under drought in large amount contributes to osmotic adjustment. It was shown that proline stabilizes proteins, membranes and sub cellular structures and protect cellular functions.The accumulation of proline in roots could have provided the root with an osmotic mechanism to maintain favourable water potential gradient for water entrance into the roots (Irigoyen *et al.*, 1992) leading to a lower droughts stress injury in the plant. In addition to acting as an osmoprotectant, proline also serves as a sink for energy to regulate redox potentials, as a hydroxyl radical scavenger, as a solute that protects macromolecules against denaturation, and as a means of reducing activity in the cell (Smirnoff, 1993).

Table 5.1: Effect of Water Stress in Rice on Proline Content (μmol l^{-1})

Rice Cultivars	Young Leaves			Old Leaves		
	Submerged	Non Submerged	Change (Per cent)	Submerged	Non Submerged	Change (Per cent)
Zayande-Rood	112	155	0.384	58	73	0.258
829	84	102	0.214	48	52	0.083
216	97	120	0.237	54	61	0.129
Average	97.67	125.67	0.286	53.3	62	0.163

Glycine Betaine

It is a quaternary ammonium compound that occurs naturally in a wide variety of plants. Its accumulation has been widely recognized as abiotic stress response where it acts as osmoprotectant by stabilizing both the quaternary structure and the highly ordered structure of membranes. It is created from choline and localized in plastids, peroxisomes and cytoplasm. Soil application increased seed germination and seedling vigour of wheat and cotton.

Polyols

Accumulation of sorbitol, mannitol and its derivatives is considered to be related to drought tolerance in many plant species. They interact with the membranes, protein complexes or enzymes and protect them by scavenging reactive oxygen species. In this respect, they became attractive candidate for screening crops under water stress.

Trehalose

It is a non-reducing disaccharide of glucose that effectively stabilizes hydrated enzymes and lipid membranes in some plants. It is a very rare compound for plant kingdom, but in the recent years, gene transfer for trehalose accumulation is one important area of scientific work during drought. However homologous genes for trehalose biosynthesis have been recently discovered in several wild and crop plants, which make them attractive candidates for gene transfer. It has a direct effect as an osmoprotectant or stabilizer of cellular structures (Figure 5.2). The observed differences can probably be attributed to pleiotropic effects caused by changes to trehalose metabolism, suggesting that this metabolic pathway plays a regulatory role in plant development. Recently, a cotton EST clone with homology to the Arabidopsis gene that encodes TPS has been found to be regulated under conditions of water stress, indicating that trehalose biosynthesis is specifically induced under drought conditions. Although the significance of this finding remains to be elucidated, it contributes towards other circumstantial evidence that trehalose metabolism in higher plants does play a role in the acquisition of drought tolerance.

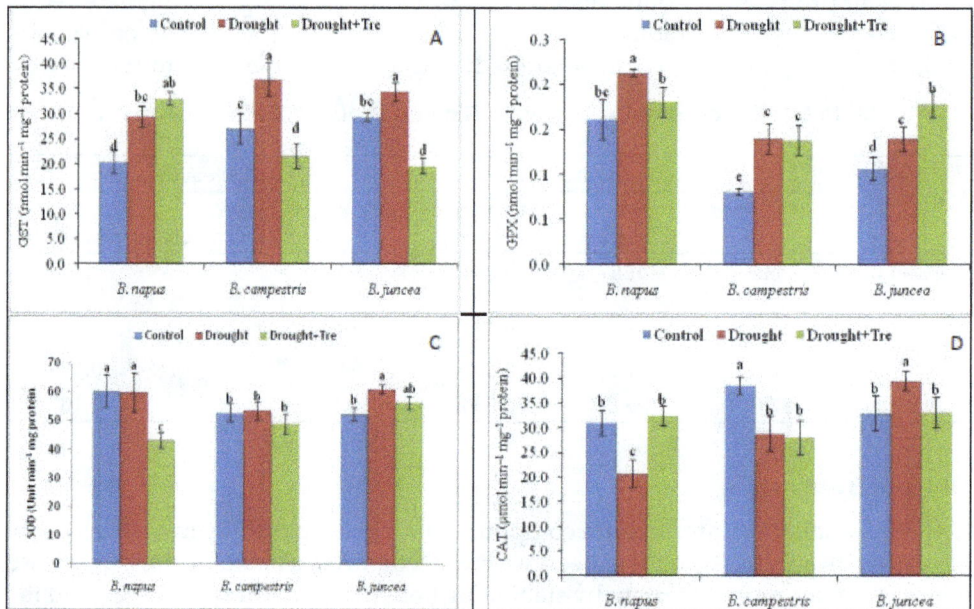

Figure 5.2: Activities of GST (A), GPX (B), SOD (C) and CAT (D) in *Brassica* Seedlings Induced by Trehalose (TRE) under Drought Stress.

Antioxidant Enzymes

ROS production during drought was reduced by scavenging enzymes by peroxidase (POX) catalase (CAT), Superoxide dismutase (SOD) and Melondialdehyde (MDA) which is fine and elaborate enough to avoid injuries of active oxygen, thus guaranteeing normal cellular function (Horvat *et al.*, 2007). The balance between ROS production and activities of antioxidative enzyme determines whether oxidative signaling and/or damage will occur (Moller *et al.*, 2007). The enzymatic components may directly scavenge ROS or may act by producing a non-enzymatic antioxidant.

Yang *et al.* (2009) exhibited that as compared with 100 per cent field capacity, at 25 per cent field capacity the increased activities of CAT, SOD, POD, APX and GR were 4.3, 103, 172, 208 and 56 per cent respectively in *P. cathayana*, whereas they were 8.1, 125, 326, 276 and 78 per cent respectively in *P. kangdingensis*. Efficient destruction of O_2^- and H_2O_2 in plant cells requires the concerted action of antioxidants. O_2^- can be dissimulated into H_2O_2 by SOD in the chloroplast, mitochondrion, cytoplasm and peroxisome. POD plays a key role in scavenging H_2O_2 which produces through dismutation of O_2^- catalyzed by SOD. CAT is a main enzyme to eliminate H_2O_2 in the mitochondrion and microbody (Shigeoka *et al.*, 2002) and thus helps in ameliorating the detrimental effects of oxidative stress. CAT is found in peroxisomes, but considered indispensable for decomposing H_2O_2 during stress. Maintaining a higher level of anti-oxidative enzyme activities may contribute to drought tolerant induction by increasing the capacity against oxidative damage (Sharma and Dubey, 2005).

Table 5.2: Effect of Drought Treatments on Antioxidant Enzymes of the Leaves of *Brassica napus* Cultivars

Drought Treatment	SOD Activity	CAT Activity	POD Activity
	unit/mg Protein		
FC	3.13 ± 2.22	2.91 ± 1.76	1.57 ±1.16
60% FC	3.44 ± 2.25	1.12 ± 5.02	1.67±1.27
30% FC	6.60 ± 5.83	0.02 ± 1.2	3.9 ±2.19
CV	11.64	12.83	6.69

Photosynthesis

The effects can be direct, as the decreased CO_2 availability caused by diffusion limitations through the stomata and the mesophyll (Flexas *et al.*, 2004, 2007) or the alterations of photosynthetic metabolism (Lawlor and Cornic, 2002) or they can arise as secondary effects, namely oxidative stress. The latter are mostly present under multiple stress conditions (Chaves *et al.*, 2009) and can seriously affect leaf photosynthetic machinery.

In addition to reduced CO_2 diffusion through the stomata, drought result in an apparent reduced CO_2 diffusion through the leaf mesophyll, *i.e.* in a reduced mesophyll conductance to CO_2 (Flexas *et al.*, 2007). Although not as straight forward as stomatal conductance measurements, estimations of g_m seem appropriate despite many assumptions involved in the most common methods used. The changes in mesophyll conductance may be linked to physical alterations in the structure of the intercellular spaces due to leaf shrinkage (Lawlor and Cornic, 2002) or to alterations in the biochemistry (bicarbonate to CO_2 conversion) and/or membrane permeability (aquaporins). Jones (1973) suggested that leaf internal diffusion conductance was depressed under water-stress conditions. However, the model used by Jones assumed that CO_2 concentration in the chloroplast was close to zero or to the compensation point, which was later shown to be untrue. Comparison of chlorophyll fluorescence with gas exchange measurements also revealed that C_c was lower than C_i, and that the difference increased under conditions of water stress or salinity (Cornic *et al.*, 1989). That water stress specifically lowers C_c below C_i was independently confirmed by measuring leaf ^{18}O (Renou *et al.*, 1990) and ^{13}C discrimination (Brugnoli and Lauteri, 1991). Still, in these early works, it was assumed that g_m was largely unaffected by stress, and that discrepancies between

C_c and C_i arose from invalid estimations of the latter due to either heterogeneous stomatal closure and/or interference of cuticular conductance (Boyer *et al.*, 1997). It was assumed that most of the mesophyll resistance to diffusion was caused by morphological and anatomical leaf traits, which are unlikely to change in response to stress, particularly in the short term. However, Genty *et al.* (1998) showed that most of the internal resistance to CO_2 diffusion was in the liquid phase inside cells instead of in intercellular air spaces, *i.e.* not so much dependent on leaf structure, and later studies specifically suggested that g_m was depressed under both salt (Centritto *et al.*, 2003) and water stress (Galmés *et al.*, 2007).

Stress Proteins

Osmotic stress induces accumulation of set of low molecular weight proteins known as stress proteins in plant tissues such as LEA (Late Embryogenesis Abundant) proteins, dehydrin, Trehalose and antioxidant enzymes. These are accumulated during seed dessication and in response to water stress. The LEA proteins were the first to be identified as genes that are expressed during maturation and dessication phase of seed development and it is recognized that these genes are also expressed in vegetative tissues during water stress.

Late Embryogenesis Abundant (LEA) Proteins

LEA were originally discovered in the late stages of embryo development *i.e.*, during maturation stage in cotton seeds (Dure *et al.*, 1981). During this stage and preceded by an increase in ABA content, gene expression and protein profiles change greatly and are associated with the acquisition of desiccation tolerance and development of capacity for seed germination. LEA proteins are accumulated in this final stage, in contrast to storage proteins which appear earlier. Moreover, their mRNAs are maintained at high levels in the dehydrated mature embryos, while transcripts of storage protein genes are completely degraded during the last embryogenesis stage (Goldberg *et al.*, 1989). LEA proteins accumulate in vegetative tissues exposed to dehydration (Bray, 1993). LEA proteins have the capacity to protect target proteins from inactivation and aggregation during water stress. A role in protein stabilization is supported by the fact that some LEA proteins preserve enzyme activity in vitro after partial dehydration (Goyal *et al.*, 2003). One mechanism of protection is the prevention of water stress induced aggregation of proteins (Kovacs *et al.*, 2008). During desiccation membrane protection is essential to preserve the cellular and organellar integrity. Some LEA proteins could contribute with sugars to H-bonding networking and protect membranes in the dry state (Hoekstra *et al.*, 2001). One consequence of drought is the increase in concentration of intracellular components, including ions. Increased ionic concentration affects macromolecular structure and function. It has been proposed that LEA proteins, because of their many charged amino acid residues might act to sequester ions (Danyluk *et al.*, 1998). A dehydrin like protein from celery is located in the vacuole and binds Ca^{++} when phosphorylated (Heyen *et al.*, 2008). LEA proteins might reduce oxidative stress in dehydrating cells by scavenging ROS and/or by sequestering metal ions that generate ROS (Tunnacliffe and Wise, 2007).

Dehydrins

It can be induced by water deficit by ABA treatment in vegetative tissues of seedlings and plants (Moons *et al.*, 1995). The high conservation of dehydrin sequences across distantly related species, and their induction by dehydration suggest that they play a key role in preventing cell damage during water stress. Their role is probably a structural one: it has been proposed that, like sucrose and other sugars which also increase in tissues subjected to water stress, they can hydrogen bond through their polar groups to macromolecules and to polar head groups of membranes replacing the hydration shell and providing the hydrophilic interactions necessary for stability in the dry state (Blackman *et al.*, 1991). As for their subcellular localization, dehydrins were first described as being mainly present in the cytosol (Mundy and Chua 1988), in the cytoplasm and chloroplasts (Schneider *et al.*, 1993) or in soluble fractions (Neven *et al.*, 1993). More recently dehydrin related proteins have also been found in nuclei. A defined role for dehydrin-like proteins in the nuclear compartment can only be hypothesized, but a protective role for nuclear proteins against denaturation and for nuclear membranes, similar to the one possibly exerted in the cytoplasm, is easily predictable. In addition, depletion of water and increase in ionic strength can be deleterious to DNA topological organization and stability, thus stabilization and protection of this macromolecule is more likely to occur. Dehydrin also suspected to have a role as antioxidative defense directly through the activity of free-radical scavengers (Hara *et al.*, 2004) or indirectly by binding toxic metals to prevent the production of ROS (Hara *et al.*, 2005). Dehydrin prevent the binding of hydroxyl and peroxyl radicals rather than superoxide anion and hydrogen peroxide (Hara *et al.*, 2004). Some amino acid residues such as Lysin, Histidin, Glycine and Serrin were assumed to have a link to radical antidote as these residues are modified when dehydrin prevent hydroxyl radicals binding. Dehydrin also reported to protect cellular components from oxidative stress (Hara *et al.*, 2009).

Hormones Involved during Drought

Phytohormones are essential for the ability of plants to adapt to abiotic stresses by mediating a wide range of adaptive responses. They often rapidly alter gene expression by inducing or preventing the degradation of transcriptional regulators via the ubiquitin–proteasome system. One of the most studied topics in the response of plants to abiotic stress, especially water stress, is ABA signaling and ABA-responsive genes.

Abscisic Acid (ABA)

ABA is a phytohormone that regulates physiological processes such as seed maturation, seed dormancy and stress adaptation. These physiological responses are triggered by the fluctuation of endogenous ABA levels in accordance with changing surroundings or developmental stimuli. Endogenous ABA levels are largely controlled by the balance between biosynthesis and catabolism. ABA synthesis is one of the fastest responses of plants to drought, triggering ABA-inducible gene expression and causing stomatal closure, thereby reducing water loss via transpiration and eventually restricting cellular growth. Numerous genes associated with ABA de novo biosynthesis and genes encoding ABA receptors

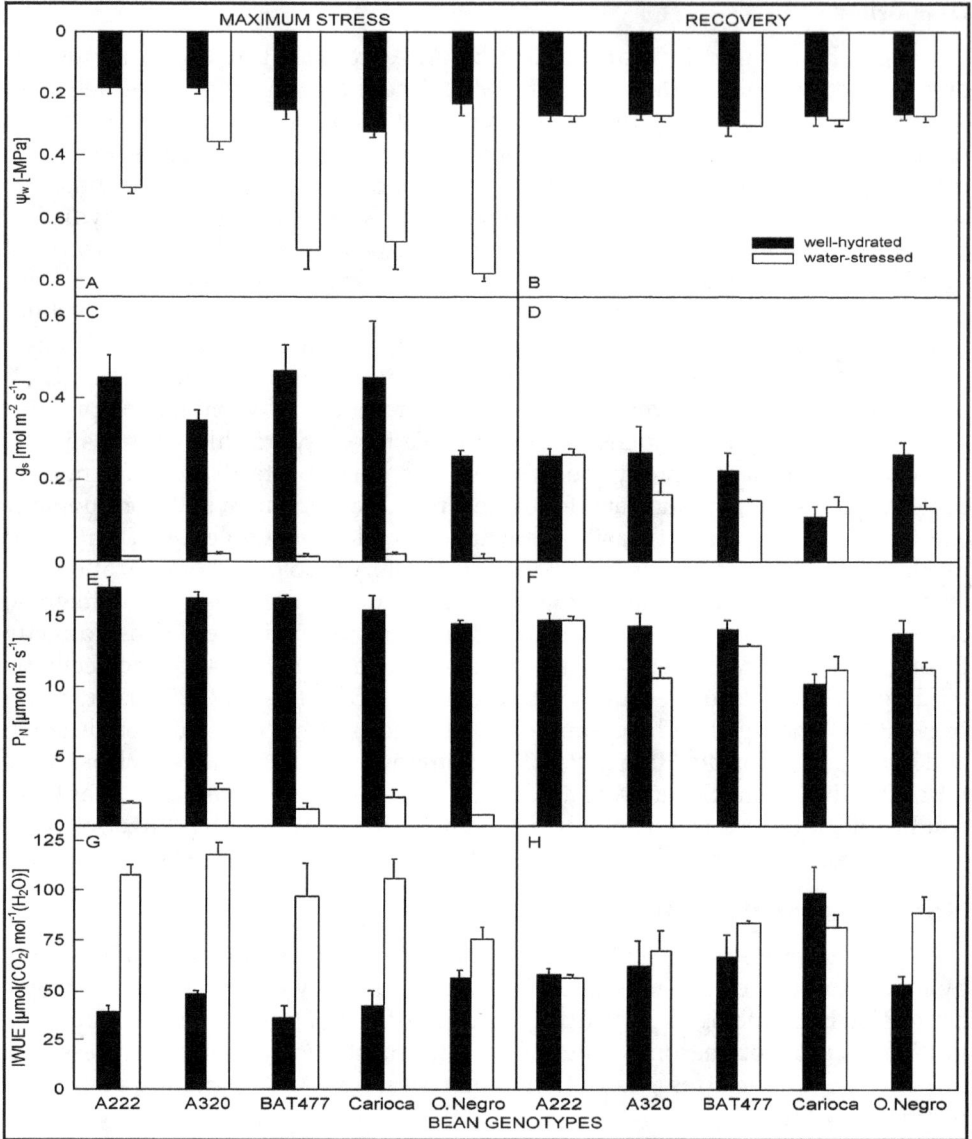

Figure 5.3: Changes in Predawn Leaf Water Ootential (Ψw - *A,B*); Stomatal Conductance (gs- *C,D*), Net Photosynthetic Rate (P- *E,F*) and Intrinsic Water Use Efficiency (IWUE - *G,H*) of Five Common Bean Genotypes (A222, A320, BAT477, Carioca and Ouro Negro) at Maximum Water Deficit (10 d of water withholding) and 2 d after Rehydration (Recovery).

and downstream signal relays have been characterized in *Arabidopsis thaliana*.It has been reported that drought tolerance was improved in transgenic plants that over expressed ABA biosynthetic genes (Luchi *et al.*, 2001). Furthermore, ABA is also involved in osmotic adjustment in some plant species (Trewavas and Jones,

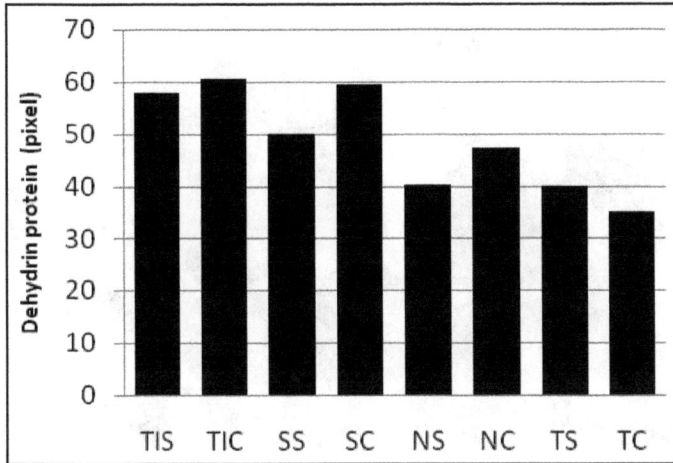

Figure 5.4: The Quantity of Dehydrin Protein in Soybean Varieties.
TC=Tanggamus control, TS= Tanggamus stress, NC=Nanti control, NS = Nanti stress,
SC=Seulawah control, SS=Seulawah stress, TC= Tidar control, TS=Tidar stress.

1991). Abdellah *et al.* (2011) reported that the increased ABA content was observed in wheat cultivar by water stress (30 per cent FC) over control. Under intense water stress, the concentrations of ABA in plants increases, which trigger a number of processes starting from decrease in turgor pressure, decline in cellular expansion and stomatal closure to reduce water loss in leaves (Thompson *et al.*, 1997).

Water stress causes ABA accumulation in stressed plants as observed by Unyayar *et al.* (2004) in sunflower. It has been hypothesized that ABA biosynthesis and signaling play an important role in a large protective effect against environmental stresses via expression of protective proteins and regulation of stomatal conductance (Goh,*et al.*, 2003). Kudoyarova *et al.* (2007) showed that the partial root zone drying increased the xylem ABA concentration 2.5 fold and decreased the cytokinins concentration of fully expanded leaves by 46 per cent in tomato plants. ABA, the universal plant stress hormone, is accumulated in roots subjected to a range of external stresses like drought. Crop varieties with high concentrations exhibit an intensified long distance ABA signaling that reduces water consumption and, in the case of grapevine, improves the quality of the berries (Hartung *et al.*, 2005).

Cytokinin **(CK)** is an antagonist to ABA, and the exposure of plants to water limiting conditions results in decreased levels of CK. The use of maturation-induced and stress-induced promoters (SARK, senescence associated receptor kinase) to drive *IPT* expression in both dicots and monocots provided an alternative approach for the induction of *IPT* and the concomitant biosynthesis of CK, without the negative effects of constitutively high CK content on plant phenology (*i.e.* owering time, plant architecture, *etc.*). *IPT* was expressed in the whole plant, its maximal expression was attained during the drought episode and the transgenic plants displayed enhanced drought tolerance and superior yields (Figure 5.5) as observed by Peleg and Blumwald (2011).

Figure 5.5: Effects of Water-stress at Pre-anthesis Stage on Growth of Rice Plants.

Brassinosteroids (BR)

BR was reported to induce the expression of stress-related genes, leading to the maintenance of photosynthesis activity, the activation of antioxidant enzymes, the accumulation of osmoprotectants, and the induction of other hormone responses.

Submergence Tolerance

Uncertainty of rainfall is a major factor affecting the rice yield in India, with flash flood affecting the plant stand seriously depending on duration of submergence stress which is considered as the third most important constraint to high yield in India (Sarkar *et al.*, 2009). Climate projections suggest that temperatures, precipitation and flooding, and sea level rise are likely to increase, with adverse impacts on crop yield and farm income in Southeast Asia (Unnikrishnan *et al.*, 2006; Wassmann *et al.*, 2009; INCCA, 2010). Rice in these areas is the major crop providing food for millions of subsistence farming families.

In rice, young rice seedlings after transplanting are particularly vulnerable to submergence stress (Joho *et al.*, 2008). The reproductive stage is the most sensitive to complete submergence, followed by the seedling and the maximum tillering stages (Reddy and Mittra, 1985). The stage most susceptible to partial submergence was the reductive division stage of the pollen mother cells followed by the heading stage, the spikelet differentiation stage and all parts of the reproductive stage (Matsushima, 1962). Flooding during the seedling stage, increasing the water depth inhibited the production of basal tillers and reduced tiller number, thereby decreasing eventual grain yield (Lockard, 1958). The reduction in yield has been attributed to a decrease in the proportion of ripened grains due to fertilization failure. Death in rice plants occurs when complete submergence lasts longer than 1-2 weeks (Palada and Vergara, 1972).

Submergence tolerance is defined as 'the ability of a rice plant to survive 10–14 d of complete submergence and renew its growth when the water subsides.

Flooding levels vary, depending on the amount and duration of rain, underlying geological formations and distance from the water. Among these factors, there are two typical kinds of flood. One is short duration over a few weeks and not very deep, termed a 'flash flood' and the other is deep flooding that lasts for a long time, called as 'deepwater flood'. Flash floods are unexpected and uncontrollable, and its flooding water level can reach 50 cm in the rainfed lowlands of the humid and semi-humid tropics of South and Southeast Asia. In these areas, flash floods at the seedling stage of rice cause severely reduced yields of rice grain (Hattori *et al.*, 2011).

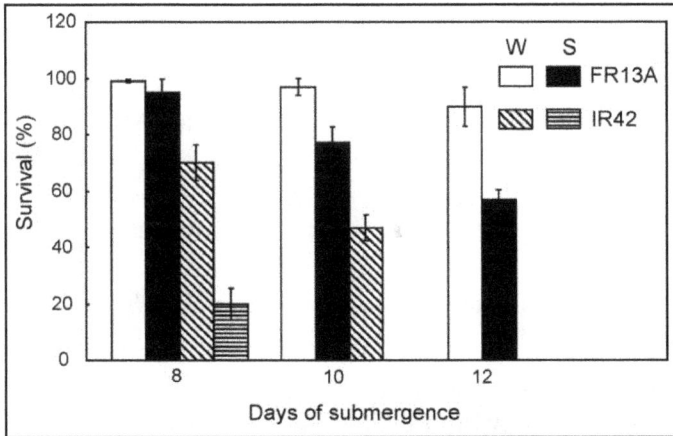

Figure 5.6: Survival of FR13A and IR42 after Submergence for 8, 10, and 12 d during the Winter (W) and Summer (S) Seasons.

The fast growth of cultivars adapted to deep water conditions seems to be mediated by the action of three different plant hormones, ethylene, GA and ABA. Under anaerobic conditions, ethylene concentration increases in plant tissue because of both increased synthesis and entrapment. This causes a reduction in ABA concentration with a concomitant increase in GA level as well as responsiveness, resulting in enhanced shoot elongation.

Pre-submergence stored carbohydrates are also reported to be associated with enhanced survival under flooded conditions, possibly by supplying the required energy for maintenance metabolism through anaerobic respiration.

Chlorophyll content of leaves decreased in all cultivars upon complete submergence, mainly because of ethylene-induced chlorophyll degradation, whereby ethylene is found to trigger gene expression and enzyme activity of chlorophyllase, the first enzyme involved in chlorophyll breakdown (Ella *et al.*, 2003) and hence reduce the potential for active CO_2 fixation during and after submergence. This finding also points to the importance of post-submergence carbohydrates for supplying the energy needed for regeneration growth.

The temperature of the floodwater ranged from 25.7 to 26.9°C in the morning and from 29.6 to 30.5°C in the evening. Carbon dioxide and oxygen dissolved in water showed opposite diurnal trends, with a higher concentration of CO_2 early in the morning than in the afternoon, whereas O_2 concentration was much higher in the afternoon than in the morning. Water depth did not seem to have significant effects on the concentration of the dissolved gases, but substantially affected light penetration. When calculated as the percentage of the total incident irradiance above the water surface, underwater light intensity decreased by about 25 percentage points just 5 cm below the water surface, and by an additional 16 percentage points with increasing depth from 5 to 30 cm and to a similar extent from 30 to 60 cm. A reduction in light intensity with increasing water depth will probably decrease underwater photosynthesis.

References

Abdellah, A., Boutraa T., Alhejely, A., 2011. The rates of photosynthesis, chlorophyll content, dark respiration, proline and abscicic acid (ABA) in wheat (*Triticum durum*) under water deficit conditions. *Int. J. Agric. Biol.*, 13(2): 215-221.

Boyer, J.S., Wong, S.C., Farquhar, G.D., 1997. CO_2 and water vapour exchange across leaf cuticle (epidermis) at various water potentials. *Plant Physiology*, 114: 185–191.

Bray, E., 1993. Molecular Responses to Water Deficit. *Pl. Phys.*, 103, 1035-1040.

Bray, E.A, 1993. Plant responses to water deficity. Trends Plant Sci., 2: 48-54.

Brugnoli, E., Lauteri, M., 1991. Effects of salinity on stomatal conductance, photosynthetic capacity, and carbon isotope discrimination of salt-tolerant (*Gossypium hirsutum* L.) and salt-sensitive (*Phaseolus vulgaris* L.) C_3 non-halophytes. *Pl. Physio.*, 95: 628–635.

Cameron, K.D., Teece, M.A., Smart, L.B., 2006. Increased Accumulation of Cuticular Wax and Expression of Lipid Transfer Protein in Response to Periodic Drying Events in Leaves of Tree Tobacco. *Pl.Phy.*, 140: 176-183.

Campbell, M.K., 1991. Biochemistry. Harcourt Brace Jovanovich College Publishers, Fort Worth, USA.

Centritto, M., Loreto, F., Chartzoulakis, K., 2003. The use of low [CO_2] to estimate diffusional and non-diffusional limitations of photosynthetic capacity of salt-stressed olive saplings. *Pl. Cell Env.*,26: 585–594.

Chaves, M.M., Flexas, J., Pinheiro, C., 2009. Photosynthesis under drought and salt stress: regulation mechanisms from whole plant to cell. *Ann. Bot.*, 103: 551–560.

Cornic, G., Le Gouallec, J.L., Briantais, J.M., Hodges, M., 1989. Effect of dehydration and high light on photosynthesis of two C_3 plants: *Phaseolus vulgaris* L. and *Elastostemarepens* (Lour.) Hall F. *Planta*, 177: 84–90.

Danyluk, J., Perron, A., Houde, M., Limin, A., Fowler, B., Benhamou, N., Sarhan, F., 1998. Accumulation of an Acidic Dehydrin in the Vicinity of the Plasma Membrane during Cold Acclimation of Wheat. *The Plant Cell Online*, 10: 623-638.

Dure, L., Greenway, S., Galau, G., 1981. Developmental Biochemistry of Cotton Seed Embryogenesis and Germination: Changing Messenger Ribonucleic Acid Populations as Shown by *in vitro* and *in vivo* Protein Synthesis. *Biochemistry*, 20: 4162-4168.

Ella, E.S., Kawano, N., Yamauchi, Y., Tanaka, K., Ismail, A.M., 2003. Blocking ethylene perception enhances ooding tolerance in rice seedlings. *Func. Plant Biol.*, 30: 813–819.

Farrant, J.M. 2000. A comparison of mechanisms of desiccation tolerance among three angiosperm resurrection plant species. *Plant Ecol.*,151: 29-39.

Flexas, J., Bota, J, Loreto, F., Cornic, G., Sharkey, T. D., 2004. Diffusive and metabolic limitations to photosynthesis under drought and salinity in C_3 plants. *Plant Biol.*, 6: 1-11.

Galmes, J., Medrano, H., Flexas, J., 2007. Photosynthetic limitations in response to water stress and recovery in Mediterranean plants with different growth forms. *New Phytologist*175: 81–93.

Genty, B., Briantais, J.M., Baker, N.R., 1998. The relationship between the quantum yield of photosynthetic electron-transport and quenching of chlorophyll fluorescence. *Biochimicaet Biophysica Acta*, 990: 87–92.

Goh, G.H., Nam, H.G., Park, Y.S., 2003. Stress memory in plants: a negative regulation of stomatal response and transient induction of rd22 gene to light in abscisic acid-entrained Arabidopsis plants. *Plant J.*, 36: 240-255.

Goldberg, R.B., Barker, S.J., Perez-Grau, L., 1989. Regulation of Gene Expression during Plant Embryogenesis. *Cell*, 56: 149-160.

Goyal, K., Tisi, L., Basran, A., Browne, J., Burnell, A., Zurdo, J., Tunnacliffe, A., 2003. Transition from Natively Unfolded to Folded State Induced by Desiccation in an Anhydrobiotic Nematode Protein. *The J. Bio. Chem.*, 278: 12977-12984.

Grime, J.P., 1979. Plant Strategies and Vegetation Process, Wiley, New York.

Hara, M., 2009. The Multifunctionality of Dehydrins: An Overview. *Pl. Signaling and Beh.*, 5(5): 503-508.

Hara, M., Fujinaga, M., Kuboi, T., 2004. Radical Scavenging Activity and Oxidative Modification of Citrus Dehydrin. *Pl. Phy. Biochem.*, 42(7-8): 657-662.

Hara, M., Fujigana, M., Kuboi, T., 2005. Metal Binding by Citrus Dehydrin with Histidine-Rich Domains. *J. Expt. Bot.*, 56(420): 2695-2703.

Hartung, W., Schraut, D., Jiang, F., 2005. Physiology of abscisic acid (ABA) in roots under stress - a review of the relationship between root ABA and radial water and ABA flows. *Aus. J. Agric. Res.*, 56: 1253-1259.

Hattori, Y., Nagai, K., Ashikari, M., 2011. Rice growth adapting to deepwater. *Curr. Opinion in Pl. Bio.*, 14:100–105.

Heyen, B.J., Alsheikh, M.K., Smith, E.A., Torvik, C.F., Seals, D.F., Randall, S.K., 2002. The Calcium-Binding Activity of a Vacuole-Associated, Dehydrin-Like Protein Is Regulated by Phosphorylation. *Pl. Physio.*, 130: 675-687.

Hoekstra, F.A., Golovina, E.A., Buitink, J., 2001. Mechanisms of Plant Desiccation Tolerance. *Trends in Pl. Sci.*,6: 431-438.

Horvatet, E., Pal, M., Szalai, G., Paldi, E., Janda, T., 2007. Exogenous 4-hydroxybenzoic acid and salicylic acid modulate the effect of short-term drought and freezing stress on wheat plants. *Biol. Plant.*, 51: 480-487.

INCCA (Indian Network for Climate Change Assessment): Climate change and India: a 4x4 assessment, a sectoral and regional analysis for 2030s. *Ministry of Environment and Forests, Government of India* 2010.

Irigoyen, J.J., Emerich, D.W., Sanchez-Diaz, M., 1992. Water stress induced changes in concentrations of proline and total soluble sugars in nodulated Alfala (*Medicago sativa*) plants. *Physiol. Plant.*, 84: 67-72.

Iuchi, S., Kobayashi, M., Taji, T., Naramoto, M., Seki, M., Kato, T., Tabata, S., Kakubari, Y., Yamaguchi-Shinozaki, K., Shinozaki, K., 2001. Regulation of drought tolerance by gene manipulation of 9-cisepoxycarotenoid dioxygenase, a key enzyme in abscisic acid biosynthesis in Arabidopsis. *Plant J.*, 27: 325-333.

Joho, Y., Omsasa, K., Kawano, N., Sakagami, J., 2008. Growth responses of seedlings in *Oryza glaberrima* Steud. to short-term submergence in Guinea, West Africa. *Japan Agricultural Research Quarterly* 42(3): 157–162.

Jones, H.G. 1973. Moderate-term water stresses and associated changes in some photosynthetic parameters in cotton. *New Phytologist*, 72: 1095–1105.

Kessler, B. 1961. Nucleic acids as factors in drought resistance of higher plants. *Recent Adv. Bot.* p. 1153-1159.

Kovacs, D., Agoston, B., Tompa, P., 2008. Disordered Plant LEA Proteins as Molecular Chaperones. *Pl. Sign. Beh.*, 3:710-713

Kudoyarova, G.R., Vysotskaya, L.B., Cherkozyanova, A., Dodd, I.C., 2007. Effect of partial rootzone drying on the concentration of zeatin-type cytokinins in tomato xylem sap and leaves. *J. Exp. Bot.*, 58(2): 161-168.

Lawlor, D.W., Cornic, G., 2002. Photosynthetic carbon assimilation and associated metabolism in relation to water deficits in higher plants. *Pl. Cell Environ.*, 25: 275-294.

Lockard, R.G., 1958. The effect of depth and movement of water on the growth and yield of rice plants. *Malayan Agric. J.*,41(4): 266-281.

Matsushima, S., 1993. Researches on the requirements for achieving high yields in rice. Matsuo T, Ishii R, Ishihara K, Hirata H, editors. Science of the Rice Plant, Vol 2. Physiology. Nobunkyo, Tokyo, pp 737–747

McKersie, B.D., Leshem, Y., 1994. Stress and stress coping in cultivated plants, *Kluwer Academic Publishers*, Netherlands.

Moller, I.M., Jensen, P.E., Hansson, A., 2007. Oxidative modifications to cellular components in plants. *Annu. Rev. Plant Biol.*, 58: 459-481.

Palada, M.C., Vergara, B.S., 1972. Environmental effects on the resistance of rice seedlings to complete submergence. *Crop Sci.*, 12:209-212.

Peleg, Z., Blumwalad, E., 2011. Hormone balance and abiotic stress tolerance in crop plants. *Curr. Opinion in Pl. Bio.*, 14:290–295.

Reddy, M.D., Mittra, B.N., 1985. Effects of complete plant submergence on vegetative growth, grain yield and some biochemical changes in rice plants. *Pl. Soil,*87: 365-374.

Renou, J.L., Gerbaud, A., Just, D., Andre, M., 1990. Differing substomatal and chloroplastic concentrations in water-stressed wheat. *Planta*, 182: 415–419.

Sarkar, R.K. and Panda, D., 2009. Distinction and characterisation of submergence tolerant and sensitive rice cultivars, probed by the fluorescence OJIP rise kinetics. *Func. Pl. Biol.*, 36: 222–233.

Sharma, P., Dubey, R.S., 2005. Drought induces oxidative stress enhances the activities of antioxidant enzymes in growing rice seedlings. *Pl. Growth Regul.,*46: 209-221.

Shigeoka, S., Ishikawa, T., Tamoi, M., Miyagawa, Y., Takeda, T., Yabuta, Y., Yoshimura, K., 2002. Regulation and function of ascorbate peroxidase isoenzymes. *J. Exp. Bot.*, 53: 1305-1319.

Smirnoff, N., 1993. The role of active oxygen in the response of plants to water deficit and dessication. *New Phytologist*, 125: 27 – 58.

Tambussi, E.A., Bartoli, C.G., Beltrano, J., Guiamet, J.J., Araus, J.L., 2000. Oxidative damage to thylakoid proteins in water-stressed leaves of wheat (*Triticum aestivum*). *Physiol. Plant.*, 108: 398-404.

Thompson, S., Wilkinson, S., Bacon, M.A., Davies, W.J., 1997. Multiple signals and mechanisms that regulate leaf growth and stomatal behaviour during water deficit. *Physiol. Plant.*, 100: 303-313.

Trewavas, A.J., Jones, H.G., 1991. An assessment of the role of ABA in plant development. In: Davies, W.J., Jones, H.G. (eds.), Abscisic Acid: Physiology and Biochemistry, pp: 169-188. BIOS Scientific Publishers Limited, Oxford, UK.

Tunnacliffe, A., Wise, M., 2007. The Continuing Conundrum of the LEA Proteins. *Naturwissenschaften*, 94, 791-812

Unnikrishnan, A.S., Kumar, R.K., Fernandes, S.E., Michael, G.S. and Patwardhan, S.K., 2006. Sea level changes along the Indian coast: Observations and projections. *Curr Sci.*, 90: 362–368.

Unyayar, S., Keles, Y., Unal, E., 2004. Proline and ABA levels in two sunflower genotypes subjected to water stress. *Bulg. J. Plant Physiol.*, 30: 34-47.

Wassmann, R., Jagadish, S.V.K., Heuer, S., Ismail, A.M., Redona, E., Serraj, R., Singh, R.K., Howell, G., Pathak, H., Sumfleth, K., 2009. Climate Change Affecting Rice Production: The Physiological and Agronomic Basis for Possible Adaptation Strategies. *Adv. Agron.*, 101: 60–122.

Yang, F., Xu, X., Xiao, X., Li, C., 2009. Responses to drought stress in two poplar species originating from different altitudes. *Biol. Plant.*, 53: 511-516.

6

Physiological Traits to Enhance Water Use Efficiency in Field Crops

M.K. Meena, B.M. Chittapur, M.R. Umesh and M.C. Naik

Water drives most of the physiological processes function, those are involved in plant growth and significant relationships exist between crop performance and many water-relations related traits such as leaf water potential, water uptake by roots, stomatal conductance, transpiration efficiency, osmotic adjustment, *etc.* Understanding of these relationships has permitted the identification of efficient tools that are used in plant selection for adaptation to water limited environments; including canopy temperature (CT) which is related to root depth and hydration status, and carbon isotope discrimination which –when measured on non-water stressed tissue– is related to intrinsic transpiration efficiency. These tools have also been applied in identifying QTLs for drought adaptation. Knowledge of water relations traits has also been used to identify complementary parents in breeding for improved adaptation of crops to water limited environments. Both CT and leaf conductance can be used as surrogates for measuring photosynthetic rate, and have application in breeding for irrigated environments, especially where yield is source limited, by heat stress, for example, water relations traits such as leaf water potential, relative leaf water content, root characteristics, and osmotic adjustment are generally too time consuming to be applied in routine breeding but are useful experimentally as accurate indicators of stress levels in field crops.

Since water drives most of the physiological processes involved in plant growth and development, measurement of plant water status can be a powerful means of assessing the adaptive potential of crop cultivars. Screening methodologies for water status involve measurement of physiological traits that integrate plant–

water relations such as: stomatal conductance (Gs), canopy temperature, carbon isotope discrimination, and relative water content, as well as direct measurement of leaf and soil water potential (Ψ leaf; Ψ soil), osmotic adjustment, and roots. The theoretical basis for these traits and their application in physiological point of view is discussed below.

Canopy Temperature

Many plant traits and environmental variables play roles in the energy balance of the plant canopy affecting its temperature. The main environmental variables are: (i) incident radiation which warms the plants directly –and which is obviously mitigated by cloud cover, (ii) soil moisture in the active root zone which determines the potential for transpiration rate and therefore evaporative cooling, (iii) wind which can increase transpiration rate through reducing boundary layers around plant structures –that would otherwise insulate the plant from atmospheric effects–, and(iv) relative humidity of the air which influences the vapor pressure deficit (VPD) between plant organ sand the air –warm, dry air being most conducive to evaporative cooling (Idso *et al.*, 1977). In terms of plant characteristics that determine genotypic differences in CT –in a given environment– the most important traits are: (i) the vascular system –of leaves, shoots and roots– which determines the capacity for transpiration,(ii) stomatal aperture which regulates transpiration rate and may be influenced by hormonal signals *i.e.*, from roots (Davies *et al.*, 2005), (iii) root depth which determines access to water –especially under drought–(Lopes and Reynolds, 2010), (iv) metabolism which if constrained for any reason (*e.g.*, by heat stress) will cause feedback inhibition of CO_2 fixation and therefore influence stomatal aperture, and (v) source–sink balance since a strong demand for assimilates (*e.g.*, from a large number of fast growing grains) will result in increased CO_2 uptake associated with larger stomatal conductance. Many studies have confirmed that CT is associated with crop yield (Blum *et al.*, 1982) as well as a range of physiological traits including stomatal conductance(Amani *et al.*, 1996), plant water status (Blum *et al.*, 1982), and deep roots. As such, CT is a versatile measurement that can complement breeding because it is highly integrative of the many physiological functions necessary to ensure adaptation to a given environment.

Stomatal Conductance

There is a strong relationship between stomatal conductance (Cs) and CT since stomatal conductance has a direct effect on transpirational cooling (Amani *et al.*, 1996). Therefore, both traits are affected by many of the same environmental and physiological factors. Under well-watered conditions, stomatal regulation maintains optimal levels of internal CO_2 concentration to feed the demand for CO_2 fixation from the Calvin cycle. However, under soil water deficit there will be a trade-off with the need to maintain a functional water status of leaves. Therefore, under such conditions differences in the VPD between canopy and atmosphere as well as chemical signals synthesized in dehydrating roots, mediate stomatal aperture and therefore water flux to the atmosphere. The closure of stomata may increase leaf temperature depending mainly on the radiation load on the canopy but will result in a better water economy or increased transpiration efficiency. Instantaneous assessment can be performed using leaf porometers, IR thermometers or infrared gas analyzers, while the use of isotopic discrimination gives an integration of the Gs over a growth period. Hand-held porometers assess the quantification of variability in Cs although that assessment is not as quick and integrative as CT due to a restriction in the number of leaves that can be measured. Therefore, under uniform canopy conditions, a single CT measurement provides a faster and more accurate estimation of the rate of transpiration and Gs of the whole plot. As mentioned, a close relationship exists between CT and Cs at moderate to high VPD conditions (Amani *et al.*, 1996). Oxygen isotopic composition in leaf and grain materials has also been recommended as an integrative way to estimate Cs under irrigated conditions.

Leaf Water Potential and Relative Water Content

Water status in plants can be defined by two main components: the content of water *per se*, and the energy status. One of the fundamental methodologies to assess energy status of plant water is water potential (Ψ). The Ψ is the water energy status at a determined time resulting from the combination of osmotic potential (Ψs), matric potential (Ψm) and pressure potential (Ψp). In most cases, Ψm is minimized and leaf water potential (LWP) can be expressed as: $\Psi = \Psi s + \Psi p$.

LWP values under irrigated conditions are typically between -0.5 and -1.1 MPa for crops. Mechanisms like leaf elongation or apical growth and spikelet formation begin to be affected by values of LWP around -1.0 MPa (Barlow *et al.*, 1977; Munns and Weir, 1981). Values of -1.2 MPa or lower water potential are associated with a reduced maximum leaf area, number of spikelets per spike and grains per unit land area. A decrease in transpiration rate of leaves and spikes was found in crops and barley as a consequence of low LWP. A decrease to minimum values of transpiration was observed in flag leaves and spikes when plants were at -2.8 and -3.1 MPa, respectively. Stress conditions have been found to induce embryo abortion, as well as affecting structures associated with both pollen sterility and embryo survival, when low LWP occurs during the critical period for grain yield determination. In addition, complete stomatal closure during grain-filling was observed in crops plants when exposed to conditions of -3.1 MPa (Frank *et al.*, 1973). RWC indicates the hydration status of plant tissue. It is expressed in relative terms as a percentage of maximum water content at full turgor. RWC is related to Ψ although both are dependent on the growth stages and water history of the plant (Hsiao, 1973). RWC in crops shows genetic variability and intermediate heritability, and also can be measured at low cost, however, it is time consuming to measure (Khan *et al.*, 2010). In the case of bread crops, additive gene effects were identified under different drought tolerant lines (Dhanda and Sethi, 1998).

Root System

Root systems determine the potential volume of soil that can be mined for water and nutrients. The interaction between environmental and genetic factors strongly determines the development of effective root systems. Kirkegaard *et al.* (2007) indicated that access to sub soil water (deeper than 1.2 m) after anthesis may increase grain yield by improving marginal water use efficiency. In crops like cereals, pulses, and oil seeds the root growth occurs during the period between germination and anthesis. In crops, root growth rate varies from 0.5 to 3.0 cm day^{-1} in primary and adventitious roots (Barley, 1970; Evans *et al.*, 1975). Vigorous and deeper root systems are recommended as one of the target traits for breeding cereals for improved water productivity in resource scarce conditions. However, extensive root systems also have higher respiration costs for plants. While genotypic variability exists in root length under favorable conditions, the benefits of a more extensive root system are more important when plants are exposed to drying soils (Blum, 2005). An alternative way to increase rooting depth and root distribution is to increase the period of root growth achievable by earlier sowing or the use of genotypes with delayed flowering. Disadvantages of these approaches are the possible negative effect of temperature (*e.g.*, frost in early sowing, or heat stress in delayed flowering) on grain yield. Methodologies for root trait evaluation under field conditions involve the use of coring methods (Nissen *et al.*, 2008) or uprooting plants to estimate root mass and architecture. Soil coring methods are a tedious and impractical job for large populations. Misinterpretation of root trait data in field studies may be related to: (i) heterogeneity of growth in the soil,(ii) physical and chemical impedance to root growth,(iii) the plasticity of roots in response to soil physical conditions, and (iv) the occurrence of biotic infections (*e.g.*, nematodes and fungal diseases).

Osmotic Adjustment

Osmotic adjustment is the process by which plants accumulate organic solutes in their cells to minimize water loss and maintain cell function as well as turgidity of cell under drought conditions. The value of Osmotic adjustment for stress tolerance has been demonstrated in C3 as well as C4 species (Blum *et al.*, 1999). The capacity of plants to accumulate solutes is one of the more important adaptive responses to drought at the cellular level. Osmotic adjustment has been identified as a mechanism to maintain grain yield under stressed conditions (*e.g.*, drought, salinity) by allowing root growth and maintaining water and nutrient capture (Turner, 1986). Osmotic adjustment is an inappropriate trait for field based evaluation due to confounding effects generated by genotypic differences in soil water exploration by roots. However, measurement of Osmotic adjustment in leaves under controlled conditions in a glasshouse can identify genetic variation. Blum *et al.* (1999) found that crops show a large expression of Osmotic adjustment in comparison to other species in response to drought treatments.

Carbon Isotope Discrimination

Carbon isotope discrimination (Δ13C) provides an integrative assessment of leaf transpiration efficiency in C_3 species (Farquhar *et al.*, 1982). It involves biochemical and physiological processes such as stomatal aperture, internal conductance of gases in plant tissue, and external environmental conditions that modify gas exchange and CO_2 fixation (Farquhar *et al.*, 1989). Carbon isotope discrimination (Δ13C) is a measure of the 13C/12C ratio in plant material relative to the value of the same ratio in the air on which plants feed, and its positive values reflect the C_3 discrimination

(Δ) against the heavier stable carbon isotope (13C) during photosynthesis. This favored discrimination of 13C is positively associated with carbon dioxide levels in the intercellular air spaces of the photosynthetic tissues and, given a constant tissue-to air VPD, is also positively associated to water uptake but negatively associated with TE. Greater overall stomatal aperture allows increased rates of leaf gas exchange, allowing the plant to favor 12C but with higher water losses. Carbon isotope discrimination has been identified as a useful trait to measure the changes in TE and CO_2 concentration at the inter cellular level of the leaf (Farquhar *et al.,* 1982) and also has been successfully used in breeding programs for improving productivity under water scarce conditions. In fact, Carbon isotope discrimination has been used as a selection tool and commercial varieties have already been released based on selection for high TE. Under favorable environments, high Carbon isotope discrimination (measured ingrains or biomass) is positively associated with grain yields. Genotypes with high Carbon isotope discrimination usually transpire more water to produce grain yield under well-watered conditions at the risk of using water with low efficiency. Therefore, selecting genotypes with low Carbon isotope discrimination (conservative water use) may help to improve productivity by increasing water use efficiency under rainfed conditions. Variability in the response of grain yield to CID can be observed under stressed conditions and it therefore depends on the phenological stage of the crops and the level of stress explored. In general, positive relationships were reported for CID and grain yield under drought conditions (Araus *et al.,* 1998). Under drought conditions, the high Carbon isotope discrimination can be interpreted as a higher capacity to sustain Cs, probably related to greater access to water at depth. Therefore, when crop growth is determined by access to water, high Carbon isotope discrimination can be associated with high yield. Alternatively, under environments where yield depends on the stored water, low Carbon isotope discrimination may be used to detect better genotypes (Richards *et al.,* 2001) due to conservative water use behavior.

Conclusion

There are different physiological approaches/traits for enhance water use efficiency in field crops ; the choice of which will be a function of breeding targets, scale of operation, and resources available. Within well-watered environments the main breeding targets are likely to be yield potential and/or heat tolerance. For these situations water use efficiency, photosynthetic capacity and efficiency are important traits to improve. Since photosynthesis is a time consuming trait to screen for using IR GAS analysis, both CT and Cs offer high-through put surrogates. Both traits show good association with photosynthesis and performance under hot, irrigated environments. Selection for drought targets should include traits that indicate a better use or capture of water, including root traits. However, current methodologies for screening root traits are still tedious and time consuming. On the other hand, CT estimates the relative water uptake capacity of roots much more efficiently than root. The selection of genotypes with more conservative water use (high TE) can be achieved by analyzing CID (of non-stressed tissue). In addition–though estimates are only feasible under controlled conditions– the evaluation of OA for determining differences in dehydration tolerance potential is also a useful trait for adaptation to limited 'stored soil water 'conditions.

References

Amani, I., Fischer, RA., Reynolds, MP. (1996) Canopy temperature depression association with yield of irrigated spring crops cultivars ina hot climate. *Journal of Agronomy and Crop Science* 176, 119–129.

Araus, JL., Amaro, T., Casadesus, J., Asbati, A., Nachit, MM. (1998) Relationships between ash content, carbon isotope discrimination and yield in durum crops. *Australian Journal of Plant Physiology* 25, 835–842.

Barbour, MM., Fischer, RA., Sayre, KD., Farquhar, GD. (2000) Oxygen isotope ratio of leaf and grain material correlates with stomatal conductance and grain yield in irrigated crops. *Australian Journal of Plant Physiology* 27, 625–637.

Barley, KP. (1970) The configuration of the root system in relation to nutrient uptake. *Advances in Agronomy* 22, 159–201.

Barlow, EWR., Munns, R., Scott, RK., Reisner, AH. (1977) Water potential, growth and polyribosome content of the stressed crops apex. *Journal of Experimental Botany* 28, 909–916.

Blum, A. (2005) Drought resistance, water-use efficiency, and yield potential: are they compatible, dissonant, or mutually exclusive? *Australian Journal of Agricultural Research* 56, 1159–1168.

Blum, A., Mayer, J., Gozlan, G. (1982) Infrared thermal sensing of plant canopies as a screening technique for dehydration avoidance in crops. *Field Crops Research* 5, 137–146.

Blum, A., Shipiler, L., Golan, G., Mayer, J. (1989) Yield stability and canopy temperature of crops genotypes under drought stress. *Field Crops Research* 22, 289–296.

Blum, A., Zhang, JX., Nguyen, HT. (1999) Consistent differences among crops cultivars in osmotic adjustment and their relationships to plant production. *Field Crops Research* 64, 287–291.

Davies, W., Kudoyarova, G., Hartung, W. (2005) Long-distance ABA signalling and its relation to other signalling pathways in the detection of soil drying and the mediation of the plant's response to drought. *Journal of Plant Growth Regulation* 24, 285–295.

Dhanda, SS., Sethi, GS. (1998) Inheritance of excised-leaf water loss and relative water content in bread crops (*Triticum aestivum*) *Euphytica* 104, 39–47.

Evans, LT., Wardlaw, IF., Fischer, RA. (1975) Crops. In: '*Crop physiology some case histories*' Evans, LT. (Ed.) Cambridge University Press, London,p. 374.

Farquhar, GD., Ehleringer, JR., Hubick, KT. (1989) Carbon isotope discrimination and photosynthesis. *Annual Review of Plant Physiology and Plant Molecular Biology* 40, 503–537.

Farquhar, GD., O'Leary, MH., Berry, JA. (1982) On the relationship between carbon isotope discrimination and the intercellular carbon dioxide concentration in leaves. *Australian Journal of Plant Physiology* 9, 121–137.

Frank, AB., Power, JF., Willis, WO. (1973) Effect of temperature and plant water stress on photosynthesis, diffusion resistance, and leaf water potential in spring crops. *Agronomy Journal* 65, 777–780.

Hsiao, TC. (1973) Plant responses to water stress. *Annual Review of Plant Physiology* 24, 519–570.

Idso, SB., Jackson, RD., Reginato, RJ. (1977) Remote-sensing of crop yields. *Science* 196, 19–25.

Jackson, RD., Reginato, RJ., Idso, SB. (1977) Crops canopy temperature: A practical tool for evaluating water requirements. *Water Resources Research* 13, 651–656.

Khan, HR., Paull, JG., Siddique, KHM., Stoddard, FL. (2010) Fababean breeding for drought-affected environments: A physiological and agronomic perspective. *Field Crops Research* 115, 279–286.

Kirkegaard, JA., Lilley, JM., Howe, GN., Graham, JM. (2007) Impact of subsoil water use on crops yield. *Australian Journal of Agricultural Research* 58, 303–315.

Munns, R., James, RA., Sirault, XRR., Furbank, RT., Jones, HG. (2010). New phenotyping methods for screening wheat and barley forbeneficial responses to water deficit. *Journal of Experimental Botany* 61, 3499–3507

Munns, R., Weir, R. (1981) Contribution of sugars to osmotic adjustment in elongating and expanded zones of wheat leaves during moderate water deficits at two light levels. *Australian Journal of Plant Physiology* 8, 93–105.

Nissen, T., Rodrigues, V., Wander, M. (2008) Sampling soybean roots: A comparison of excavation and coring methods. *Communications inSoil Science and Plant Analysis* 39, 1875–1883.

Richards, RA., Rebetzke, GJ., Watt, M., Condon, AG., Spielmeyer, W., Dolferus, R. (2001) Breeding for improved water productivity in temperate cereals: phenotyping, quantitative trait loci, markers and the selection environment. *Functional Plant Biology* 37, 85–97.

Royo, C., Villegas, D., Garcia del Moral, LF., Elhani, S., Aparicio, N., Rharrabti, Y., Araus, JL. (2002) Comparative performance of carbon isotope discrimination and canopy temperature depression as predictors of genotype differences in durum wheat yield in Spain. *Australian Journal of Agricultural Research* 53, 561–569.

Turner, N. (1986) Adaptation to water deficits: a changing perspective. *Australian Journal of Plant Physiology* 13, 175–190.

7

Hydrophilic Polymers: As Water Holding Capacity Enhancer for Maximizing Crop Yield in Agriculture

M.K. Meena, U.K. Shanwad and M. Chandranaik

During the 20th century, the main emphasis of agricultural development all over the world was the increasing productivity per unit area of land used for crop production to feed the ever-increasing population. At present time the use of hydrophilic polymers in agricultural and horticultural sector is gaining popularity. This has provided solutions to the problems of the present day agricultural and horticultural sector which is to maximize the land and water productivity without threatening the environment and the natural resources. The uses of hydrophilic polymers potentially influence soil permeability, density, structure, texture, evaporation and infiltration rates of water through the soils. Many arid and semi-arid regions are facing the problems of uncertain and inadequate rain fall. Spatially diversified soil characteristics, shortage of large agricultural lands and underprivileged condition of farmers do not allow them to adopt advantageous and economical application of traditional irrigation methods as well as micro-irrigation techniques (drip and sprinkler irrigation).

Though, not much research in India has been undertaken on the use of hydrophilic polymers in agriculture as well as in horticultural crops, the researchers world over (specifically western countries of the world) have extensively worked on utilizing hydrophilic polymers for increasing water use efficiency and enhancing crop yield. Various studies have strongly recommended that soil conditioning with hydrophilic polymers could be an innovative facet in the field of agriculture as well as in horticulture, which works as miniature water storage reservoirs. Research evidences suggest that problems associated with traditional micro irrigation and the factors which are catalyst in practicing efficient irrigation techniques can be taken care of by conditioning the soil with hydrophilic polymers. Better water management can be attained with the application of hydrophilic polymers and considerable water saving can be done without compromising the crop yield. In this regard detailed information about the soil application of hydrophilic polymers in the field of agriculture and horticulture especially for the small and marginal farmers living under arid and semi arid regions, provided highlighting latest research findings inclination towards the various parameters like improving soil health, water retention capacity of soil and effects on plant growth and development.

Hydrophilic Polymers in Agriculture as Biocides, Herbicides and Molluscicides

Synthetic polymers play important role in agricultural uses as structural materials for creating a climate beneficial to plant growth *e.g.* mulches, shelters or green houses; for fumigation and irrigation, in transporting and controlling water distribution. However, the principal requirement in the polymers used in these applications is concerned with their physical properties; such as transmission, stability, and permeability or weather ability; as inert materials rather than as active molecules.

The polymeric biocide has many advantages and its potential benefits include: it allows lower amounts than conventional biocides to be used as it releases the required amount of active agent over a long period, number of applications is reduced because of long period of activity by a single application, it eliminates the time and cost of repeated over applications because less active materials is needed, reduction of toxicity, it eliminates the need for widespread distribution of large amount of biocide levels in the surrounding environment, reduction of evaporation and degradation loses by environmental forces or leaching by rain in to the soil or waterways due to the macromolecular nature, it extends the duration of activity of less or non-persistent biocides which are unstable under an aquatic environment by protecting them from environmental degradation and hence, enhances the practical applicability of these materials, reduction of phytotoxicity by lowering the high mobility of the biocides in the soil and hence reduces its residue in the food web, extension of herbicide selectivity to additional crops by providing a continuous amount of herbicide at a level sufficient to control weeds but without injury to the crop.

The main problem with the use of less persistent conventional herbicides that have greater specificity is the use of excess amounts than that actually required to control the herb because they are unstable in an aquatic environment and of the need to compensate the amount wasted by the environmental forces of photodecomposition, leaching and washing away by rain. They are also highly toxic to farm workers and expensive on multiple applications which are required because of their lower persistence. On the other hand, the applications of large amounts of persistent herbicides are undesirable because of their frequent incorporation into the food chain. The major drawback to the economic use of these polymeric herbicides is the large amount of inert control that must be employed as a carrier for the herbicides and the disposal of the herbicide residual materials which may be harmful for the soil and the plant. Hence the use of excessive amounts of such bioactive polymers usually necessary for herb control is inevitable.

Basics and Science behind Hydrophilic Polymers

Mohammad *et al.* (2010) have experimentally concluded that hydrophilic polymer materials are hydrophilic networks that can absorb and retain huge amounts of water or aqueous solutions. They can uptake water as high as 100,000 per cent. Common hydrophilic polymers are generally white sugar-like hygroscopic materials that can be used for agricultural purpose. Francesco *et al.* (2008) have found that in agricultural field, polymers are widely used for improving irrigation efficiency; hydrophilic polymer material has smart delivery systems. It can help the agricultural industry to combat viruses and other crop pathogens. Functionalized polymers were used to increase the efficiency of pesticides and herbicides, allowing lower doses to be used and to indirectly protect the environment through filters or catalysts to reduce pollution and clean-up existing pollutants.

Water Absorption Capacity of Hydrophilic Polymers

The use of hydrophilic polymers, particularly under green house condition has shown that they have great potential to hold water and release slowly for crop growth and development. Polymeric soil conditioners are known since the 1950s. Hydrophilic polymers work as water conservative reservoirs, which works near the root mass zones of the plant; On watering, these polymers expand to around 200-400 times of its original volume, further irrigation water or rainwater can be collected, stored, and then released gradually for crop requirements over a fairly long duration between water applications. The authors also further conclude that polymers when mixed with the soil structure create good air permeability in soil, improve water absorption property of soil and fertilizer conservation capacity hence economy in irrigation can be attained. It also decreases loss of fruits and vegetables caused by the insects by 10-30 per cent. Further, Fonteno and Bilderback (1993) have also noted that hydrophilic polymers can be used to improve agricultural areas as it can absorb and store upto 400 times their own weight of water. Similar experiments by Abd El-Rehim *et al.* (2004) reveals that the water retention of sandy soils may improve considerably increasing the plant performance on those soils. On small scales, these products might be useful to enhance rainfall retention in the soil, especially on slopes that allow little runoff. The polymers absorb water as it infiltrates through the soil.

Hydrophilic Polymers for Water Conservation in Agriculture

The presence of water in soil is essential for vegetation. Liquid water ensures the feeding of plants with nutritive elements, which makes it possible for the plants to obtain a better plant growth rate. It seems to be interesting to exploit the existing water potential by reducing the losses of water and also by ensuring better living conditions for vegetation. Taking into account of hydrophilic polymer (luquasorb), the possibilities of its application in agricultural field has increasingly been investigated to alleviate certain agricultural problems like water stress condition.

Table 7.1: Influence of Hydrophilic Polymer (Luquasorb) on per cent Seedling Establishment in Tomato

Treatment	Days after Transplanting				
	20	40	60	80	At Harvest
HP@0.50 g/plant	97.7 (81.28)	97.7 (81.28)	94.4 (76.31)	94.4 (76.31)	94.4 (76.31)
HP@ 0.75 g/plant	97.7 (81.28)	97.7 (81.28)	95.5 (77.75)	95.5 (77.75)	95.5 (77.75)
HP@ 1.00 g/plant	97.7 (81.28)	97.7 (81.28)	95.5 (77.75)	95.5 (77.75)	95.5 (77.75)
HP@ 1.25 g/plant	98.9 (83.98)	98.9 (83.98)	96.6 (79.37)	96.6 (79.37)	96.6 (79.37)
HP@ 1.50 g/plant	100.0 (90.00)	100.0 (90.00)	97.7 (81.28)	97.7 (81.28)	97.7 (81.28)
HP@ 1.75 g/plant	100.0 (90.00)	100.0 (90.00)	98.9 (83.98)	98.9 (83.98)	98.9 (83.98)
Control	96.6 (79.37)	96.6 (79.37)	93.3 (75.00)	93.3 (75.00)	93.3 (75.00)
Mean	98.5 (83.98)	98.5 (83.98)	96.0 (78.46)	96.0 (78.46)	96.0 (78.46)
CD (0.05)	NS	NS	2.8	2.8	2.8

Figures in the parenthesis are arcsine transformed values.

HP= Hydrophilic polymer; NS =Non Significant (Meena *et al.*, 2011).

Fidelia and Chris (2001) have studied that improved water retention capacity of soil can be achieved by the amendment of hydrophilic polymers with various percentage of dosage in to the soil structure by weight. Water retention capacity of soil can be increased by 50 per cent to 70 per cent. Experimental investigations on soil carried out by Yangyuoru, *et al.* (2006) conditioned with different treatments of polymers suggests that extensive retention of water by the hydrophilic polymers was observed which would otherwise lost due to evaporation and percolation. The study suggested that soil moisture increased by 6.20 - 32.80 per cent with HP application, while soil bulk density was reduced by 5.50–9.40 per cent relative to the control, especially with a moderate water deficit when the relative soil moisture contents were about 40–50 per cent. Further, experimental studying of saturated soil water content, saturated soil hydraulic conductivity and soil water diffusivity of the soil with hydrophilic polymers application, it was found that saturated water volumetric content increased significantly up to 0.186cm^3, which is meaningful for improving agricultural water use efficiency in the arid and semi-arid areas. Both saturated hydraulic conductivity and diffusivity had a significant reduction mainly because hydrophilic polymers repeatedly absorbed and desorbed water. The water absorbing and desorbing capacity of hydrophilic polymers showed a downward trend with time and out side water condition. Under a relatively stable water condition, such capacity of hydrophilic polymers reduced more slowly. Therefore, the saturated water volumetric content of the mixed hydrophilic polymers and soil samples decreased gradually but were still higher than or close to that of the ordinary soil, and the hydraulic conductivity and diffusivity increased gradually but still respectively lower than those of the ordinary soil.

Soil Application of Hydrophilic Polymer and Seedling Establishment and Survival

Seed germination and seedling development are critical phases in early growth and establishment of any plant species, which unequivocally depends on moisture availability in the soil. The Hydrogel hydrophilic amendments improved seedling growth and establishment by increasing water retention capacity of soil and regulating the plant available water supplies. Wallace and Wallace (1986) observed the significant increase in the rate of emergence of tomato and lettuce seedlings with increasing concentration of polyacrylamide (hydrophilic) polymer in glasshouse. Similarly, Saleh and Hussein (1987) studied the beneficial effects of incorporating polymeric material (Aquastore) into sandy soil and reported that the seedling emergence of wheat cultivars was earlier and associated with a higher germination percentage. Significant increase in seedling survival was also reported in cucumber (Al-Harbi *etal.*, 1999), barley and wheat (Akhter *et al.*, 2004) and in wheatgrass (Mengold and Shelay, 2007), when soil was treated with hydrogel as compared to control.

Increasing Irrigation Efficiency by Soil Application of Hydrophilic Polymers

Among the resources increasingly involved in man-made agricultural activities, water is certainly one of the most precious sources affecting plant growth and development. Nearly three quarters of the water used all over the world are being consumed for agricultural needs in each and every country. Hence, there is an urgent necessity to optimize its use. Among the recent initiatives to achieve better use of water supply is the introduction of hydrophilic polymer which when applied into the soil absorbs and retains large quantities of water and releases absorbed water, thereby allowing the plants to have water available at wilting point. Hossein *et al.* (2010) suggests that with the use of hydrophilic polymers in irrigation practice, water stress significantly alter in decreasing the number of leaves per plant, chlorophyll content, seed yield and water use efficiency. Whereas the application of super absorbent polymer moderated the negative effect of deficit irrigation, especially in high rates of polymer (2.25 and 3 g/kg of soil). Polymer have the best effect to all characteristics of the plants in all levels of water stress treatment and it is strongly suggested that the irrigation period of cultivation can be increased by application of polymer. In terms of water conservation and optimize water use efficiency where water scarcity is a common problem, hydrophilic polymers can be used as a water conservator in agriculture as suggested by Sannino (2008). Another study on the effects of hydrophilic polymers by Shooshtarian, *et al.* (2011) on the physical characteristics of soil and plant species (flowers and ground cover plants, turf grasses, trees and shrubs) it was found that reaction and amount of irrigation efficiency enhanced after using HP for green spaces within arid and semi arid regions. Flannery and Busscher (1982) studied the use of synthetic polymer (Permasorb) in potting soils to improve water holding capacity in ryegrass and reported that permasorb (6.4g/l) significantly reduced the watering frequency by increasing water holding capacity of soil. The mean number of watering was significantly reduced (20.3) in soil treated with 6.4g/l Permasorb as compared to control (32.5).

Table 7.2: Influence of Hydrophilic Polymer (Luquasorb) on Yield and Yield Parameters of Tomato

Treatment	Fruits/Plant	Fruit Yield (kg/plant)	Fruit Yield (t/ha)	Fruit Volume (cm³/plant)
HP @ 0.50 g/plant	80.2	4.65	28.5	4772.0
HP @ 0.75 g/plant	84.3	4.77	31.3	4934.7
HP @ 1.00 g/plant	85.3	4.94	32.7	5130.3
HP @ 1.25 g/plant	88.7	4.96	34.2	5231.0
HP @ 1.50 g/plant	91.5	5.19	35.9	5836.3
HP @ 1.75 g/plant	92.8	5.61	36.6	6111.0
Control	78.5	4.45	26.9	4270.3
Mean	85.9	4.94	32.3	5183.7
CD (0.05)	7.30	0.20	2.03	411.46

Meena *et al.*, 2011.

Soil Application of Hydrophilic Polymers to Reduce Drought Stress

Drought continued to be a recurring phenomenon worldwide and scarcity of water imposed by drought is affecting plant growth and productivity. Although sometimes, water is present in the soil, becomes unavailable to the plant growth and development due to various reasons. To grow and to fulfill the plant functions, they need not only fertile subsoil, but appropriate amount of water, too. The amount and distribution of rainfall largely influence the success of rainfed crops. An effective and planned utilization of available water or rainfall has therefore, become one of most essential factors, in Indian agriculture specific to vegetables such as tomato. In this direction hydrophilic polymers which retain more moisture and nutrients with potentiality of releasing them the use as and when required would be advantageous for proper plant establishment and helps to obtain quality produce of tomato (Sendur kumaran *et al.*, 2000).

In the study as the effect of super absorbent polymer and different levels of irrigation on characteristics of the plants it was observed that the effect of interaction between super absorbent and water stress was significant and a major decrease was observed with increase in stress level (Atiyeh and Ebrahim, 2013). Further the authors studied other parameters such as the highest plant height, number and surface area of the leaf, fresh and dry weight and they found that diameter of plant was related to irrigation after 4.0 days with 3.0 per cent polymer application and the lowest was related to irrigation after 12 days with no application of polymer. The study indicated that using hydrophilic polymers controlled relationship soil, water and plant, decrease water stress; drought stress cause molecular damage to plants, either directly or indirectly through formation of reactive oxygen species (ROS). Alteration of antioxidant enzymes activities is an element in the defense process.

Changes in antioxidant enzymes activities and seed yield in the plants were investigated by Nazarli *et al.* (2011) under drought stress and super absorbent synthetic polymer application and suggested that application of polymers could be advantageous against drought stress, and could protect plants in drought stress conditions. After adding HP to the soil of pots in deepness of root development and applying drought stress in four different leaf stage the results indicated that water stress significantly decreased the number of leaves per plant, chlorophyll content, dry weight, relative water content and Water Use Efficiency, whereas the application of HP can also compensate the negative effect of drought stress, especially in high rates of polymer application (0.2 (per cent) and 0.4(per cent) (g/kg)).It is strongly suggested by Leila *et al.* (2012) that the irrigation intervals could be increased by application of HP. Property of high water retention capacity and protection against drought was also observed by Nazarli *et al.* (2011). As referred above, drought stress leads to production of Oxygen radicals, which result in increased lipid perioxidation and oxidative stress in the plant, but with the use of HP could reserve different amount of water in itself and ultimately increase the soil ability of water retention and preserving and at last in water deficiency as concluded in the experimental study of Tohidi, *et al.* (2009). Considerable decrease in drought stress was also observed during another study by Javad *et. al.* (2011) in silage corn.

Hydrophilic Polymers to Enhance Fertilizer Efficiency

Robiul *et al.* (2011) concluded that optimum use of fertilizer, herbicides and germicides is also considered to be one of the major constraints in designing efficient irrigation technology. While making effort in arid and semiarid regions of northern China, there was an increasing interest in using reduced rate of chemical fertilizer alongwith HP for field crop production by evaluating the effectiveness of different rates of HP, it was observed that application by 11.2 per cent under low 18.8 per cent under medium and 29.2 per cent under high rate with only half amount (150 kg ha^{-1}) of fertilizer compared with control plants, which received conventional standard fertilizer rate (300 kg ha^{-1}). At the same time plant height, stem diameter, leaf area, biomass accumulation and relative water content as well as protein and sugar contents in the grain also increased significantly following HP treatments. The authors also suggested that the application of HP at 15 kg ha^{-1} plus only half the amount of conventional fertilizer rate (150 kg ha^{-1}) would be a more appropriate practice for sustainable crop production under arid and semiarid conditions and the regions with the similar ecologies; moreover, polymers are safe and non-toxic, it also reduce excessive nutrient loss from soil thereby preventing pollution of agro-ecosystem.

Biodegradability of Hydrophilic Polymers in Soil

While studying about the soil amendment elements their post consequences must be known to the users as it may be absorbed by the root and ultimately penetrates in to the fruits and any crops and may cause noxious effect on the consumers. Studies carried out by Fidelia, *et al.* (2012) have proven that HP is sensitive to the action of ultraviolet rays, which by breaking bonds; degrade the polymer into oligomers(molecules of much smaller size). These polyacrylates thus becomes much more sensitive to the aerobic and anaerobic processes of microbiological degradation, therefore degrade naturally in soils (up to 10 to 15 per cent per year), in water, carbon dioxide, and nitrogen compounds. The polyacrylates is too voluminous to be absorbed into the tissues and walls of plants that it has no potential for bioaccumulation. It is a ideal solution for containers, hanging plants, and houseplants, and it has also shown its effectiveness in large-scale farming, especially at the time of germination and development of the root net work due to good aeration of the soil.

Soil Application Hydrophilic Polymers and Yield of Crops

In water stress condition, yield level can be sustained or enhanced by using new advanced initiatives like hydrophilic polymer. The differences in yield levels with the use of hydrophilic polymer and farmers practices showed greatest potential of achieving higher productivity in agricultural as well as horticultural crops. In agricultural system, hydrophilic polymers have been used either as soil amendments or seed coatings. In either case, results have been inconsistent. Tomato, radish and eggplant have shown increased yields, when grown in polymer media (Matro, 1963 and King and Jenson, 1973). Increased fresh weights of chrysanthemum and bean grown in polymer-amended soil have been reported (Jensen and Eikhof, 1971). Baasiri *et al.* (1986) studied the influence of Aquastock (polymer) on yield of cucumber and reported that the cucumber yield was significantly increased from 0 to 2 kg/m³, though further yield increase up to 4 kg/m³, but the increase over 2 kg was non-significant, when polymer was applied to a depth of 20 cm into the soil. They found similar trend in fruit number. Silberbush *et al.* (1993) studied the effects of hydrophilic polymer (Agrosoak®) on water storage and availability in maize and tomato grown in sand dunes and concluded that agrosoak compensated for low water application and increased dry weight of plants increasing yields which is probably due to moisture stored by hydrophilic polymer. In addition to increasing yield in cotton and maize, the time required to produce yield was significantly reduced in the presence of polymer (El-Sayed *et al.*, 1995). They also reported that the yield increased significantly with an increase in hydrogel polymer concentration in both species. The maximum yield production in presence of polymer was ranged from 1.8-3.3 times the level measured in the absence of hydrogel. The effects of hydrophilic gel polymer on the yield of barley studied by Volkamar and Chang (1995) showed that grain yield of barley increased by 15 per cent and biomass by23 per cent by a polymer @ 1.87 g/kg soil, which was due to either more grains per spike or larger grains. El-Hady and Wanas (2006) studied the effects of hydrogel on water and fertilizer use efficiency in cucumber under water stress condition and concluded that at same irrigation amount, higher the application rate of hydrogel, higher is the marketable yield produce. On the other hand, reducing irrigation amount from 100 per cent to 85 per cent of crop water requirement has caused an increase in marketable yield of cucumber relative to that of control plants equal to 38.6, 54.2 and 78.3 per cent when 2, 3 and 4 g of hydrogel crystals were incorporated, respectively.

Soil Application Rate of the Hydrophilic Polymers

Considering the above discussion it can be imagined that more the hydrophilic polymers mixed in the soil, more would be the water retention and improved soil moisture level of soils and releasing the same slowly to crop plant during the crop growth period. To justify the cost of HP material and to be achieving benefit cost ratio in irrigation practice; optimum quantity of the hydrophilic polymers would require to be understood. Various studies have suggested to use proper concentration of hydrophilic polymers indifferent proportion basically depending upon the type of soil structure and other properties of soil, type of atmospheric condition, under which it is being practiced, type of crop to be irrigated and mostly the quality of irrigation water to be utilized for the irrigation. Various researchers have suggested different rate of hydrophilic polymers to be applied as soil conditioner are listed in the Table 7.3.

Table 7.3

Sl.No.	Name of the Researchers	Suggested Dosage for Hydrophilic Polymers (HP)	Recommendation for Usage
1.	Huttermann *et al.* (1995)	0.4 per cent to 0.6 per cent (g/kg)	To increase the Pinus seedling survival
2.	Volkamer and Chang(1995)	1.87 g/plant	To increase the root volume and biomass
3.	Shooshtarian *et al.* (2011)	4.0 to 6.0g/kgof soil	For crops growing in arid and semi arid areas of the world
4.	Sendur K *et al.* (2001)	1.50g/plant	To increased processing quality of tomato
5.	Hossein *et al.* (2010)	2.25 to 3.0g/kg of soil	For all level of water stress treatment and improved irrigation period
6.	Flannerand Busscher (1982)	6.4g/l of water	To reduced the watering frequency by increasing water holding capacity of soil
7.	Altarawneh, *et al.* (2012)	0.2, 0.4 and 0.8 per cent soil	To delay permanent wilting in sandy soil
8.	El-Hady and Wanas (2006)	2.0 to 4.0 g/plant	To save irrigation frequency upto15 per cent
9.	Liyuan and Yan (2013)	0.5 to 2.0 g/pot	To improve relative water content and increased leaf water use efficiency
10.	Leila *et al.* (2012)	0.2 to 0.4 per cent (g/kg)	To moderate the adverse effects of drought stress on horticultural crops

Sl.No.	Name of the Researchers	Suggested Dosage for Hydrophilic Polymers (HP)	Recommendation for Usage
11.	Atiyeh and Ebrahim (2013)	3.0 per cent (g/kg)	To decrease water stress
12.	Meena, *et al.* (2011)	1.75g/plant in soil	To improve and maintained relative water content and delay permanent wilting symptoms in tomato
13.	El-Sayed *et al.* (1995)	5-20 per cent (g/kg)	To increase the plant height in cotton and maize
14.	Meena, *et al.* (2011)	1.75g/plant in soil	To reduce the irrigation frequency upto 50 per cent in tomato
15.	Meena, *et al.* (2011)	1.75g/plant in soil	To increase the seedling establishment upto 100 per cent in tomato

Conclusion

Since irrigation water is a limiting factor in the country; it is important to improve the water use efficiency of the plants. The use of water retaining polymers has potential for horticultural and other crops. Throughout human civilization history, agriculture has been a main source of food, fuel and fiber. Opportunities have arisen through external events and trends that impacted patterns of production and utilization. When working in this field, we always deal with water, aqueous media and bio-related systems. Thus, we increasingly walk in a green area becoming greener via replacing the synthetics with the bio-based materials, *e.g.*, polysaccharides and polypeptides. Uncertainty of rainfall, increased temperature in arid and semi-arid regions is prominent throughout the world which encourages efficient water conservative irrigation technology of Hydrophilic polymers. The technique has a great water absorption capacity by its own weight which helps in improving soil moisture capacity and hence reduces water stress on the plant during prolonged drought stress condition and during irrigation intervals. Quantity of total water required for the irrigation is also reduced by 20 to 50 per cent when the soil conditioning by hydrophilic polymers in different proportion is adopted. A hydrophilic polymer is safe, bio degradable, non toxic and inert with increased shelf life and lasts for years (15 to 20 per cent degradation/year is observed) in the soil actively. Moreover, better aeration in the root zone enhances germination, root development and microbial activities. Reduced use of fertilizers and pesticides adds to the overall benefits and the quality of yield. This technology is widely unheard of to the farmers due to various reasons and therefore calls for special attention. Availability of agricultural hydrophilic polymers from the market is difficult for the ordinary farmers. These polymers were developed to improve the physical properties of soil in view of: Increasing their water-holding capacity, Increasing water use efficiency, enhancing soil permeability and infiltration rates, reducing irrigation frequency, reducing compaction tendency, stopping erosion and water run-off and increasing plant performance especially in structure -less soils in areas subject to drought.

References

Akhter, J., Mahmood, K., Malik, K. A., Mardan, A., Ahmad, M., Iqbal, M. M., 2004. Effects of hydrogel amendment on water storage of sandy loam and loam soils and seedling growth of barley, wheat and chickpea. *Plant Soil Environment*, **50**(10): 463-469.

Al-Harbi, A. R., Al- Omran, Shalaby, A. A., Choudhary, M. I., 1999. Effeficacy of a hydrophilic polymer declines with time in greenhouse experiments.*Hortic. sci.*34:223-224.

Alessandro Sannino, 2008. Application of Superabsorbent Hydrogels for the optimization of water resources in agriculture, The 3rd International Conference on Water Resources and AridEnvironments (2008) and the 1st Arab Water Forum,

Altarawneh, A., Kreuzig, R., Batarseh, M., Bahadir, M., 2012. Impacts of Soil Amended Superabsorbent Polymers on the Efficiency of Irrigation Measures in Jordanian Agriculture, The 4th International Congress Eurosoil, Bari, Italy.

Atiyeh Oraee and Ebrahim Ganji Moghadam, 2013. The effect of different levels of irrigation with superabsorbent (S.A.P) treatment on growth anddevelopment of Myrobalan (*Prunus cerasifera*) seedling, African J. Agril. Res., 8(17): 1813-1816.

Baasiri, M., Ryan, J., Mucheik, M., Harik, S.N., 1986, Soil application of a hydrophilicconditioner in relation to moisture, irrigation frequency and crop growth. *Commun. inSoil Sci. Plant anal.*, **17**(6): 573-589.

Dhumal, K. N., 1993, Effect of Jalshakti on growth and yield of some vegetables under water stress conditions. *J. Maharashtra Agric. Univ.*, **18**: 307-311

El-Hady O.A., Wanas, Sh.A., 2006. Water and Fertilizer Use Efficiency by Cucumber Grown under Stress on Sandy Soil Treated with Acrylamide Hydrogels, J. Applied Sci. Res., 2(12): 1293-1297.

El-Sayed, H., Kirkwood, R.C., Graham, H.R., 1995, Studies on the effects of salinity and hydro gel polymer treatments on the growth, yield production, solute accumulation in cotton and maize. *J. King Saudi Univ. Agric. Sci.*, 7(2): 209-227.

Flannery, R.L., Buscher, W. J., 1982, Use of a synthetic polymer in potting soils to improve water holding capacity. *Commun. in Soil Sci. Plant anal.*, 13(2): 103-111.

Fidelia, N. Nnadi, 2012. Super Absorbent Polymer (HP) and Irrigation Water Conservation, Irrigation and Drainage Systems Engineering, Indian Irr. Drainage Sys. Engg. 1:1, 2012

Fidelia Nnadi, Chris Brave, 2011. Environmentally friendly superabsorbent polymers for water conservation in agricultural lands, J. SoilSci. Envl. Manag. 2(7): 206-211

Fonteno, W.C., Bilderback, T.E., 1993. Impact of Hydrogel on Physical Properties of Course- Structured Horticultural Substrates, Journal American Society of Horticultural Sciences 118: pp. 217-222,

Abd El-Rehim, H. A., Hegazy, E.S.A., Abd El-Mohdy, H.L., 2004. Radiation synthesis of hydrogels to enhance sandy soils water retentionand increase plant performance, *J. App. Polymer Sci.*, 93: 1360-71

Francesco Puoci, Francesca Iemma, Umile Gianfranco Spizzirri, Giuseppe Cirillo, Manuela Curcio, Nevio Picci, 2008. *American J. Agril. Bio. Sci.*, 3 (1): 299-314,

Hossein Nazarli, Mohammad Reza Zardashti, Reza Darvishzadeh, Solmaz Najafi, (2010). The Effect of Water Stress and Polymer onWater Use Efficiency, Yield and several Morphological Traits of Sunflower under Greenhouse Condition,*Nat Sci. Biol.* 2 (4), 53-58,

Huttermann, A., Zommordi, M., Reise, K., 1990, Addition of hydrogels to soil forprolonging the survival of pinus seedlings subjected to drought. *Soil Tillage Research*, 50: 295-304.

Javad Khalili Mahalleh, Hossein Heidari Sharif Abad, Gorban Nourmohammadi, Farrokh Darvish, Islam Majidi Haravan and EbrahimValizadegan, 2011. Effect of superabsorbent polymer (Tarawat A200) on forage yield and qualitative characters in corn under deficit irrigation condition in Khoy Zone (Northwest of Iran), *Adv. Env. Bio.*, 5(9): 2579-2587.

Jensen, M.H., Eikhoff, R.H., 1971, A hydrophilic polymer as a soil amendment. *Proc.10th Nat. Agric. Platics Conf.* 10:69-79.

King, P.A, Jenson, M.H., 1973 The influence of insolublbisised poly ethylene in the soil water matrix effects on vegetable crops, *Proc. 11th Nat. Agric. Platics Assoc.* 11: 106-116.

Leila Keshavars, Hasan Farahbakhsh and Pooran Golkar, (2012). The effects of drought stress and super absorbent polymer on morpho-physiological traits of pearmillet. (*Pennisetum glaucum*). *Int. Res. J. App. Basic Sci.* 3 (1): 148-154.

Liyuan Yan, Yan Shi, 2013. Effects of Super Absorbent Resin on Leaf Water Use Efficiency and Yield in Dry-land Wheat, Advance Journal ofFood Science and Technology 5(6): 661-664.

Yangyuoru, M., Boateng, E., Adiku, S.G.K., Acquah, D., Adjadeh, T.A., Mawunya, F., 2006. Effects of Natural and Synthetic Soil Conditionerson Soil Moisture Retention and Maize Yield, West Africa Journal of Applied Ecology (WAJAE) –ISSN: 0855-4307,

Matro, G.J., 1963. A study of the moisture status of tomato transplants treated with certain proprietary compounds. *M.S. thesis, ratager state univ. U.S.A.*

Meena, M.K., Nawallagatti, C.M., Chetti, M.B. 2011. Impact of hydrophilic polymer on irrigation requirement and biophysical parameters in tomato. *Int. J.Agril. Sci.*,7(2):424-428.

Nazarli, H.F. Faraji, Zarsashti, M.R., 2011. Effect of Drought Stress and Polymer on Osmotic adjustment and Photosynthetic Pigments of Sunflower. Cercetãri Agronomice în Moldova, 44:1 145

Nazarli, H., Zardashti, M.R., Darvishzadeh, R., Mohammadi, M., 2011. Change in activity of antioxidative enzymes in young leaves of sunflower (*Helianthus annuus* L.) by application of super absorbent synthetic polymers under drought stress condition, African J. Crop Sci. 5(11): 1334-38.

Robiul Islam, Xuzhang Xue, M., Sishuai Mao, Xingbao Zhao, Egrinya Eneji, A., Hu, Yuegao, 2011. *African J. Biotech.* 10(24): 4887-4894.

Saleth, R. Maria, 1996. Water Institutions in India: Economics, Law and Policy, Commonwealth Publishers, New Delhi,

Shooshtarian, S., Abedi-Kupai, J., Tehrani Far, A., 2011. Evaluation of application of Superabsorbent polymers in green space of arid and semi-arid regions with emphasis on Iran. *J. Biodiv. Ecol. Sci.*,1(4): 9287.

Silberbush, M., Adar, E., De-Malach, Y., 1993, Use of hydrophilic polymer to improvewater storage and availability to crops grown in sand dunes II. Tomato irrigatedsprinkling with different water salinities. *Agricultural Water Management*, **23**:303-313.

Sendur Kumaran, S., Natrajan, S., Muthvel, I., Sathiayamurthy, V.A., 2001. Standardization of hydrophilic polymers on growth and yield of tomato. *J. Madras Agric.*, **88**(1-3): 103-105

Tohidi, H.R., Shirani-Rad, A.H., Nour-Mohammadi, G., Habibi, D., Marshhadi-Akbar-Boojar, M., 2009. Effect of super absorbent application on antioxidant enzyme activities in Canola (*Brassica napus* L.) cultivars under water stress conditions. *American J. Agril. Bio. Sci.*, 4(3): 215-223.

Volkamar, K.M., Chang, C., 1995, Influence of hydrophilic gel polymers on waterrelations, growth and yield of barley and canola. *Can. J. Plant Sci.*, **75**: 605-611.

Waham Ashaier Laftah, Shahrir Hashim, Akos N. Ibrahim 2011. Polymer Hydrogels: A Review, *Polymer-Plastics Tech. Engg.*, 50: 1475–1486.

8

Recent Developments in Water Management in Rice

B.G. Masthana Reddy

Rice is the staple food of over 3 billion people in the world (FAOSTAT). Rice cultivation is the principal activity and source of income for more than 100 million households in developing countries in Asia, Africa and Latin America. World food production during 2013-14 was 473.2 m. tonnes and by 2020 the requirement would be around30-50 per cent of the current demand.

Rice is an important crop of India grown on an area of 44 million ha with a production of 106 million tonnes. In TungaBhadra command it is grown on an area of 3.5 Lakh ha. Since, 2000, world rice production has been almost stagnant with no substantial increase. The world population continues to grow steadily, while land and water resources are declining and rice need to be produced under such situations to feed the growing population.

Rice Ecosystems

Depending on water source and depth of water, rice is cultivated mainly in four eco systems.

1. Irrigated Lowland Rice

It is most extensive and productive ecosystem. Land is prepared by intensive pudding and transplanting of 25-30 days age seedlings is done. However, direct wet seeding is also practiced. Standing water of 2-5 cm is maintained. Midseason drainage is practiced. It account for 58 per cent of total area, and about 45 per cent irrigation potentials used and it contributes to 75 per cent of total rice production.

Water requirement includes land preparation (150 to 250 mm) seepage and percolation (200-700 mm) and midseason drainage (50-100 mm). Total water requirement varies from 900-2250 mm.

2. Rainfed Lowland Rice

Rice area lack water supply and water control for irrigation. Area is subjected either for drought or flood. Salinity is a problem in coastal area where sea water submerge in rice area. Bunding of field to retain water is done. Crop is established by direct seeding or transplanting. Fields remained submerged for most part of the crop. Yields are highly variable depending on salinity stress.

3. Rainfed Upland Rice

Grown under aerobic conditions without irrigation and puddling. Practices include field preparation in summer and seeds are broad cast or drill sown is lines. Drought, weed competition and low external input limit the yields. Only one crop is raised during rainy season.

4. Deep Water Rice

Deep water rice is sown directly at on aerobic rice and is subjected to water stagnation at later stages of crop growth. Low external inputs and lower yields are characteristic.

Major water uses of low land rice are

☆ Transpiration losses

☆ Evaporation

☆ Combined seepage and percolation

☆ Puddling and drainage.

On an average it requires about 1432 l of evapo-transpired water to produce 1kg of rough rice (Bouman, 2009). When accounted for ET, seepage and percolation losses on field scale it requires about 2500 litres irrespective of sources.

Rice Water Productivity

Water productivity represents the weight of grains produced per unit of water.

Depending upon the water balance components involved water productivity can be defined as:

1. Grain yield per unit of water evapotranspired (WP_{ET})
2. Grain yield per unit of total field water supply(WP_{IR})

Tuong (1999) and Bouman and Toung (2001) reported that at field level WP_{ET} under low land rice varies from 0.4 to 1.6g grain kg^{-1} water and WP_{IR} is about 0.2 to 1.2 g grain kg^{-1} water.

Moisture Stress Effects at different Growth Stages

Water requirement of rice is low at seedling stage. Immediately after transplanting sufficient water should be provided to facilitate early rooting. After that shallow water depth facilitates tiller production and root anchorage. Rice is most sensitive to water stress from 20 days before heading to 10days after heading. Panicle primordial development, booting, seedling and flowering are critical stages for moisture. Drought at this stage causes panicle sterility caused by impeded panicle formation, heading, flowering or fertilization. The ripening period which includes milk, dough, yellowish and grain ripening requires less water.

Different Water Management Practices in Rice

1. Flooding

Water is delivered to fields surrounded by bunds on at all sides. Usually a depth of 5-10 cm standing water is maintained throughout to crop growth period. This is practiced whether water is available in plenty or unlimited water such as casual supplies. Conveyance and application losses are more. The overall efficiency is around 30-35 per cent water requirement on average is around 1300-1500 mm.

2. Saturated Soil Culture (SSC)

In SSC the soil is kept as close to saturation as possible, thereby reducing the hydraulic head of ponded water which decreases the seepage and percolation flows. It means that a shallow irrigation is given to obtain about 1cm of ponded water a day or so after disappearance of ponded water. Bouman and Tuong (2001) found that water input decreased on average by 23 per cent from continuously flooded check with no significant reduction in yield.

3. Raised Beds

Effective in keeping soil around saturation. Rice plats are grown on beds (120 cm wide) and water in the furrows (30 cm wide) is kept close to surface of beds. Compared to flooded rice water saving was 34 per cent and yield losses were 16-34 per cent.

4. Alternate Welting and Drying

In alternate wetting and drying (AWD) irrigation water is applied to obtain flooded conditions after a certain number of days have passed after disappearance of ponded water. The number of days of non-flooded soil before irrigation is applied can vary from 1to10 days. Advantages of AWD include improved rooting system, reduced lodging, periodic soil aeration and better control of certain diseases. Bouman and Toung (2001) reported that 92 per cent of AWD treatments resulted in yield reduction of 0-70 per cent compared to flooded rice. They reported that AWD resulted in increased water productivity with respect to total water input, because reduction in water input is larger than reduction in yield. In lowland rice areas with heavy soils and shallow ground water table in China and the Philippines Lampayan *et al.* (2005) reported that in AWD water inputs decreased by 15-30 per cent without significant yield reduction. However in areas with loamy and sandy soils with deeper ground water table in India, Sharma *et al.* (2002) reported reduction in water inputs by 50 per cent but observed yield losses up to 20 per cent compared to flooded check.

5. System of Rice Intensification (SRI)

Alternate wetting and drying (AWD) is the water management practice adopted in SRI. Important characteristics of SRI includes Transplanting of 8-12 days old seedlings, transplanting of single seedling, wider spacing (25 cm x 25 cm) weeding by churning soil through rotary weeder and use of compost. Studies at Agricultural Research Station Gangavati, Karnataka revealed that due to adoption of AWD as water management practice SRI resulted in 34 per cent savings in water compared to traditional continuous submergence.

6. Direct Seeded Rice (DSR)

Direct seeded rice is the method of cultivating rice wherein dry seeds are directly sown into dry soil or pre germinated seeds are broad cast onto the wet puddled soil. The crop is irrigated without submergence during the growth period. Advantages of DSR includes

☆ Nursery raising is eliminated

☆ Puddling is not done and considerable saving in land preparation

☆ Transplanting costs are eliminated

☆ Less requirement of seeds

☆ Since AWD method of irrigation is practiced savings in water up to 30 per cent is possible

☆ Yields are similar to transplanted rice

Studies at Agricultural Research Station, Gangavati, Karnataka revealed that water requirement of DSR to be about 950 mm as compared to 1300 mm into the call of puddle transplanted.

7. Micro Irrigation in Rice

Micro irrigation is practiced under conditions where water is not sufficient to grow rice under submergence conditions. There are two important micro irrigation system in rice

i. Sprinkler Irrigation

Sprinkler irrigation is a method of applying irrigation water which is similar to rainfall where water is sprayed into the air through sprinklers so that it breaks up into small water drops which fall to the ground. Irrigation water requirement is 20-50 per cent less than flooded rice and depends on soil type, rainfall and water management. In Brazil, Stone *et al.* (1999) reported that using new aerobic rice varieties and using sprinklers at row spacing of 20cm and N supply at 90 kg/ha a grain yield of 5-6 t/ha had been recorded.

ii. Drip Irrigation

Drip irrigation is a method that allows water to drip slowly to the roots of the plants either on to the soil surface or directly into the root zone though a network of valves, pipes, tubing and emitters. At Madurai in sandy clay soil drip irrigation scheduled at 150 per cent evaporation replenishment from USWB pan evaporimeter produced a grain yield of 5.62 t ha^{-1}. Vijaylaxmi (2010) reported scheduling drip at 200 per cent PE recorded comparable yield as that of flood irrigation 5days after disappearance of ponded water. Similarly Medleys and Wilson (2008) observed 17.3 per cent higher yield (8.36 t ha^{-1}) with subsurface drip in comparison to flooded rice at Texas.

Advantages of Micro Irrigation

Potential water savings under sprinkle, surface and subsurface drip by scheduling irrigation to water crop water requirements (ETc), elimination of seepage, percolation and runoff losses.

1. Elimination of free water evaporation
2. Higher capacity to capture incident rainfall
3. Direct seeding in dry soil eliminates completely water resources for puddling
4. Controlled frequency of application of water
5. Fertilizers and insecticides can be applied trough fertigation
6. Easier harvesting
7. Favour multiple cropping and conservation agriculture
8. Automation is possible, and
9. Enhanced water productivity

Large scale adoption of drip irrigation has not been possible owing to high system cost and lot of research is yet to be carried out to reduce the system cost before it is adopted on field scale.

9

Water Productivity and Tillage, Crop Residue Management

M.N. Thimmegowda and B.K. Ramachandrappa

Rainfed agriculture in India covers 58 per cent of the cultivated area, supports 40 per cent human population and 2/3rd of livestock. 87 per cent coarse cereals, 85 per cent food legumes, 72 per cent oilseeds, 65 per cent cotton and 44 per cent rice are contributed to food bowl in the country by the rainfed ecosystem. The fragile ecosystem is characterized by diverse climate, soils and cropping systems, poor economic status of farmers, application of low external inputs and low investment (Ramachandrappa *et al.*, 2014). In sub-Saharan Africa, more than 95 per cent farm land, 90 per cent in latin America, 60 per cent in south Asia, 65 per cent in East Asia are under rainfed ecosystem (Wani *et al.*, 2009).

Resource degradation is an important aspect of rainfed production system because of aberrant rainfall and frequent droughts. The rainfed lands suffer from a number of biophysical and socio-economic constraints which affect productivity of crops and livestock. These include low and erratic rainfall, land degradation and poor productivity, low level of input use and technology adoption, low draft power availability (Mayande and Katyal, 1996), inadequate fodder availability, low productive livestock (Singh, 1988) and resource poor farmers and inadequate credit availability. Nearly 73 m ha is affected by water erosion (61 per cent of the total degraded area) characterized with shallow soils, poor water holding capacity, sub-surface hard pan, surface crusting, sealing and cracking, very low soil organic matter and wide spread nutrient deficiencies, low productivity and biomass return to soil.

Intensive tillage practices employing inversion implements such as mould board plough results in loss of surface crop residue and subsequent loss of soil organic carbon from soil aggregates. The burning of fossil fuels for generating energy for draft power in tillage operations also contributed for the greenhouse gasses and climate change under intensive agriculture system. Crop residues in India are either used for feeding livestock or burnt on the field. The residues especially in sugarcane, cotton, partly in rice, wheat *etc.* are burnt on the field and exasperating the greenhouse gasses in to the atmosphere contributing to the climate change. The combination of intensive tillage, imbalanced fertilization and poor crop residue management resulted in soil quality deterioration leading to low productivity in rainfed regions (Roldan *et al.*, 2003).

The climate change impacts were witnessed with vagaries of weather in terms of temperature and rainfall anomalies, which are having larger impacts on rainfed production system directly and in the long range on soil health through rapid degradation of organic matter. Practices such as zero/no tillage or reduced tillage, green manuring, crop residue recycling *etc.* proved effective in improving soil fertility and productivity in many parts of the world (Unger, 1990) under both irrigated and rainfed ecosystems. The impacts are more profound in rainfed ecosystem than irrigated situations. Perhaps, conservation agriculture is gaining momentum in this context.

Conservation agriculture (CA) defined as minimal soil disturbance (no-till, NT) and permanent soil cover (mulch) combined with rotations, is a recent agricultural management system that is gaining popularity in many parts of the world. It is an approach to managing agro-ecosystems for improved and sustained productivity, increased profits and food security while preserving and enhancing the resource base and the environment. The key features for conservation agriculture are:

☆ Minimum soil disturbance by adopting zero/minimum tillage and reduced traffic for agricultural operations

☆ Leave and manage the crop residues on the soil surface

☆ Adopt spatial and temporal crop sequencing/rotation, green manures, intercrops to derive maximum benefits from inputs and minimize adverse environmental impacts.

Conservation Tillage

The history of tillage dates back to many millennia when humans changed from hunting and gathering to more sedentary and settled agriculture mostly in the Tigris, Euphrates, Nile, Yangste and Indus river valleys. Reference to ploughing or tillage is found from 3000 BC in Mesopotamia (Hillel, 1998). Lal (2007) explained the historical development of agriculture with tillage being a major component of management practices. With the advent of the industrial revolution in the nineteenth century, mechanical power and tractors became available to undertake tillage operations; today, an array of equipment is available for tillage and agricultural production. There is no doubt that this list of tillage was beneficial to the farmer, but at a cost to him and the environment, and the natural resource base on which farming depended.

Tractors consume large quantities of fossil fuels that add to costs while also emitting greenhouse gases majorly CO_2 and contributing to global warming when used for ploughing. Animal-based tillage systems are also expensive since farmers have to maintain and feed a pair of animals for a year for this purpose. Animals also emit methane, a greenhouse gas 21 times more potent for global warming than carbon dioxide (Grace *et al.*, 2003). Zero-tillage reduces these costs and emissions. Farmer surveys in Pakistan and India show that zero-till of wheat after rice reduces costs of production by US$60 per hectare mostly due to less fuel (60–80 l ha^{-1}) and labour (Hobbs and Gupta 2004).

Since the 1930s, the farming community has been advocating a move to reduced tillage systems that use less fossil fuel, reduce run-off and erosion of soils and reverse the loss of soil organic matter. The first 50 years was the start of the conservation tillage (CT) movement and, today, a large percentage of agricultural land is cropped using these principles. Baker *et al.* (2002) defined CT as the collective umbrella term commonly given to no-tillage, direct-drilling, minimum-tillage and/or ridge-tillage, to denote that the specific practice has a conservation goal of some nature. Usually, the retention of 30 per cent surface cover by residues characterizes the lower limit of classification for conservation-tillage, but other conservation objectives for the practice include conservation of time, fuel, earthworms, soil water, soil structure and nutrients. Thus residue levels alone do not adequately describe all conservation tillage practices.

Conservation tillage is a set of practices that leave crop residues on the surface which increases water infiltration and reduces erosion. It is a practice used in conventional agriculture to reduce the effects of tillage on soil erosion. However, it still depends on tillage as the structure forming element in the soil. Nevertheless, conservation tillage practices such as zero tillage practices can be transition steps towards Conservation Agriculture.

Lal (2007) reported the benefits of zero tillage *viz.*, erosion control, water conservation, soil fertility enhancement, carbon sequestration *etc.* are directly attributed to the amount of crop residue mulch and application of dung/manure as soil amendment. Camara *et al.* (2003) studied the long term effects of tillage, N and rainfall on winter wheat yield and found that despite beneficial effects on soil properties, conservation or low tillage tended to be less productive for wheat than mould board ploughing due to lack of downy brome weed control in low tillage. David Kwaw-Mensah and Mahdi Al-Kaisi (2006) observed that maize yield and above ground biomass response to tillage systems were not significantly different for all N rates and sources. Although, much efforts have gone into such studies in temperate regions, systematic studies in rainfed semi-arid tropical regions are rare especially in developing countries because of difficulties in controlling weeds, less water infiltration in compacted soil and non-availability of appropriate seeding equipments (Sharma *et al.*, 2008).

The experiments on conservation tillage at AICRP for dryland agriculture at Bengaluru center indicated no significant beneficial effect of deep tillage on yield of crops other than pigeonpea. The yields of finger millet were reduced due to deep tillage. Deep tillage using mould board plough reduced the weed incidence. Pigeonpea and maize crops have responded to deep tillage, the weed growth was low in deep tillage treatment and moisture storage was higher. Across the crops, light disc appears to be appropriate to get maximum grain yield. The tillage treatments did not significantly influence the pH and electrical conductivity of soils (Ramachandrappa *et al.*, 2013).

Tillage and current agricultural practices result in the decline of soil organic matter due to increased oxidation over time, leading to soil degradation, loss of soil biological fertility and resilience. Although this SOM mineralization liberates nitrogen and can lead to improved yields over the short term, there is always some mineralization of nutrients and loss by leaching into deeper soil layers. This is particularly significant in the tropics where organic matter reduction is processed more quickly, with low soil carbon levels resulting only after one or two decades of intensive soil tillage. Zero-tillage, on the other hand, combined with permanent soil cover, has been shown to result in a build-up of organic carbon in the surface layers (Lal, 2005). No-tillage minimizes SOM losses and is a promising strategy to maintain or even increase soil C and N stocks (Bayer *et al.*, 2000).

Higher bulk densities and penetration resistance have been reported under zero-tillage compared with conventional tillage (Gantzer and Blake, 1978). This problem is greater in soils with low-stability soil aggregates (Ehlers *et al.*, 1983). Bautista *et al.* (1996) working in a semi-arid ecosystem found that zero-tillage plus mulch reduced bulk density. The use of zero-till using a permanent residue cover, even when BD was higher, resulted in higher infiltration of water. Minimum tillage was found superior with respect to nutrient availability.

Residue Management

In tropical countries like India, after harvest, the crop residue is removed from the soil surface and used as fodder for livestock or fuel for domestic cooking.Unger *et al.* (1988) reviewed the role of surface residues on water conservation and indicated that this association between surface residues, enhanced water infiltration and evaporation led to the adoption of CT after the 1930's dust bowl problem. Bissett and O'Leary (1996) showed that infiltration of water under longterm (8-10 years) conservation tillage (zero and sub-surface tillage with residue retention) was higher compared to conventional tillage on a grey cracking clay and a sandy loam soils.

Kumar and Goh (2000) concluded that crop residues of cultivated crops are a significant factor for crop production through their effects on soil physical, chemical and biological functions as well as water and soil quality. They can have both positive and negative effects.

The energy of raindrops falling on a bare soil result in destruction of soil aggregates, clogging of soil pores and rapid reduction in water infiltration with resulting runoff and soil erosion. Mulch intercepts this energy and protects the surface soil from soil aggregate destruction, enhances the infiltration of water and reduces the loss of soil by erosion (Dormaar and Carefoot, 1996). NT plus mulch reduces surface soil crusting, increases water infiltration, reduces run-off and gives higher yield than tilled soils (Thierfelder *et al.*, 2005). Similarly, the surface residue, anchored or loose, protects the soil from wind erosion (Michels *et al.*, 1995). The dust bowl is a useful reminder of the impacts of wind and water erosion when soils are left bare.

Surface mulch helps reduce water losses from the soil by evaporation and also helps moderate soil temperature. This promotes biological activity and enhances nitrogen mineralization, especially in the surface layers (Hatfield and Pruegar, 1996). This is a very important factor in tropical and sub-tropical environments but has been shown to be a hindrance in temperate climates due to delays in soil warming in the spring and delayed germination (Swanson and Wilhelm, 1996). Fabrizzi *et al.* (2005) showed that NT had lower soil temperatures in the spring in Argentina, but NT had higher maximum temperatures in the summer, and that average temperatures during the season were similar.

Karlen *et al.* (1994) showed that normal rates of residue combined with zero-tillage resulted in better soil surface aggregation, and this could be increased by adding more residues. A cover crop and the resulting mulch or previous crop residue help reduce weed infestation through competition and not allowing weed seeds the light often needed for germination. There is also evidence of allelopathic properties of cereal residues in respect to inhibiting surface weed seed germination (Jung *et al.*, 2004). Weeds will be controlled when the cover crop is cut, rolled flat or killed. Farming practice that maintains soil micro-organisms and microbial activity can also lead to weed suppression by the biological agents (Kennedy, 1999).

Soil microbial biomass (SMB) has commonly been used to assess below-ground microbial activity and is a sink and source for plant nutrients. Amendments such as residues and manures promote while burning and removal of residues decrease SMB (Alvear *et al.*, 2005). Increased SMB occurs rapidly in a few years following conversion to reduced tillage (Alvarez and Alvarez, 2000). Increased SBM increased soil aggregate formation, increased nutrient cycling through slow release of organically stored nutrients and also assisted in pathogen control (Carpenter-Boggs *et al.*, 2003). Ground cover promotes an increase in biological diversity not only below ground but also above ground; the number of beneficial insects was higher where there was ground cover and mulch and these help keep insect pests in check.

Crop production in the next decade will have to produce more food from less land by making more efficient use of natural resources and with minimal impact on the environment. Only by doing this, food production will keep pace with demand and the productivity of land be preserved for future generation. Conservation tillage and ground cover with more than 30 per cent can be considered conservation agriculture helps for reducing soil erosion, runoff and evaporation, improve soil physical, chemical and biological properties of soil in long term to enhance soil productivity under changed climatic situations of rainfed ecosystem.

References

Alvarez, C.R., Alvarez, R., 2000, Short term effects of tillage systems on active soil microbial biomass. *Biol. Fertil. Soils.*31: 157–161. doi:10.1007/s003740050639.

Alvear, M, Rosas, A, Rouanet, J.L., Borie, F., 2005, Effects of three soil tillage systems on some biological activities in an Ultisol from southern Chile. *Soil Tillage Res.* 82: 195–202.doi:10.1016/j.still.2004.06.002.

Baker, C.J., Saxton K.E., Ritchie W.R., 2002, *No-tillage seeding: science and practice.* In CAB International 2ndEdn.Oxford, UK.

Bautista, S., Bellot, J., Ramon-Vallejo, V., 1996, Mulching treatment for post fire soil conservation in semi-arid eco-systems. *Arid Soil Res. Rehabil.*, 10: 235–242.

Bayer, C., Mielniczuk, J., Amado. T.J.C., Martin-Neto, L., Fernandes, S.V., 2000, Organic matter storage in a sandy loam Acrisol affected by tillage and cropping systems in southern Brazil. *Soil Tillage Res.*, 54: 101–109. doi:10.1016/S0167-1987(00)00090-8

Bissett, M.J., O'Leary, G.J., 1996, Effects of conservation tillage on water infiltration in two soils in south-eastern Australia. *Aust. J. Soil Res.*,34: 299–308. doi:10.1071/SR9960299.

Carpenter-Boggs, L., Stahl, P.D., Lindstrom, M.J., Schumacher, T.E., 2003, Soil microbial properties under permanent grass, conventional tillage, and no-till management in South Dakota. *Soil Tillage Res.*71:15–23. doi:10.1016/S0167-1987(02)00158-7.

Dormaar, J.F., Carefoot, J.M., 1996, Implication of crop residue and conservation tillage on soil organic matter. *Canadian J. Pl. Sci.* 76: 627–634.

Ehlers, W., Kopke, U., Hess, F., Bohm, W., 1983, Penetration resistance and root growth of soils in tilled and untilled loess soil.*Soil Tillage Res.*,3: 261–275. doi:10.1016/0167-1987(83)90027-2

Fabrizzi, K.P., Garcia, F.O., Costa, J.L., Picone, L.I., 2005, Soil water dynamics, physical properties and corn and wheat responses to minimum and no-tillage systems in the southern Pampas of Argentina. *Soil Tillage Res.* 81: 57–69. doi:10.1016/j.still.2004.05.001.

Gantzer, C.J., Blake, G.R., 1978, Physical characteristics of LaSeur clay loam following no-till and conventional tillage. *Agron. J.*,70: 853–857.

Grace, P.R, Harrington L, Jain M.C., Robertson, G.P., 2003, *Long-term sustainability of the tropical and subtropical rice–wheat system: an environmental perspective.* In: Improving the productivity and sustainability of rice–wheat systems: issues and impact Ladha J.K, Hill J, Gupta R.K, Duxbury J, Buresh R.J ASA special publications 65 2003. pp. 27–43. Eds. Madison, WI:ASA.

Hatfield, K.L., Pruegar, J.H, 1996, Microclimate effects of crop residues on biological processes.*Theor. Appl. Climatol.*, 54: 47–59. doi:10.1007/BF00863558.

Hillel, D., 1991, *Out of the earth: civilization and the life of the soil.* New York, NY

Hillel, D., 1998, *Environmental soil physics.*Academic Press, San Diego, CA

Hobbs, P.R., Gupta, R.K., 2004, *Resource conserving technologies for wheat in rice–wheat systems.*In: Improving the productivity and sustainability of rice–wheat systems: issues and impact Ladha J.K, Hill J, Gupta R.K, Duxbury J, Buresh R.J paper 7, ASA special publications vol. 65 pp. 149–171. Eds. Madison, WI, US.

Jung, W.S, Kim, K.H, Ahn, J.K, Hahn, S.J., Chung, I.M., 2004,Allelopathic potential of rice (*Oryza sativa* L.) residues against *Echinochloacrusgalli.Crop Protect.* 23: 211–218. doi:10.1016/j.cropro.2003.08.019.

Kennedy, A.C., 1999. Soil microorganisms for weed management. *J. Crop Prod.* 2, 123–138. doi:10.1300/J144v02n01_07.

Kumar, K., Goh, K.M., 2000. Crop residues and management practices: effects on soil quality, soil nitrogen dynamics, crop yield and nitrogen recovery. *Adv. Agron.* 68, 198–279.

Lal, R., 2007. Managing world soils for food security and environmental quality. *Adv. Agron.* 74: 155–192.

Lal, R., 2005. Enhancing crop yields in the developing countries through restoration of the soil organic carbon pool in agricultural lands. *Land Degrad. Dev.* 17, 197–209. doi:10.1002/ldr.696.

Mayande, V.M and Katyal, J.C., 1996. Low Cost Improved Seeding Implements for Rainfed Agriculture. *Technical Bulletin. 3*, CRIDA, Hyderabad p. 26.

Michels, K., Sivakumar, M.V.K., Allison, B.E., 1995., Wind erosion control using crop residue. I. Effects on soil flux and soil properties. *Field Crops Res.* 40, 101–110. doi:10.1016/0378-4290(94)00094-S.

Ramachandrappa, B.K., Shankar, M.A., Dhanapal, G. N., Sathish, A., Jagadeesh, B.N., Indrakumar, N., Balakrishna Reddy, P.C, Thimmegowda, M.N, MaruthiSankar, G.R., CH. Srinivasa Rao, Murukanappa, 2014. *Four Decades of Dryland Agricultural Research for Alfisols of Southern Karnataka (1971-2010)*, Pub: Directorate of Research, UAS, Bangalore. p 308.

Ramachandrappa, B.K., Shankar, M.A., Sathish, A., Alagundagi, S.C., Surakod, V.S., Thimmegowda, M.N., Shirahatti, M.S., Guled, M.B., Khadi, M.B., Jagadeesh, B.N., Ch. Srinivasa Rao, 2013. *Rainfed technologies for Karnataka*AICRP for Dryland Agriculture, UAS, Bangalore and Dharwad. p 115.

Singh, R.P., 1988. Dryland Agriculture Research in India. Pages 136-164 in 40 years of Agricultural Research and Education in India, ICAR, New Delhi, India,

Swanson, S.P., Wilhelm, W.W., 1996. Planting date and residue rate effects on growth, partitioning and yield of corn. *Agron. J.* 88, 205–210.

Thierfelder, C., Amezquita, E., Stahr, K., 2005. Effects of intensifying organic manuring and tillage practices on penetration resistance and infiltration rate. *Soil Tillage Res.* 82: 211–226. doi:10.1016/j.still.2004.07.018.

Unger, P. W., Langdale, D. W., Papendick, R. I., 1988. Role of crop residues—improving water conservation and use. *Cropping strategies for efficient use of water and nitrogen*, vol. 51 (ed. W. L. Hargrove), pp. 69–100. Madison, WI:American Society of Agronomy.

10

Role of Liquid Organic Manures in Fertigation and Crop Production

Satyanarayana Rao, B.K. Desai and R. Venkanna

Sustainable agriculture is the broad spectrum of production method that is supportive to environment. In rainfed and irrigated eco-systems, water, nutrient and tillage are the main critical factors responsible for enhancing the productivity of soil, water and crop. Improvement in water use efficiency in agriculture is essential because of irrigation sources are declining, energy costs make irrigation more expensive to deliver, world demand for food, feed, and fiber is increasing and production is being pushed into more arid environments. Higher water use efficiency is achieved through agronomic (Varieties, time and method of planting, row spacing and row orientation, seed rate/plant population, fertilizer, irrigation, cropping systems *etc.*) and engineering measures (water harvesting, water saving irrigation, modernization of irrigation system).

Among various production methods, a keen awareness has been created on the adoption of organic farming as a remedy to manoeuvre the ill effects of modern chemical agriculture. Organic farming gives major emphasis on recovery and maintenance of soil fertility and other management practices for sustainable yield. However, under arable production systems, organic manures suffer from the drawback of slow release of nutrients, which may cause significant reduction in crop yield and net farm income. This could be overcome by use of judicious combination of various sources of organic manures which includes compost, green manures, crop residues, bio-fertilizers, liquid organic manures. In general, in crop production, combined application of organic manures or/and chemical fertilizers alongwith liquid organic manures (either through foliar application or soil application or alongwith irrigation water (micro irrigation system) not only enhance the crop productivity and nutrient and water use efficiency but also provide nutrients in a more synchronized system and release the nutrients as per the need of crop to sustain higher productivity (Kanwar *et al.*, 2006). It is imperative to use both slow and fast nutrient releasing organic manures in an integrated manner in organic production system. Some of promising liquid organic manures can be categorized as follows.

Table 10.1: Approximate Quantity of Dung and Urine Produced per Head Annually by different Animals

Animal	Average Weight of Animal (kg)	Total Urine (kg)	Total Dung (kg)	Total Excreta (kg)
Cow	272	2,187	5,137	7,324
Bullock	362	2,935	6,849	9,784
Sheep	45	188	382	570
Pig	90	456	639	1,095
Horse	634	2,283	9,132	11,415

Source: Yawalikar *et al.* (1996).

Table 10.2: Excreta Produced per Head by Various Animals

Animal	Dung Produced per Year (cartloads)*	Urine Produced per Year (kerosene tins)**
Bullock	14.0	162
Cow	10.0	121
Horse	18.0	126
Sheep	0.8	10
Pig	1.3	25

* 1 cartload carries about 500 kg dung; **kerosene tin contains about 16 litres urine.

Organics Fermented at Farm Level

1. Beejamruth
2. Jeevamruth
3. Panchagavya
4. Cow urine
5. Bio-digester solution
6. Amrit pani
7. Humic acid
8. Sea weed

Organic Fertilizers

1. Organic NPK
2. Organic PK
3. Bio-phos
4. Bio-zinc
5. Bio-magnesium
6. Liquid bio-fertilizers

Panchagavya, jeevamruth and Beejamrutha are eco-friendly organic preparations made from cow products. Panchagavya is a plant growth stimulant and enhances the biological efficiency of the crops. While, jeevamruth activates biological activity of the soil, Beejamrutha protects the crop from soil borne and seed borne pathogens apart from improving the germination. All these fermented liquid organics contain macro nutrients, essential micronutrients, vitamins, essential amino acids and growth promoting factors like IAA, GA and beneficial organisms. The liquid manures are applied by various means through soil, foliage and irrigation water (fertigation).

Table 10.3: Approximate Quantity (kg) of Plant Nutrients Excreted per Head Annually by Various Animals

Animal		*N*	P_2O_5	K_2O
	Dung	20.71	10.20	5.10
Cow	Urine	21.87	Traces	29.52
	Total	42.28	10.20	34.62
	Dung	27.21	13.60	6.80
Bullock	Urine	29.17	Traces	39.56
	Total	56.37	13.60	46.36
	Dung	2.84	1.89	1.71
Sheep and goat	Urine	2.52	0.09	3,93
	Total	5.36	1.98	5.64
	Dung	3.74	3.40	2,72
Pig	Urine	1.81	0.45	2.04
	Total	5.55	3.85	4.76
	Dung	49.89	27.21	36.28
Horse	Urine	30.61	Traces	28.35
	Total	80.50	27.21	64.63

Source: Yawalikar *et al.* (1996).

An experiment conducted with different liquid formulations indicated that use of Beejamrutha, cow urine, panchagavya and liquid biofertilizers under Palekar method and organic method revealed that there was an increase in paddy yields 5 to 11 per cent (Devakumar, *et al.*, 2008). The use of liquid organics *viz.*, panchagavya, cow urine and bio-digester solution through foliar nutrition during crop critical stages apart from other organic manures application well ahead of sowing of crops, meet out the nutrients demand of the crop plants in organic production system. Sharada (2013) opined that application of compost (37.5 per cent), vermicompost (37.5 per cent) and Glyricidia (25 per cent) equivalent to 100 per cent RDN alongwith 3 per cent foliar spray of panchagavya during critical stages has resulted in on par yields with that of chemical fertilizers (RDF) in both the crops in greengram-rabi sorghum cropping system under rainfed conditions. Patil *et al.* (2012) reported that among various liquid organic manures, 3 per cent Panchagavya spray twice at flower initiation and 15 days later resulted in on par yield (2189 kg ha^{-1}) with 10 per cent cow urine spray (2114 kg ha^{-1}) but found superior over 10 per cent bio-digester solution spray (1734 kg ha^{-1}). Kiran (2014) found higher chickpea yield with panchagavya, vermiwash and cow urine sprays alongwith the application of FYM + Vermicompost and vermicompost.

Use of Liquid Organics through Fertigation

Fertigation is the application of water soluble solid fertilizer or liquid fertilizer through drip irrigation system. Fertigation has become an attractive method of nutients application in modern intensive agriculture systems. Water and nutrient are the main factors of production in irrigated agriculture and are the major inputs in contributing higher productivity. Saha *et al.* (2005) found that aloe vera responded differently to organic and inorganic sources under drip irrigation system. At equivalent nutrient level $N_{80}P_{40}K_{30}$, organic sources of fertilizer treatments was superior to inorganic sources thus showing prospects of growing organic aloe vera on commercial basis. Mahendra *et al.* (2013) opined that drip fertigation of 100 per cent RDF as WSF and drip fertigation of 100 per cent RDF + liquid bio-fertilizers+ Humic acid) were best practices to obtain higher pod yield, net returns, WUE and water productivity in bhendi cultivation as compared to traditional method of applying fertilizers through surface irrigation. Jayakumar *et al.* (2014) while studying the drip fertigation effects on yield and soil fertility at Coimbatore, Tamil Nadu noticed that application of 150 per cent RDF as drip fertigation combined with bio-fertigation of liquid formulation of azophosmet @ 250 ml ha^{-1} (10^{12} cells/ ml) registered the higher seed cotton yield of 3395 kg ha^{-1} and was significantly superior to control. Bio-fertigation significantly increased seed cotton yield and a progressive increase in seed yield was noticed with increasing levels of NPK fertilizer application. Sabreen *et al.* (2015) observed highest sesame yield (533 kg/ fed),water use efficiency (0.307 kg seeds/m^3) of irrigation water at 75 per cent ET irrigation levels with liquid poultry manure (4 t ha^{-1}) as compared to 50 and 100 per cent ET levels with 1 and 3 tonnes of liquid poultry manure.

Organic bio-stimulants hold great promise for the future of agriculture in general and organic farming in particular. These bio-stimulants aid in plant growth and metabolism, increase root biomass and promote soil organisms which in turn improves the crop productivity.

References

Devakumar, N., Rao, G.G.E., Nagaraj, S., Shubha, S., Imran Khan, Goudar, S.B., 2008. Organic Farming and Activities of Organic Farming Research Centre, Bangalore.pp.18-37.

Jayakumar, M., Surendran, U., Manickasundaram, 2014. Drip fertigation effects on yield, nutrient uptake and soil fertility of Bt Cotton in semiarid tropics. *Int. J. Pl. Prod.* 8(3): 1735-8043.

Kanwar, K.S., Paliyal, Manjinder Kaur Bedi, 2006. Integrated management of green manure, compost and nitrogen fertilizer in a rice-wheat cropping sequence. *Crop Res.*, 31(3): 196-200.

Patil, S. V., Halikatti, S. I., Hiremath, S. M., Babalad, H. B., Sreenivas, M. N., Hebsur, N. S., Somanagouda, G., 2012. Effect of organics on growth and yield of chickpea (*Cicer arietinum* L.) in *Vertisols. Karnataka J. Agric. Sci.,* 25(3): 326-331.

Sabreen Kh. Pibars, Mansour, H. A., Imam, H.M., 2015. Effect of Organic Manure Fertigation on Sesame Yield Productivity under Drip Irrigation System. *Global Advanced Res. J. Agric. Sci.,*4(8) :378-386.

Sharada, 2013. Studies on nutrient management practices through organics in greengram-rabi sorghum cropping system. *M.Sc. (Agri.) Thesis,* UAS, Raichur.

11

Irrigation Water Management in Commercial Crops

M.Y. Ajayakumar, Y.M. Ramesh and
Manjunatha Bhanuvally

Cotton (*Gossypium hirsutum*)

In the past five years, Indian cotton scenario has changed dramatically, largely due to the adoption of *Bt* cotton. The number of Bt hybrids released for commercial cultivation till date has crossed 1000 with more than 35 seed companies and public sector institutions. The first true breeding cotton variety was released in 2002 by the Indian Council of Agricultural Research (ICAR). This provides an opportunity to the farmers to save their own seed without losing the efficacy of *Bt* gene. The area under Bt cotton reached 10.0 million hectares in 2013-14 constituting nearly 81 per cent of the total cotton area in India. As a result, the production also reached 5.5 million tonnes. All these are indicators of the extraordinary impact and acceptance of *Bt* technology in cotton by the Indian farmers. This is quite comparable to the success of dwarf varieties of wheat and rice during the Green Revolution period. Several studies have established considerable economic benefits of *Bt* cotton cultivation to the farmers of all strata. Another significant development relates to creation of enabling environment by the Government of India. The Ministry of Environment and Department of Biotechnology simplified the regulatory procedures leading to expeditious commercial release, especially of events with well-established bio-safety record.

Cotton as such being the most important commercial crop of India (10.3 m ha with a production of 29.5 million bales of lint in 2009-10) contributes to around

60 per cent of the raw material to the textile industry and provides employment to nearly 60 million peoples with productivity of 494 kg ha⁻¹. Further impetus in improvement of Indian average cotton productivity (less than that of the world, 725 kg ha⁻¹) is possible through efficient and optimal use of precious on farm inputs *i.e.*, water and nutrient. Management of water and nutrients plays a key role in breaking of the undesired tempo in productivity plateau reached after major enhancement by introduction of *Bt* cotton and occupying more than 85 per cent area. Cotton is one of the identified crops for promotion of drip irrigation.

Production Practices of *Bt*-Cotton

Sowing Time

Normally *Bt* cotton hybrids are sown starting from last week of May to the last week of July. However, early sowing within 15th of June is more ideal sowing time in order to increase the yield.

Spacing and Seed Rate

Spacing	Seed Rate (kg acre⁻¹)
90 cm × 45 cm	1.20
90 cm × 60 cm	0.90
120 cm × 60 cm	0.675

FYM Requirement

The well decomposed FYM @10t ha⁻¹ should be incorporated before 2-3 weeks of sowing or 10 t compost/ha or 1 ton of vermi-compost ha⁻¹.

Nutrient Requirement

For hybrid Bt-Cotton recommended fertilizers are 180: 90: 90 kg NPK ha⁻¹. During basal, 50 per cent of the recommended N, entire dose of P and K quantity should be applied. Remaining quantity of N should be applied in three equal splits at 50, 80 and 110 days after sowing. Soil application of 25 kg $MgSO_4$ha⁻¹ 10 kg $ZnSO_4$ and 10 kg $FeSO_4$ ha⁻¹was recommended at the time of sowing.

Leaf Reddening

Reasons

☆ Acute nitrogen and magnesium deficiencies

☆ Imbalanced fertilization

☆ Improper irrigation or moisture stress

☆ When crop enters into cold season

☆ Sudden fall in night temperature

☆ Insect induced reddening

Management

★ Balanced nutrition to correct N and Mg deficiencies

★ Timely water management

★ Proper sowing time (June first week)

★ Spraying 1 per cent 19:19:19 (10 grams in 1 litre of water) + 1 per cent $MgSO_4$ (10 grams in 1 litre of water)= 3 times during flowering (60-65 DAS), boll initiation stage (80-85 DAS) and boll development stage (100-105 DAS)

★ Spraying micronutrient mixture (Either Bio-20 or Multiplex 5 grams in 1 l of water)

Weed Management

★ 50-60 days are critical stages

★ Pre-emergent application of 3.33 l ha^{-1} of Pendimethalin 30EC or 1.75 l Pendimethalin 38.7 CS or Diuron 80WP on the day of sowing or on the next day. While applying the herbicide make sure that there should be sufficient moisture in the field.

★ Post-emergent application of 1.25 litre Pyrithiobac Sodium 10 per cent EC during 20-25 days after sowing or when the weeds are at 3-4 leaf stage.

★ Two times inter-cultivation

★ One or two times hand weeding

Water Management

★ Irrigate the crop once in 15-20 days in case of black soils and 10-12 days in sandy loam soils

★ Critical stages are: Flowering, boll initiation and boll development stages

★ Making drainage channel in case of black soils is necessary

★ Irrigating the alternate rows will save the water upto 25-40 per cent

Modern Techniques of Water Management in *Bt* Cotton

★ Drip Irrigation

★ Fertigation

Advantages of Drip Irrigation

★ Enhanced water utility

★ Better crop growth and yield

★ Superior fibre quality

★ Reduced salinity

★ Higher fertilizer use efficiency

★ Reduced weed competition

☆ Saving of labour

☆ Enhancement of yield

☆ Timely sowing can be done

☆ Most suitability to long duration crops like cotton

☆ More economic returns and more suitability to light soils

Shankaranarayanan *et al*. (2011) reported that existing drip system with fertigation resulted in higher seed cotton yield (2.71 t ha⁻¹) as compared to traditional system of irrigation *i.e.* flooding. However, low cost poly tube drip and low cost micro tube system recorded on par seed cotton yield with that of existing drip system (2.62 and 2.53 tha⁻¹, respectively). But higher B: C ratio was recorded with low cost poly tube system.

Table 11.2: Yield Parameters, Yield and Economics as Influenced by Low Cost Drip Systems

Treatments	Seed Cotton Yield (t ha⁻¹)	Boll weight (g)	Seed Cotton Yield (g plant⁻¹)	Cost of Culti-vation (Rs. ha⁻¹)	Gross Returns (Rs. ha⁻¹)	Net Returns (Rs. ha⁻¹)	B: C Ratio
Existing Drip system	2.71	4.9	148.9	32865	67625	34760	2.06
Low cost micro tube system	2.53	5.1	137.8	27244	63367	36123	2.33
Low cost poly tube drip	2.62	4.9	143.3	27190	65500	38310	2.41
Ridges and furrows	2.41	5.0	128.6	25550	60300	34750	2.36
CD (P=0.05)	018	NS	11.2	-	-	-	-

Sugarcane (*Saccharum officinarum* L.)

Sugarcane is an old energy source for human beings and, more recently, a replacement of fossil fuel for motor vehicles, was first grown in South East Asia and Western India. Around 327 B.C. it was an important crop in the Indian sub-continent. It was introduced to Egypt around 647 A.D. and, about one century later, to Spain (755 A.D.). Since then, the cultivation of sugarcane was extended to nearly all tropical and sub-tropical regions. Portuguese and Spaniards took it to the New World early in the XVI century. It was introduced to the USA (Louisiana) around 1741.

Worldwide sugarcane occupies an area of 20.42 million ha with a total production of 1333 million metric tons. Sugarcane area and productivity differ widely from country to country. Brazil has the highest area (5.343 million ha), while Australia has the highest productivity (85.1 tha⁻¹). Out of 121 sugarcane producing countries, fifteen countries (Brazil, India, China, Thailand, Pakistan, Mexico, Cuba, Columbia, Australia, USA, Philippines, South Africa, Argentina, Myanmar, Bangladesh) account for 86 per cent of area and 87.1 per cent of production. Out of the total white crystal sugar production, approximately 70 per cent comes from sugarcane and 30 per cent from sugar beet. Sugarcane area and productivity differ widely from country to country. Brazil has the highest area (5.343 m ha), while Australia has the highest productivity (85.1 ha^{-1}).

It is a renewable, natural agricultural resource because it provides sugar, besides biofuel, fibre, fertilizer and myriad of byproducts/co-products with ecological sustainability. Sugarcane juice is used for making white sugar, brown sugar (Khandsari), Jaggery (Gur) and ethanol. The main byproducts of sugar industry are bagasse and molasses. Molasses, the chief by-product, is the main raw material for alcohol and thus for alcohol-based industries. Excess bagasse is now being used as raw material in the paper industry. Besides, co-generation of power using bagasse as fuel is considered feasible in most sugar mills.

Climatic Requirement

Mean total radiation received in 12 months of growth has been estimated to be around 6350 MJ/m^2. About 60 per cent of this radiation was intercepted by the canopy during formative and grand growth periods. The total dry matter production showed linear relationship with the intercepted PAR and the test of correlation yielded R_2 value of 0.913 (Ramanujam and Venkataramana, 1999). However, the energy conversion rate in terms of dry matter production per unit of intercepted radiation showed a quadratic response with per cent light interception indicating that the rate of energy conversion increased linearly up to 50 per cent light interception and beyond this level, the rate of photosynthetic conversion of solar radiation gets reduced. In sugarcane crop canopy the upper 6 leaves intercept 70 per cent of the radiation and the photosynthetic rate of the lower leaves decreased due to mutual shading. Therefore, for effective utilization of radiant energy a LAI of 3.0 - 3.5 is considered optimum. Areas having short growing period benefit from closer spacing to intercept high amount of solar radiation and produce higher yields. But in areas with long growing season wider spacing is better to avoid mutual shading and mortality of shoots. Rough estimates show that 80 per cent of water loss is associated with solar energy, 14 per cent with wind and 6 per cent with temperature and humidity. High wind velocities exceeding 60-km/hour are harmful to grownup canes, since they cause lodging and cane breakage. Also, winds enhance moisture loss from the plants and thus aggravate the ill effects of moisture stress.

Effect of Climate on Sugarcane Yields and Sugar Recovery

The sugarcane productivity and juice quality are profoundly influenced by weather conditions prevailing during the various crop-growth sub-periods. Sugar recovery is highest when the weather is dry with low humidity; bright sunshine hours, cooler nights with wide diurnal variations and very little rainfall during ripening period. These conditions favour high sugar accumulation.

Irrigation Water Management

Sugarcane being a long duration crop producing huge amounts of biomass is classed among those plants having a high water requirement and yet it is drought tolerant. It is mostly grown as an irrigated crop. The plant crop season being 12-18 months in India, 13-14 months in Iran, 16 months in Mauritius, 13-19 months in Jamaica, 15 months in Queensland (Australia) and 20 - 24 months in Hawaii.

Moisture Extraction Pattern

Most root biomass for sugarcane is found close to the surface and then declines approximately exponentially with depth. Typically, approximately 50 per cent of root biomass occurs in the top 20 cm of soil and 85 per cent in the top 60 cm. The percentage of roots in the 0-30 cm horizon was 48-68 per cent ; from 30 to 60 cm, 16-18 per cent ; 60 to 90 cm, 3-12 per cent ; 90 to 120cm, 4-7 per cent ; 120 to 150 cm, 1-7 per cent ; and 150 to 180 cm, 0-4 per cent. Thus the moisture extraction pattern from different soil layers follows the root biomass distribution.

Root growth responds strongly to the soil environment, creating plasticity in the form and size of the root system. The size and distribution of the root system is strongly affected by the distribution and availability of soil water, causing differences in the capacity of crops to exploit deeper soil reserves. Root distribution of sugarcane crop raised on loamy soil irrigated at 7, 14 and 21 days interval. Roots of a 12-month old plant crop were more deeply distributed under less frequent irrigation presumably in response to drying of the surface. Deeper rooting reduces the vulnerability of crops to soil water deficits by providing increased capacity for uptake of deep reserves of soil water. It also aids in reducing lodging. Hence, drip irrigated cane should be scheduled irrigations at less frequency during the initial 2 to 3 months period to promote deeper rooting. Nutrient supply has also been shown to similarly affect the rooting patterns. High soil strength causes slower root growth with reduced branching and thickened roots. High water markedly affects root distribution, with a majority of studies showing that rooting ceases within approximately 0.1 m of static water tables. Restricted root growth above shallow water tables does not necessarily reduce crop growth, as capillary rise can supply the crop with water and instances of water uptake from within the saturated zone have been observed. A risk of water stress does result from the lack of root penetration in soils with high water tables if ground water height falls rapidly, leaving the root system restricted to dry soil.

Physiological Characteristics to be Considered for Efficient Water Management

A liberal water supply reduces the cane yield and/or sugar yield, while mild water stress enhances the yield, Excessive watering at tillering should be avoided since it coincides with active root development and hinders nutrient uptake due to poor O_2 diffusion. Length of the cane determines the sink available for sugar storage since there is no secondary thickening of the stem in sugarcane A drying off or cut out period of 4-6 weeks prior to harvest ensures an optimum sugar yield. Reduction of water during the ripening to flower stage helps to control flowering or arrows.

Field Irrigation Schedule

The goal of an efficient irrigation scheduling programme is to provide knowledge on correct time and optimum quantity of water application to optimize crop yields with maximum water use efficiency and at the same time ensure minimum damage to the soil. Thus, Irrigation scheduling is the decision of when and how much water to apply to a cropped field. Its purpose is to maximize irrigation efficiencies by applying the exact amount of water needed to replenish the soil moisture to the desired level and make efficient use of water and energy.

Therefore, irrigation scheduling for sugarcane involves precise estimation of depth of water to apply at each irrigation, and the interval between irrigation, for each soil-plant-climatic condition. With drip irrigation, intervals of irrigation are usually daily irrespective of the evaporative demand of the atmosphere. The crop evapotranspiration under standard conditions, denoted as is the evapotranspiration from disease free, well-fertilized sugarcane crop, grown in large fields, under optimum soil water conditions, and achieving full production under the given climatic condition. The amount of water required to compensate the evapotranspiration loss from the cropped field is defined as crop water requirement. Although the values for crop evapotranspiration and crop water requirement are identical, crop water requirement refers to the amount of water that needs to be supplied, while crop evapotranspiration refers to the amount of water that is lost in evaporation + transpiration.

The irrigation water basically represents the difference between the crop water requirement and effective precipitation. The irrigation water requirement also includes additional water for leaching of salts and to compensate for non-uniformity of water application. Adequate soil moisture throughout the crop-growing season is important for obtaining maximum yields because vegetative growth including cane growth is directly proportional to the water transpired. Depending on the agro-ecological conditions, cultivation practices adopted and crop cycle (12-24 months) seasonal water requirements of sugarcane are about 1300mm to 2500 mm distributed throughout the growing season. The amounts of water required to produce 1.0 kg cane, dry matter and sugar are 50-60, 135-150 and 1000-2000 g, respectively. The transpiration coefficient of sugarcane is around 400. This means 400 m³ of water is required to produce one ton of dry matter.

Sugarcane Water Requirements in Various Countries

Sl.No.	Country	Water Requirement (mm)
1.	Australia	1522 (Drip)
2.	Burundi, Central Africa	1327 to 2017 (Furrow)
3.	Cuba	1681 to 2133 (Plant)
4.	Hawaii	2000 to 2400 (24 months)
5	Jamaica	1387
6	Mauritius	1670 (Drip)
7	Philippines	2451 (Furrow)
8	Pongala, South Africa	1555
9	Puerto Rico	1752
10	South Africa	1670

Sl.No.	Country	Water Requirement (mm)
11	Subtropical India	1800 (Furrow)
12	Taiwan	1500 to 2200 (Furrow)
13	Tropical India	2000 to 2400 (Furrow)
14	Venezuela	2420 (Furrow)
15	Thailand	2600 (Furrow)

Water Supply and Cane Yield

Frequency and depth of irrigation should vary with growth periods of the cane. The relationship between relative yield decrease (1-Ya/Ym) and relative evapotranspiration deficit for the individual growth sub-periods is shown in the picture. During the initial germination, field emergence and establishment of young seedlings the crop requires less water, hence light and frequent irrigation water applications are preferred. The water supply must be just sufficient to keep the soil moist with adequate aeration. If the soil is allowed due to infrequent and less water application, the germinating buds get desiccated leading to a lower and delayed germination.

On the other hand over irrigation leads to bud rotting due to lack of aeration, fungal attack and reducing soil temperature. Thus both under and over irrigation are detrimental for germination, resulting in low stalk population per unit area. During the early vegetative period (formative) the tillering is in direct proportion to the water application. An early flush of tillers is ideal because this furnishes shoots of approximately same age. Any water shortage during tillering phase would reduce tiller production; increase tiller mortality and ultimately the stalk population-an important yield component. However, excess irrigation during tillering phase is harmful particularly in heavy soils, since it coincides with active root development, which may be hampered by anaerobic condition created in the soil as a result of over irrigation.

The yield formation or grand growth period is the most critical period for moisture supply in sugarcane. This is because the actual cane yield build-up or stalk growth takes place in this period. The production and elongation of inter-nodes, leaf production on the stalk and its expansion, girth improvement, ultimately the stalk weight takes place in this period. It is also the period for production of sugar storage tissues. Therefore, crop reaches its peak water requirement in this stage. With adequate water supply to maintain a sheath moisture content of 84-85 per cent in the leaf sheaths, 3,4,5 and 6 from the top during this period of active growth produces longest inter-nodes with more girth (thick cane) and the total cane weight is greater. On the other hand water deficits in yield formation period reduce stalk elongation rate due to shortening of inter-nodes resulting in less cane weight and the effect is well marked on yield at harvest.

When the crop is in the ripening period, a farmer may also have a just planted crop on his farm in most situations. Therefore, the tendency of the farmer will be to provide sufficient water to the new (young) crop and neglect the grown up crop that is to be harvested. This situation is particularly true under limited water availability situations. If the grown up crop is not irrigated as required it experiences severe water deficits and there could be cane breakage, pith formation, significant reduction in cane weight, increase in fibre content and deterioration in juice quality. The situation is further aggravated if the harvesting is delayed. Thus both the farmer and the factory would suffer. Therefore, even for the grown up crop, reasonable amount of water with restricted supply is necessary to obtain good cane yield.

Sheath Moisture Content

The sheath moisture or relative water content determined by crop logging technique. It has long been used to control water application to commercial sugarcane crop, more particularly during ripening phase, when gradual increase in water stress is used to stimulate sugar storage in the stem. A ripening log is used to compare measured and desired sheath water content during approximately 12 to 24 weeks (depending upon the crop duration) prior to harvest. Sheath water content is measured on a periodic basis, and irrigation intervals and amounts are varied to produce a gradual decline of sheath water content, from about 83 per cent at the beginning of ripening to about 75 per cent at harvest.

In Hawaii and Taiwan sheath water content has been found to be a good indicator of stem sugar content. Similar methods involving other tissues are in use in Mexico, South Africa, India and Zimbabwe. Irrigation water is often limited and costly input. Therefore determination of optimum amount of water over a crop period to achieve higher water use efficiency assumes significance. Several experiments conducted world over have indicated that the relationship between cane yield and seasonal crop water use under a given climatic condition is linear. When irrigation plus rainfall is greater than the crop water requirement, anaerobic soil conditions or N losses may reduce crop growth rates and cane yield.

Drip Irrigation

Drip irrigation in sugarcane is a relatively new innovative technology that can conserve water, energy and increase profits. Thus, drip irrigation may help solve three of the most important problems of irrigated sugarcane - water scarcity, rising pumping (energy) costs and depressed farm profits. Whether or not drip will be successful depends on a host of agronomic, engineering and economic factors. "Drip irrigation is defined as the precise, slow and frequent application of water through point or line source emitters on or below the soil surface at a small operating pressure (20-200 kPa) and at a low discharge rate (0.6 to 20 LPH), resulting in partial wetting of the soil surface.

Surface Drip

The application of water to the soil surface as drops or a tiny stream through emitters placed at predetermined distance along the drip lateral is termed as surface drip irrigation. It can be of two types - online or integral type surface drip system. Integral dripline is recommended for sugarcane.

Subsurface Drip (SDI)

The application of water below the soil surface through emitters moulded on the inner wall of the dripline, with discharge rates (1.0 - 3.0 LPH) generally in the same range as integral surface drip irrigation. This method of water application is different from and not to be confused with the method where the root zone is irrigated by water table control, herein referred to as sub-irrigation. The integral dripline (thin or thick-walled) is installed at some predetermined depth in the soil depending on the soil type and crop requirements. There are two main types of SDI - one crop and multicrop.

Agronomic Advantages of Drip over Sprinkler and Furrow Irrigation

Adoption of drip irrigation (surface or subsurface) system in sugarcane cultivation is technically feasible, economically viable and beneficial in many ways:

1. Higher water application uniformity
2. Deceased energy costs due to reduced pumping time to irrigate a given design area
3. Saving in water up to 45 to 50 per cent contributing to higher water use efficiency
4. Saving in fertilizer (25 to 30 per cent) due to fertigation consequently improved fertilizer use efficiency *i.e.*, agronomic efficiency, physiological efficiency and apparent recovery fraction
5. Less weed growth and saving in labour due fewer weed control, fertigation and plant protection operations
6. Less pest and disease incidence due to better field sanitation
7. Optimum soil-water air relations contributing to better germination, uniform field emergence and maintenance of optimum plant population
8. Early harvesting and more ratoons
9. Day/Night irrigation scheduling is possible
10. Facilitates growing of crop on marginal soils due to frequent irrigations and fertigation
11. High frequency irrigation, micro-leaching and higher soil water potential enables use of saline water for irrigation and higher cane and sugar yields

Effective drip technology requires a more intense application of crop, soil, climatic, engineering, and economic factors than is usually present with furrow irrigation. New management perspectives and skills are required planting configuration, land preparation, drip design features, irrigation scheduling, fertigation, operation and maintenance of the system. The new management practices induced with drip technology seem to have significantly helped increase cane and sugar yields.

Both surface and subsurface drip irrigation system were technically feasible in sugarcane under diverse conditions. Available of dripline types for application in sugarcane are thick-walled for surface drip irrigated cane, thin-walled for subsurface drip irrigated cane, non-pressure compensated driplines for levelled land, pressure compensated driplines for undulated topography *etc.* Subsurface drip irrigation is superior over surface drip in terms of water availability, uniformity, water use, water use efficiency, cane yield and quality, management *etc.* Paired row or dual row or pineapple planting configuration with variable spacings depending upon the soil texture with one dripline for every two rows was found to be technically feasible, economically viable and potentially profitable in comparison to rectangular single row planting configuration with dripline for every crop row. Driplines can be successfully buried before planting cane without waiting for planting and germination of cane. Cane germination and field emergence will be adequate to give satisfactory plant stand both under surface and SDI systems without any supplementary use of surface furrow or overhead sprinkler system for germination.

Chilli

Chilli, in the matured green form is considered as a vegetable while fully matured red coloured dry fruits after drying are considered as a spice. Chilli as a vegetable belonging to the solanaceae family because of its pungency due to the chemical constituent – 'Capsaicin' is used in various culinary purposes as well as in the raw form also. Andhra Pradesh is the leading and largest chilli growing state in the country followed by Karnataka, Maharashtra and Tamil Nadu. The productivity is more in Andhra Pradesh and Punjab as it is cultivated under irrigated conditions in these states. But with efficient irrigation systems like micro irrigation especially drip system with other precision farming technologies will help in boosting up the productivity even in areas where it is being cultivated as a rainfed crop.

It is a source of Vitamin A and C which can be cultivated in all the three seasons under South Indian condition in areas having a temperature range between 20 and 30°C.

Improved Varieties: NP-46A, Guntur, Byadgi, Pusa, Jwala, Gauribidanur and hybrids available in the market.

Soil Type

Red sandy loam soils, well drained medium black to mixed clay soils are also suitable, with a pH of 6.0 to 7.0

Climatic Requirement

Temperature – (Day : 24°C – 34°C, Night 18°C – 21°C) relative humidity : 60-65 per cent.

Season

June – July

October – November

January – February

Seed/Seedling requirement

For one acre 50 grams of seeds are sufficient to raise healthy, vigorous and plantable seedlings needed for planting in one acre (7,756 seedling/acre). One good seed is sown per plug of plastic pro trays filled with sterilized and enriched coco-peat. 82 trays of 98 plug capacity each are required.

Nursery

Pro-tray method: Sterilized coco-peat as nursery media is enriched with *Azospirillum* or *Azotobacter* or microbial consortium, *Pseudomonas fluorescens* and *Trichoderma harzianum* @ 10g. each/tray. Trays filled with above media are protected by keeping in a nylon net house. Trays are watered lightly twice a day early morning and late evening hours. Seeds germinate in 7-8 days after sowing. 19:19:19 water soluble fertilizer @ 1g/1 is sprayed on 15th day after sowing. 25-30 days old seedlings are preferred for transplanting in the main field.

Land Preparation

Land is ploughed well and the soil is brought into a fine tilth. Raised beds of size 15-20 cm. Height, 100 cm. width and required length leaving a space of 45cm. as working space between the beds are prepared.

FYM and Neem Cake Application

Prior to sowing application of 10tons/acre well decomposed FYM is recommended. Further, One tonne of FYM should be treated with *Azospirillum* or *Azotobacter* or *Pseudomonas fluorescens* and *Trichoderma harzianum* @ 1kg each and mixed with the remaining FYM and applied to the beds. Neem cake @ 250kg/acre is applied to the beds treated with bio agents.

Recommended Dose of Nutrients (RDF): 105: 52.5:52.5 kg NPK ha^{-1}

Basal Fertilizer Application

Apply 20 per cent of RDF 21:10:10 kg NPK ha^{-1} (38 kg ammonium sulphate + 25 kg single super phosphate + 7 kg muriate of potash/acre) mix well and level the beds properly.

Laying of Drip Line

16 mm class-2 in-line drip laterals, having emitters with a discharge capacity of 4 lph spaced at 40 cm are laid in two lines at a distance of 60 cm, 5,600 meter length of lateral is required per acre. The discharge of emitters and leaks in the pipe checked before covering with the mulch film. 70 kg of mulch film is required to cover 1 acre area when 1m wide bed with a inter-bed spacing of 45cm is followed.

Plastics Mulch Application

UV stabilized, either black alone or bi colored (grey on top and black at bottom) poly ethylene mulch films available in rolls of 1.2-1.5 meter width and length of 400/600/1000 meters usually of 30 microns (120 guage) thick are laid manually with three persons, one holding the plastic roll with the help of pipe, another person taking/spreading the film on bed and hold tightly on other side and the soil with spades is put on sides of the film by the third person. The film can be laid mechanically also by using mulch laying machine. Drip irrigation laterals should be placed below mulch sheet before laying.

Spacing, Plant Population and Transplanting

At 60 cm x 60 cm sapcing, plastic mulch film covering 1m wide bed is punched by using a 2" dia PVC or GI pipe exactly at a distance of 60cm. to accommodate 7,756 seedling/acre in a double row. Transplant 25-30 days old seedlings at the centre of the hole. Contact of seedlings with the mulch film has to be avoided while planting.

Irrigation

Irrigate the crop daily depending on the stage of the crop and evaporation transpiration demand. Irrigation water should be provided @ 8 litres/m^2 (first month) and 12 litres/m^2 (subsequent months) *i.e.* for 30 minutes during first month and 45 minutes during later months.

Fertigation

For injecting the fertilizer separate ventury system/fertilizer tank can be used. Fertigation has to be scheduled twice a week starting from 21 days after transplanting. Fertigation has to be stopped one week before final harvest. Generally 5 months duration crop requires 15 weeks of fertigation *i.e.* 30 fertigations.

Water Soluble Fertilizers

34 : 17 : 17 kg NPK/acre and for **Fertigation (80 per cent RDF)**

Foliar Nutrition

Apply foliar nutrition having major, secondary and micronutrients depending on the soil conditions, crop growth given through foliar application atleast three times starting from 45 days after transplanting at an interval of 15 days @ 5 g/litre of water.

Harvesting and Yield

Well matured tender green coloured fruits should be harvested two times within a week while fully matured ripe fruits with uniform colour development are harvested for dry chilli marketing. Harvested fruits are graded according to size and market demand, packed in unit packages and sent to market. The expected yield of 12 tons/acre excellent quality chilli fruits can be obtained by following precision farming techniques.

Fertigation Schedule (For one acre)

Crop Stage	Fertilizer Scheduling (weeks)	Quantity of Fertilizer (kg/week/acre)			
		19:19:19	Urea	MAP	SOP
Pre-vegetative	3	No fertigation			
Vegetative	4	3 (12)	3 (12)	2 (8)	0 (0)
Vegetative, Flowering and harvesting	6	5 (30)	5 (30)	1 (6)	1 (6)
Vegetative, Flowering and harvesting	5	2 (10)	1 (5)	0 (0)	2 (10)
	2	No fertigation			
Total quantity of Fertilizers	15 (30 fertigation)	52 kg	47 kg	14 kg	16 kg

Plant Protection

Sl.No.	Pest/Disease	Chemical for Foliar Apray	Concentration per litre of Water
1.	Damping-off	Captaf drenching	2 ml
2.	Powdery Mildew	Wettable Sulphur/Hexaconozole	1.5 g/1.5 ml
3.	Pytophthora Blight	Mancozeb + Metalaxyl (Pre-mixed)	1.5 to 2.0 ml
4.	Leaf Minors and Aphids	Monocrotophos/Abomectin	1.0 ml/0.5 ml
5.	Anthracnose	Carbendezim	1 g
6.	Thrips	Imidacloprid	0.5 ml
7.	Root rot	Carbendezim	2.0 g
8.	Leaf curl (white flies)	Acetamaprid	0.5 g
9.	Leaf Spot	Mancozeb	2.0 g

References

www.sugarcanecrops.com.

https://giongmia.files.wordpress.com/2007/01/sugarcane-crop-ebook.pdf

Ramanujam, T.; Venkataramana, S., 1999. Radiation interception and utilization at different growth stages of sugarcane and their influence on yield. *Indian J. Pl. Phy.*, 4(2): 85-89.

12

Precision Micro-Irrigation and Water Measurement Methodologies

R.H. Rajkumar, J. Vishwanatha, S.R. Anand and A.V. Karegoudar

Rice is the important crop of Thunga Bhadra Project (TBP) command area, though only 8.6 per cent (29,032 ha) of the TBP command was earmarked for paddy cultivation, now it has been increased to more than 70 per cent (2, 55,366 ha) and in all these cases, rice is traditionally grown by transplanting under puddled fields. For this puddling operation, farmers in this area are going for intensive tillage under continuous ponded water nearly 10 cm throughout the season, which serves to break down soil aggregates, reduced macro-porosity, disperse the clay fraction, and farming dense zone of compaction at depth. The head reach farmers of this command are using excessive irrigation water and fertilizers with unscientific method, which is leading to wastage of precious natural resource *i.e.* water. Due to this, the tail end farmers are not getting sufficient water in time and ultimately the lands of low laying area are becoming degraded by waterlogging and salinity and it is estimated that around 96,125 ha land has been affected in TBP command. In this command area, around 20 – 30 per cent of the total water required for the rice culture is being dedicated by the farmers to nursery rising, puddling and transplanting. Therefore, every drop of water received at the farmer's field by way of rainfall, canal irrigation or pumped from aquifer, is valuable and needs to be used effectively. Looming water shortages coupled with costly labour inputs have made these farmers to think in modification in transplanting cropping to direct seeded rice system with precise method of irrigation. The water availability during *Kharif* and late *rabi* is a biggest problem in tail end region. The DSR will facilitate better establishment of second crop in *rabi* for tail end farmers. The extent of water saving will facilitate expansion of irrigated area and cropping intensity in the command area.

Micro irrigation such as surface drip, subsurface drip and sprinkler irrigation methods under DSR in command area has added advantages mainly through saving in water and fertilizers with increased water and nutrient use efficiency. The potential water saving under micro irrigation is mainly due to moisture based irrigation scheduling leading to reduced seepage, percolation and evaporation losses. With micro-irrigation in DSR are can get more water and fertilizer use efficiency with attaining the same or higher productivity of the rice without compensating yield and also reduction in usage of water could ultimately results in reduction or avoiding development of waterlogged saline soils.

Precise Micro-Irrigation

Drip Irrigation System - Components and their Function

A drip irrigation system for DSR consists essentially of mainline, sub mains, lateral, drippers, filters and other small fittings and accessories like valves, pressure regulators, pressure gauge, fertilizer application components *etc.*

1. Filter

It is the heart of drip irrigation. A filter unit cleans the suspended impurities in the irrigation water so as to prevent blockage of holes and passage of drip nozzles. The type of filtration needed depends on water quality and emitter type. A two stage filter unit is usually needed.

a) Gravel Filter (Sand Filter)

These filters are effective against inorganic suspended solids, biological substances and other organic materials. This type of filter is essential for open reservoir, when algae growth take place. The dirt is stopped and accumulated inside the media in the filter. Gravel filter consist of small basalt gravel or sand (usually 1-2 mm dia) placed in cylindrical tank, made of metal. Water enters form the top and flows through the gravel while leaving the dirt in the filter. The clean water is discharge at the bottom. The filter is cleaned by reversing the direction of flow. Pressure gauges are fitted at the inlet and cutlet of the filter. When the dire accumulates, the pressure difference between the inlet and outlet increase and when the pressure difference is more than 0.5 to 1.0 kg/cm^2 (5-10 m), then filters must be cleaned by opening the clover or back washing, Automatic self cleaning filter are also available.The flow rate of the filters may be 10, 15, 20, 25, 30, 40, 50 cu m/hr and the tank diameter may range from 10-50 cm depending on the capacity of the system.

b) Screen Filter

1. These are installed with or without gravel filter, depending upon quality of water. The screens are usually cylindrical shape and are made of non-corrosive metal or plastic material. Screens filters are specified as below:

2. By the diameter of inlet and outlet (range from ¾ "to 4" inches)

3. By the recommended range of flow rate (ranges from 3, 5, 7, 10, 15, 20,30, 40, m³/hr).

4. By the size of holes in the screen (in mm, micron or in mesh *i.e.* the number of holes per square inch).

5. As a approximation, 20, 40, 80, 100, 120, 150 and 200 mesh (0.15, 0.1 and 0.08 mm) respectively. The most common mesh selected for drip irrigation is 100 to 200 meshes (0.15 to 0.08 mm dia).

6. By the total surface area of the filter (in sq. m) or the active or net filter area, which is usually about 1/3 of the total filter surface are.

7. By the cleaning methods: manual or automatic. The head loss across the filter should not be more than 3 m and otherwise needs cleaning. The filters are cleaned by flushing the screen with a stream of water. After cleaning the screen is checked for tears and the gasket should be checked and replaced when necessary.

8. With relatively clean water, screen filters can be used alone.

c) Disc Filter

The filtration elements are grooved plastic disc, which are piled together around a telescopic core, acceding to the desired degree of filtration. Both sides of the discs are grooved and the grooves cross each other when piled up and frightened together. The housing is made of plastic or metal and comes in many different sizes mainly 3 ¾ to 3". The water passes through the filter from the outside to the inside. There is no danger of filter tearing. The filtration is affected in two stages: the larger outer surface operates as a screen filters and collects the larger particles. The grooves inside the disc allow the adhesion of fine particles, mainly organic matter. The filter element can be cleaned easily. When opening the core, the discs are released and can easily be rinsed under running water. The pressure drop is slightly higher than screen filter but disc filters have better cleaning capacity than screen filter. The water flow should be on a tangent to the disc to allow them to spin freely.

Automatic Backwash Filtration Systems

The Filtration Process

Water flows through the inlet valve via of the first filter where it is filtered. The filtered water then flows through the Second station outlet which will flush the filter, same action vise versa for first filter.

The Backwash Process

Command based on time settings is sent from the controller to (two) separate components in the filter:

1. First station valve Enters FORWARD mode

2. Second station valve Enters backwash mode (entrance closed, drain opens).

Water flows via the FIRST Station filter. It enters the first station filter (which is open), Purifies water in the Sand media. and enters in second station outlet the water carries away impurities toward the inlet valve which is already closed inlet and drain open to pass away the impurities. At the end of the backwashing process (20 seconds) the backwash command is withdrawn, the First station will be cleaned in same principle.

2. Main Line

The main line conveys the water from filtration system to the sub main. They are normally made of rigid PVC pipes in order to

Automatic Backwash Filtration Systems Installed at Primary Filtration Unit.

minimize corrosion and clogging. Usually they are placed below the ground *i.e.* 60 to 90 cm (2 to 3 ft), so that they will not interfere with cultivation practices. Their diameter is based on the system flow capacity. The velocity of flow in mains should not be greater than 1.5 m/s and the frictional head loss should be less than 5m/1000 m running length of pipeline.

3. Submain

The Submain conveys the water from mainline to the laterals. They are also buried in ground below 2 to 2.5 ft and made of rigid PVC. The diameter of Submain is usually smaller than main line. There may be number of submain from one mainline depending upon the plot size and crop type.

4. Laterals

Laterals are small diameter flexible pipes or tubing made of low density polyethylene (LDP) or liner low density polyethylene (LLDPE) and of 12 mm, 16mm, and 20 mm size. Their colour is black to avoid the algae growth and effect of ultra- violet radiation. They can withstand the maximum pressure of 2.5 to 4 kg/cm^2. They are connected to submain at predetermined distance. The pressure variation between two extreme points of lateral should not be more than 15-20 per cent and discharge variation should not be more than 10 per cent. On slopping ground, the laterals are placed along the contour with 1 per cent extra length for sagging purpose.

5. Emitters or Drippers

It is the main component of Drip irrigation system for discharging water from lateral to the soil. *i.e.* to the plants. There are various types and size of drippers, based on different operating principles. They are made of plastic, such as polythene or polypropylene. Their discharge range is between 1-15 ph. Each dripper has its own characteristics, advantages and disadvantages which determines its use. The drippers can be classified according to working principle, discharge, type, structure, working pressure, designation, durability, regulated and non regulated discharge.

The main principle when planning a dripper is to achieve the minimum discharge with maximum size of water passage. The large water passage is essential to minimize clogging and provide the minimum discharge for cheapest set-up. Therefore, an emitter is necessary, (a hole in a pipe is not a dripper). Emitters may be on the lateral or inside to lateral, accordingly they are called on line or inline emitters.

6. Controls Valves (Ball Valves)

These are used to control the flow through particular pipes. Generally, they are installed on filtration system, mainline, and on all submain. They are made up of gunmetal, PVC cast iron and their size ranges from ½" to more than 5".

Solenoid/Electrical Valves

Solenoid Valves Installed in the Field for Automation Purpose.

Generally in drip irrigation system, for irrigating the area PP Ball Valves are used. But in atomized drip irrigation system, solenoids/electrical valves are used for controlling though controllers. These valves are connected by cables to controller, cables are inside PVC/Conduits pipe which are laid adjacent to main or sub main lines.

7. Flush Valve

It is provided at the end of each sub main to flush out the water and dirts.

8. Air Release cum Vacuum Breaker Valve

It is provided at the highest point in the main line to release the entrapped air during the start of the system and to break the vacuum during shut off. It is also provided on submain if submain length is more.

9. Non Return Valve

It is used to prevent the damage of pump from flow of water hammer in rising main line.

10. Pressure Gauge

It is used to indicate the operating pressure of the drip system.

11. Gromate and Take-off

These are used to connect the lateral to submain. A hole is punched with hand drill of predetermined size in submain. Gromate is fixed into the hole. Take off is pressed into the hole. Take off is pressed into the gromate with take of punch upto the step provided. Gromate acts as a seal. The sizes are different for 12 mm, 16mm, and 20 mm lateral.

12. End Caps (End Sets)

They are used to close the lateral ends, Submain ends or mainline ends. Sub mains and mains are preferably provided with flush valve. They are convenient for flushing the line.

13. Fertilizing System

It is used to add the chemical irrigation water; however, fertigation is not free of hazards. Chemicals added to water may be toxic to human beings and animals so, safeguard must be taken to prevent back flow of irrigation water into the water source, which might be used for drinking purpose. Only water-soluble fertilizers should be used to minimize the clogging hazard.

14. Water Meter

In drip automation system, irrigation and fertigation will be by time base. For quantity base irrigation and fertigation water meter is necessary. In quantity base irrigation and fertigation only, EC monitoring and pH control will be possible with the help of acid quantity.

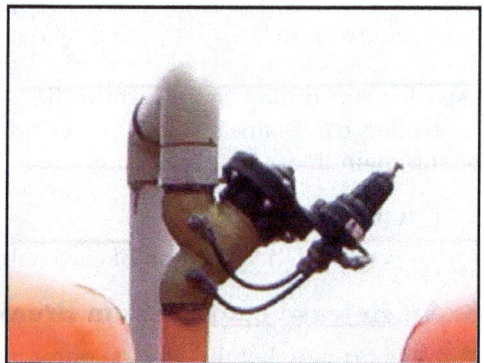

Water Meter to Measure Quantity of Water Applied in Drip Irrigation.

15. Pressure Relief Valve

This instrument is very important and it acts as safety equipment for drip system which can be installed before sand filter in the drip head unit. In case, the output discharge/head of the pump is more than the required discharge/head, the excess pressure can be released through this equipment. The required out going pressure/head will be set in the pressure relief valve, the excess discharge/head automatically will be released or bypassed to water source. From this equipment the pipe damage or head unit damage can be minimized greatly.

Further, the development of automation in micro irrigation is much more precise method of irrigation and fertilizer management. Automation can help to overcome the uncertainty and limited hours of power and labor. In addition, analysis and manipulation of the basic crop data with judicious water management could be attained by automation technique in micro-irrigation for DSR under command area.

Micro-irrigation Automation

To manage the crop effectively, Irrigation and fertigation automation is very important

Why Automation is required.........?

☆ To avoid over/under irrigation in drip irrigation

☆ To manage uncertainty of power/electrical supply

☆ To overcome problem of limited hours power/electrical supply

☆ To give proportionate fertigation

☆ To overcome labour management problem

☆ To analyze and manipulate the basic crop data

☆ To overcome problem of maintenance of Hi –Tech application in modern agriculture

☆ To reduce failure of drip irrigation in large scale projects

☆ To get accurate readings of climate data

☆ To save and distribute the successful crop data to other farmers.

The benefits, salient and key features of the automation technique are listed below

Benefits of Automation

☆ Improve yield

☆ Optimize/reduces water and fertilizer input expenses

☆ Improves consistency and quality of crop

☆ Focus management on bottom line

☆ Tension free irrigation and fertigation for drip irrigation

☆ Useful where limited hours of power supply availability

☆ Useful where uncertainty of power supply availability

☆ Reduces irrigation labour cost

☆ Judicious water management of available resources

☆ Cost effective for long term basis

☆ Irrigation and fertigation is possible as per growth stages of crop

☆ Effectively useful for large drip irrigation projects

☆ Accurate and basic data available to analyze and manipulate

☆ Crop and water planning can be possible as per climatic condition

☆ Chemical injection and EC/pH control is possible through the automation.

Salient Features of Automation

☆ Designed to ensure precise fertigation

☆ Effectively increases water and nutrient uptake to plant.

☆ Provide excellent EC and pH Monitoring

☆ The highest quality components

☆ Clean and impressive design

☆ Significantly improved crop quality and yield

☆ Significantly reduced fertilizer expenses

☆ Significantly reduced labours requirement

Key Features

☆ Superior irrigation and fertigation control

☆ User friendly

☆ Large graphic display

☆ Modular Input and Output cards

☆ Memory backup key

☆ Hot keys for easy navigation

☆ PC Interface; Optional

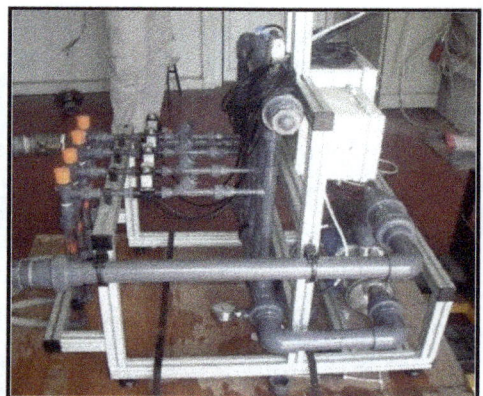

Drip and Fertilizer Automation Equipment

Weather Station and Sensors

One sensor based weather station has been recommended for the following purpose

☆ To give exact irrigation duration based on the daily current evaporation data which will be downloaded daily from the weather station

☆ To know the rainfall data

☆ To know the soil temperature

☆ To know the temperature, relative humidity and wind velocity

Tensiometers

Tensiometers are the most commonly used soil moisture sensors and could be the entry level for monitoring application. A tensiometer is a closed, water-filled tube with a porous ceramic tip at one end and a vacuum gauge at the other, which acts like an artificial plant root. As the soil dries out, it exerts suction on the water in the tensiometer and this is measured on the gauge

Advantages

☆ Easy to use

☆ Measurements taken quickly

☆ Direct indication of plant suction in kPa

☆ Inexpensive

☆ Simple to install

☆ Minimal soil disturbance no calibration required

Water Measurement Methodologies

Irrigation water flow can be measured with Orifices, flumes, notches and flow meters. The common and easy method to measure flow is flumes.

1. Cut Throat Flume

Flume has engraved, molded and in - built gauge plate for discharge measurement in lit/second and cusecs. Any non-technical person or farmer can understand discharge easily. Fiberglass reinforced plastic (FRP) is 100 per cent

Cut Throat Flume

anticorrosive material and having impact strength just like steel and completely maintenance free for years together. Moulded arrow is given to understand flow direction on the cut throat flume. Discharge table is attached with the instrument alongwith the information of calibration formula, installation procedure and ready recknor for discharge readings.

Application: Discharge measurement in laboratory, research institutes, irrigation canal and waste water from factories.

Range of Cut Throat Flume - (One cusec=28.32 l/second=0.02832 cumecs)

A. 20x90 and 30x90 section - 0.33, 0.5, 1, 2, 3 and 4 cusec

B. 40x180 and 60x180 section - 5, 6, 8, 10, 12, 15, 18 and 20 cusec

C. 60x270 and 100x270 section – 25, 30, 35, 40, 50, 60, 70 and 80 cusecs

Known Quantity of Water Applied through Cut Throat Flume to Puddle Paddy Field

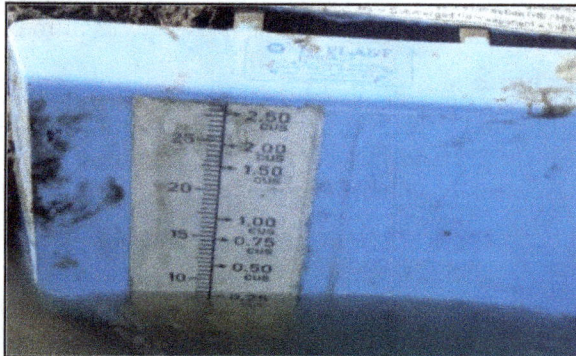

Ready Recknor to know Water Applied.

Parshall Flume

Flumes are molded flumes and available in assemble-dissemble system for easy and economical transportation and installation. Molded and engraved gauge plate is given to show water level in flume and discharge in liter/second and cusecs. These Parshall flumes are manufactured using high quality grade resins and glass fibers to sustain tremendous load of flowing water and their high velocity. Parshall flume discharge measurement system is supported with "Electronic Discharge Measuring Recorders" for 24x365 days for automatic data recording.

Application: Research institutes, laboratories, irrigation small and big canals and waste water from factories

Range of Parshall Flume at Laboratory level - 0.33, 0.5, 1 and 2 cusecs.

Irrigation and waste water: 5, 10, 20, 30, 50, 80, 100, 150, 200, 300, 400, 500, 750, 1000, 1500, 2000, 2500 and 3300 cusecs.

V-notch

FRP V-Notches are useful for seepage measurement of irrigation dam and toe drains. Discharge in V-Notches can measure from 6 lit/second to 120 liter/second. In V-notch, advantage is of head over the crest is large and thus flow remains unaffected by surface tension and viscosity. Limitation of V-Notch is small discharge range and V-plate itself become obstruction to silt and hence not suitable for field channels in irrigation and for waste water management. V-Notches are having molded and engraved gauges for measuring discharge in lit/second and levels in centimeter. Provision can be made on demand to remove V-Plate as and when required cleaning silt gathered in V-Notch.

Application: Seapage measurement of dam in foundation, inspection and top galleries and toe-drains. Limited use in waste water and laboratories.

Range of Cut Throat Flume - 15, 30, 45, 60, 90 and 120 l/s. (Above 120 l/s not recommended)

Trapezoidal Weirs

Replogle flumes are most accurate flumes and easy for installation. Due to vertical crest in downstream of flume above CBL, heavy scouvering and energy dissipation problems may come. Hence in India it is used only for laboratory and research institutes level. Replogle flumes are moulded, single piece FRP flume without joint. Gauge plate is molded and engraved for discharge and level

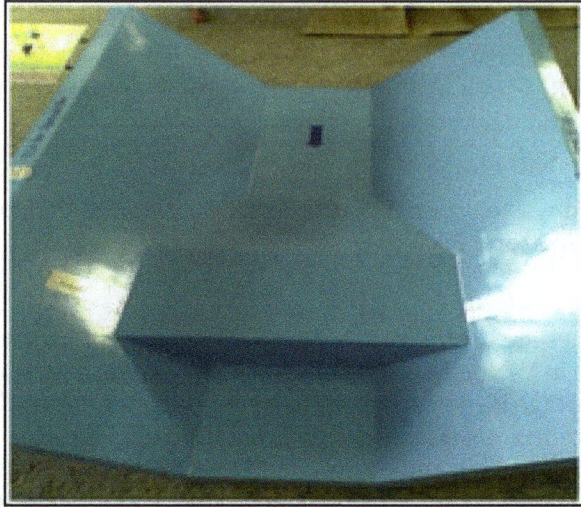

measurements. Moulded arrow is given to understand flow direction and discharge table is attached with information of calibration formula, installation procedure and ready reckoner for discharge measurement.

Application: Discharge measurement in small and big channels of irrigation and research Institutes.

Range of Cut Throat Flume - 0.33, 0.5, 1, 2 and 3 cusecs capacity. Big capacity on demand.

Orifice Meter

Orifice Meter is used at laboratory and research level for discharge measurement. Orifice meter are practically abounded due to considerable fall, excessive loss of head and silt deposition in upstream side. These orifice meters are made of FRP and having 50 mm to 120mm orifice. Discharge range is from 15 liter/second to 100 liter/second. Gauge marking in centimeter and liter/second is given on orifice plate. Discharge table is given with calibration formula for free flow condition.

Application: Laboratory, research institute, engineering and agriculture collages for demonstration

Material: Fiberglass reinforced plastic (FRP)

Size : Standard design.

H-Flume

H-Flumes are very accurate for discharge measurement in open channel. Advantage of H-Flume is wide opening at high flows to reduce backwater effect. Due to complicated geometry, it is not much popular in India. These H-Flumes are made in FRP with gauge marking at both wings of H-Flumes and shows discharges in liter/second and water level in flume for free flow condition. Moulded arrow is given to understand flow direction. Discharge table is attached with information of calibration formula, installation procedure and ready recknor.

Area of application : Laboratory, Institute, engineering and agriculture collages for demonstration and limited purpose for field irrigation

Range : 15,30,60 and 150 liter/second. Big capacity H-Flume available on demand.

Water Meter

In Drip Automation system, Irrigation and fertigation will be by time base. For quantity base irrigation and fertigation water meter is necessary. In quantity base irrigation and fertigtion only, Ec monitoring and pH control will be possible with the help of acid quantity.

Conclusion

Micro-irrigation in DSR with automation technique in command area will help in following aspects, Without compensating yield could get more water and fertilizer use efficiencies with attaining the same or higher productivity of the rice, Saving in water and fertilizer. Reduction in usage of water could ultimately result in reduction or avoiding development of waterlogged saline soils in command area, Lesser weed growth. Sufficient water availability to tail end farmers. Uniform crop growth can be attained and able to grow more with less water. Labour and power cost can be minimized.

13

Potential Crops for Salt Affected Soils in Command Areas

S.R. Anand, J. Vishwanatha, A.V. Karegoudar and R.H. Rajkumar

Sugarbeet (*Beta vulgaris* var. *Saccharifera* L.) is a sugar producing tuber crop, grown in temperate countries. With the recent development of tropical hybrids of sugar beet, it is possible to raise the crop in tropical and subtropical areas in India. Sugar beet root contains 15-20 per cent sucrose and in the process of sugar extraction, 12-14 per cent recovery is possible. Sugarbeet being a salt tolerant crop, it can also be grown in saline soils where other crops fail to grow. Sugar beet is not only the source of sugar, but also provides several by-products like ethanol, cattle feed and betaine. Like sugarcane molasses, beet molasses can also be used in ethanol production as well as in the pharmaceutical industry for vitamin B_{12} Production.

The Key Advantages of Sugar Beet in Comparison with Sugarcane

It is well known that sugarbeet has a shorter growing cycle (around 5 months), lower water requirement (about 1/3 to 1/2 of the water needed to grow sugar cane) and a slightly higher sugar (16-19 per cent sucrose) and ethanol yield per acre (2,500 to 3,000 l/acre). Finally, by processing sugar beet after or before the sugar cane crushing, factories can stretch their operations over a longer period of time and reduce their production costs (Steven Cosyn *et al.*, 2011). When compared to sugar cane sugarbeet is high yielding (30-40 MT/acre), tolerant to high temperature, less water requirement and drought tolerant, improves soil conditions and excellent in performing on saline and alkaline soils. It saves water and easy for cultivation and harvesting. It is an Industrial crop ready to be processed when the factory needs it. Just in time.

Advantages to Community at Large

1. Every acre under TSB will release an acre of farm land for food crops: potential 10 to 12 m ha.

2. Land reclaims and land conservation is possible

3. TSB uses about one third of the water actually needed for cane

4. Beet molasses is more environment friendly than cane molasses (low effluent volume, low non sugar solid and almost no colour) and can be sprayed on land as fertigation.

5. Transformation of the existing cane factories will generate industrial activity and the development of new know-how

6. Will help meet the increased need of bio-ethanol, efficiently

7. Export possibility

Benefits of New Crop Promotion

☆ Increase in the cropping intensity (*i.e.* One Sugar beet and two crops of Sweet sorghum in a year) thus increase the farm income to Rs.22,000 per acre per annum.

☆ Provides continuous and increased employment opportunities in agriculture throughout the year.

☆ Dependence on single source of raw material *i.e.* sugarcane for sugar and ethanol production is avoided.

☆ Since water requirement is less, more area could be brought under irrigation.

☆ Water saving –10,000 m^3/ha

☆ Problem of Weather abnormalities and water scarcity is avoided being a seasonal crop.

Comparative Analysis of Sugarcane and Sugarbeet

Salient Features	Sugar Beet	Sugar Cane
Duration	6 Month	12 – 13 months
Growing Season	Throughout the year	Only one season.
Soil Requirement	Grows well in Sandy loam, tolerates alkalinity.	Grows well in loamy soil
Water Requirement	400-500 mm	1800-2000 mm
Crop Management	Moderate management, Low fertilizer requirement and less	Good management and more fertilizers requirement
	pest and disease complex	

☆ Forecast of sugar deficit could be offset with sugar beet due to higher sugar productivity.

☆ Ensure continuous operation of plant (300 days) and provides opportunity for direct and indirect industrial employment continuously.

Agronomical Practice of Sugar Beet

1. Tropical varieties: Pasoda, Hi 0064, Doratea
2. Soil

   ☆ Well drained, loamy to clay loam

   ☆ pH 6.5 to 8.0 – tolerate mild salinity

   ☆ pH <6 – cannot be grown
3. Season and Climate

   ☆ Oct – Nov to March – May (sub tropical varieties)

   ☆ Optimum temperature regimes

   ☆ Germination: 20 - 25°C

   ☆ Growth and maturity: 30 - 35°C

   ☆ Sugar accumulation: 25 - 35°C
4. Crop establishment:

   ☆ Seed rate : 3.6 kg/ha (Rs.5700/ha)

   ☆ Spacing : 50 x 20 cm

   ☆ Population : 1 to 1.2 lakhs plants/ha
5. **Fertilizer :** 120 : 60 : 60 kg N, P_2O_5 and K_2O/ha (Time and Split: Not standardized)
6. **Irrigation** (Quantity and Schedule: Not standardized)

   ☆ Pre-sowing (seeds germinate in a week)

   ☆ 1st irrigation – early establishment

   ☆ Subsequent irrigation – need based

   ☆ Sensitive to water stagnation

   ☆ Stop irrigation 1 month before harvest

   ☆ Irrigation just prior to harvest

Sugar Beet as Bio-fuel Crop and its Scenario in India

In India, sugar beet was tried firstly at Indian Institute of Sugarcane Research (IISR), Lucknow in 1970, under All India Co-ordinated Research Project (AICRP) on sugar beet. Later, sugar beet cultivation was initiated by Vasantdada Sugar Institute (VSI) Pune, Maharashtra. On the basis of agronomic trials conducted at various locations in India, the sugar beet cultivation was taken up in Sriganganagar area of Rajasthan for sugar production and in Sundarban area of West Bengal for alcohol production (VSI, 2009). In Karnataka, sugar beet cultivation and processing was first initiated by leading sugar factory *viz.*, Shree Renuka Sugars in Athani and Gokak Taluka of Belgaum district and it aims to increase the processing of sugar beet to 50,000 tonnes in the year 2013-14 (Times of India, April 18th 2013). In Karnataka, research on sugar beet has been first initiated at Agricultural Research Station, Gangavathi during 2010-2012 and it was found that sowing of sugar beet during I fort night of August with a spacing of 60 x 30 cm is ideal for successful cultivation of sugar beet under saline soils (ECe ranges between 6-8 Dsm^{-1}).

Fuel Ethanol - New Horizon

Govt. has allowed 5 per cent blend with petrol in first phase and 10 per cent in second phase

☆ Current production of ethanol is only 1.3 billion litres ethanol can be produced directly from beet juice

☆ 90 litres t^{-1} of beet at 15 per cent sucrose

☆ 7200 litres ha^{-1} at 80 t yield

Simultaneously Worked with Machinery Manufacturers

☆ Development of small sized factories (typically 2000 TPD as against10,000 in EU)

☆ Process development for sugar and alcohol

☆ Showing trials and machinery to potential customers

Current Status of Demand and Supply of Ethanol in India

In January 2003, the Government of India launched the Ethanol Blended Petrol Programme (EBPP) promoting the use of ethanol for blending with gasoline and the use of biodiesel derived from non-edible oils for blending with diesel at 5 per cent blending. Due to ethanol shortage during 2004-05, the blending mandate was made optional in October 2004, and resumed in October 2006 in the second phase of EBPP with a gradual rise to 10 per cent blending. These ad-hoc policy changes continued until December 2009, when the Government came out with a comprehensive National Policy on Bio-fuels formulated by the Ministry of New and Renewable Energy (MNRE), calling for blending at least 20 per cent bio-fuels with diesel and petrol by 2017. It is estimated that India need 10,000 lakh litres of ethanol, out of which 5000 lakhs litres of existing production and there is demand of another 5000 lakhs litres per annum (www.ethanolindia.com).

Sugar Beet as an Alternative Crop under Salt Affected Soils

In Karnataka, most of the salinity-affected area is in the irrigated commands due to secondary salinization. The extent of damage in the state is reported to be 10 per cent of the total irrigated area. The severely-affected command is the Tungabhadra Project (TBP) area (96,215 ha), which alone accounts for over 32 per cent of the total command area in the project (3.62 lakhs ha). Unscientific soil water management and violation of cropping system has lead to the twin problems of water logging and soil salinity especially in the low lying areas of the command. Due to shortage of water and soil salinity, second crop of paddy in paddy- paddy cropping system is becoming not viable in low lying areas of the command. In this context, thinking of thought for alternative crops for rice- rice mono culture system of cultivation is necessary. Sugar beet being short duration crop, as well as salt tolerant crop, can fit well in cropping system. To avoid further land becoming water logged and salinity and also make use of already affected land economically, cultivation of sugar beet appears to be one option available.

Research Results in India and Abroad

Starting in 1995, agronomic trials of varieties from different gene pools in India,

☆ Planted **every month** for almost 80 months, rejecting the unsuitable ones and adding new ones

☆ Suitability data for on almost 80 cycles in Pune

☆ Extended to 40 other locations in India (four year data available now)

☆ Trials in Pakistan, Sudan, Kenya, Malawi, Thailand

Quality Comparison between Sugar Beet and Sugarcane

Character	Sugar Beet	Sugarcane
Duration (months)	6-7	10-12
Brix reading	23-24	18-20
Pol per cent	20-22	13-16
Sugar recovery (per cent)	15-16	11-12
Sugarrecovery in factory (per cent)	10-12	8-10
Yield (t ha⁻¹)	60-80	100
Water requirement	120 cm	200 cm

Performance of Sugar Beet

Planting date: 15th November

Date of harvest: 23rd April

Variety	Tuber Yield (t ha⁻¹)	Sucrose (per cent)	Leafy Matters (per cent)
Posada	72.73	15.50	9.88
Dorotea	72.41	16.70	11.35
HI 0064	61.89	15.70	15.31

Though sugarcane is an important crop of Tungabhadra command previously, it was slowly withdrawn from cultivation due to non availability of sufficient water from canal system and other local problems. Sugarbeet being a short duration crop and requiring less water than sugarcane fits well in the cropping programme particularly under saline soils. However, there is no information on its agronomic requirement particularly on saline Vertisols.

Pooled data of three years indicated that the August first fortnight sowing recorded significantly higher sugar beet root yield (39.67 t ha⁻¹) over August II fortnight (35.54 t ha⁻¹), September I fortnight (29.75 t ha⁻¹) and September II fortnight (27.26 t ha⁻¹) sowings. This could be due to water logged situation observed during the year 2010-11 and 2012-13 at early growth stages of the crop as sugarbeet is highly sensitive to water logging and hence, the sowing time should not coincide with rainy period.

Generally, in TBP command August first fortnight and September first fortnight coincides with heavy rainfall. In the years of study, the crop received 288.5 mm, 29.5 mm and 142.75mm rainfall during the year 2010, 2011 and 2012, respectively. On the contrary, the agronomic trials conducted by VSI (2009) under normal soils the sugarbeet root yield range from 60 to 75 tha⁻¹. Under saline soils, overall lower root yield could be attributed to the effect of salinity constraints in achieving root yield at par with the non saline soils. However, in another study at Digraj in Maharashtra, it was found that, the September, October and November sowing is the optimum sowing period for sugarbeet in Maharashtra.

Among different planting geometry, there was no significant difference with respect to root yield and TSS per cent. But, there was a significant difference in number of beets per plot and weight of 10 beet roots in different planting geometry. Significantly higher number of beets per plot was recorded in sowing at 45 x 20 cm (179.1) followed by sowing at 45 x 30 cm, 60 x 20 cm and 60 x 30 cm (156.0, 157.5 and 135.9, respectively). This could be due to variation in spacing and hence number of plants per plot was different. Similarly, significantly higher weight of 10 beets was recorded in sowing at 60 x 30 cm (14.91 kg) followed by sowing at 60 x 20 cm, 45 x 30 cm and 45 x 20 cm (14.30, 13.78 and 12.97 kg, respectively). The higher weight of beets could be due to larger area associated with each plant and sufficient availability of plant nutrients alongwith lesser pest and disease incidence leading to higher weight of ten beets at wider row spacing as compared to closer spacing. As no relevant references on weight of ten beets as influenced by different spacing was available, further confirmation on this line is needed.

Economics

The cost of cultivation was same for all the four dates of sowing (Rs. 27442 ha⁻¹), while, it differed with planting geometry mainly due to difference in the requirement of seeds. The closer spacing of 45 x 20 and 45 x 30 cm incurred Rs. 28803 and Rs. 28013 ha⁻¹, respectively. Whereas, wider spacing of 60 x 20 and 60 x 30 cm incurred the cost of Rs.26980 and Rs. 25973 ha⁻¹, respectively. However, significantly higher gross returns was obtained with August first fortnight (Rs.71410 ha⁻¹) followed by August second fortnight (Rs. 63978 ha⁻¹). Similarly, net returns and B:C ratio were significantly higher with August first fortnight (Rs. 45636 ha⁻¹and 2.61, respectively) followed by August II fortnight (Rs. 36536 ha⁻¹and 2.34, respectively). This could be due to lower cost of cultivation of sugar beet as compared to sugar cane and it is also getting higher price per ton (Rs. 2000-2500). However, significantly lower net returns and B:C ratio were obtained with September first fortnight and September second fortnight sowing. This could be due to lower beet yield obtained with delayed sowing. Planting geometry did not differ significantly with respect to gross returns, net returns add B:C ratio. Interaction effect was also found non-significant (Times of India April 18, 2013).

Challenges can be Overcome

There are also challenges ahead for the sugar beet crop, which can be overcome by intensive research. When moving sugar beet from one area to the other and especially when moving to the hotter areas with often suboptimal conditions of irrigation, several new diseases can appear, for which not always the proper resistances or chemical treatments are yet known. The effect of the well known pests and diseases can be more severe under hotter and higher moisture regions. Besides agronomic improvement, chemical and genetic solutions have to be developed and improved to secure stable yields needed for the application of sugar beet as a significant agricultural crop.

Studies by SESVanderHave with tropical sugar beet commenced in India in 2005; both by agricultural research institutes and sugar factories. These studies have shown that sugar beet yields with high sugar contents can be achieved. Tests are now in a semi commercial stage to introduce the sugar beet as a commercial crop. Also in other tropical and subtropical regions, sugar beet has shown excellent results. In the Australian continent, several tests have been made with very interesting results. The water use efficiency of the crop was in several cases the trigger to start these experiments. In Indonesia, some experiments have been running to use sugar beet as an off season crop for sugar production.

The idea was to produce a product comparable to red sugar of cane at the farmers' place and to collect this product when the roads are accessible after the rainy season. In the highlands of Africa and South America there are large areas with temperatures very comparable to the temperate zones, where sugar beet could appear to be very efficient in the conversion of the very intense sun energy. Most optimal are the sub-tropical regions with a wet and dry season where sufficient water is available.

Conclusion

Together with the Indian research institutes, SESVanderHave has demonstrated that growing sugar beet in a country like India is feasible from an agronomic point of view. There is large potential to apply sugar beet as a profitable crop outside the current regions of production.Three years study indicated that, among the different dates of sowing August I fortnight sowing is ideal for higher root yield of sugar beet. Further delay in sowing of sugar beet under Thunga Bhadra project (TBP) command area may lead to lower root yield as rainy period coincides with sowing time resulting in to poor germination and poor crop stand. Planting geometry did not have influence on root yield of sugar beet. Generally, wider spacing (60 x 30 cm) gives more root yield as compared to closer spacing (45 x 30 cm). It yields more under saline conditions and having good market value as that of sugar cane (2000 to 2500 per ton). Hence, Sugar beet could be more remunerative especially under saline soils of TBP area.

References

http//www.vsi.in. Vasantdada sugar institute official website.

Steven Cosyn, Klaas Van der Wounde, Xavier Sauvenier, Jean noel Evrard, 2011. Sugar beet: A complement to sugar cane for sugar and ethanol production in tropical and subtropical area. *International Sugar Journal*. **1346** (113):120-123.

Times of India daily newspaper April 18, 2013. (PTI) by Vidya Murkumpi, Managing Director, Shree Renukha Sugars ltd, Karnataka, India.

VSI, 2009. Evaluation of sugar beet varieties under different agro climatic zones, soil type and proper sowing and harvesting months in Maharashtra. *Final Research Report*, Pune, Maharashtra state. India. PP-232.

14

Laser Guided Land Levelling for Efficient Irrigation Water Use

P.S. Kanannavar, P. Balakrishnan, Y. Ravindra,
C. Rudragouda and B. Anuraj

Soil and water conservation is essential in agriculture to increase the crop productivity to meet the food demand for raising population. In agriculture, land development plays a key role because of undulating topography of the soil surface which has a major impact on the germination, water saving and crop yield. Traditional methods of land levelling are more cumbersome and not so precise and repetitive in nature. The latest technology of laser guided land levelling can address these important issues and help to accomplish yield increase and agricultural sustainability. In Karnataka during 2008-09, the study of this technology was taken up first time in University of Agricultural Sciences, Raichur. After three years of continuous study, the uniformity in soil-moisture distribution (95-98 per cent), irrigation time and water saving (25-30 per cent) and high levelling index (2-9 cm) was proved. Thereafter in order to popularise and transfer the technology to farmers, it was demonstrated to the several farmers of this region during Krishimelas held during every year, demonstrations and field level trainings. After usage of this technology by the farmers, it was observed that 25-35 per cent increase in paddy yield as well as saving in man power and 30 to 40 per cent saving in energy requirement in cultivation made the farmer to promote usage of this technology in entire Karnataka.

A significant (20-25 per cent) amount of irrigation water is lost during its application at the farm due to poor farm design and unevenness of the field. Better Land levelling saves irrigation water and facilitates field operation, conserves vital resources and increases the yield (Rickman, 2002). For an efficient irrigation system the level difference between high and low spots of a field should be as minimum as possible whereas under actual field conditions higher difference is very common. The latest technology of laser guided land levelling can address these important issues and help to accomplish yield increase and agricultural sustainability.

For the first time in Karnataka a study was initiated to find the feasibility and benefits of tractor operated laser guided land leveller in the research farms of UAS, Raichur. A comparative evaluation of the laser guided land leveller with the traditional method of levelling was carried out to know Reduced Levels' (RLs) Standard Deviation (SD), Levelling Index (LI), Soil-moisture Deviation (moisture SD) and Uniformity Coefficient (U$_c$) *etc.*

The levelling indices of both the fields before and after levelling were calculated using the formula given by Agarwal and Goel.

$$\text{Leveling index} = \frac{\sum \text{Numerical difference between the designed and existing grid levels}}{\text{Number of grid points}}$$

After rainfall of 19 mm and 70 mm at different durations, the moisture contents at grid points were measured and its distribution uniformity or Uniformity co-efficient was calculated in both the fields using *Christiansen* formula (Michael, 2011).

$$Cu = 100\left[1 - \frac{\sum X}{mn}\right]$$

where,

Cu = Uniformity Coefficient or Moisture distribution uniformity in per cent

m = Average value of all moisture contents in per cent

n = Total number of grid points

X = Numerical deviations of individual observations or grid moisturecontent from the average moisture content.

After the fields were levelled with both methods of levelling, the results revealed considerably lower value of RLs, SD when the fields were graded by and laser leveller. It indicated lower undulations in the Laser levelled fields compared to traditional levelling field. The uniformity of soil-moisture distribution on the laser levelled and traditional levelled fields were found as 93.94 per cent and 77.7 per cent, respectively. The increase and decrease in the cost of levelling and field capacity per hour in laser levelling method due to increased time requirement to achieve precise levelling with more earthwork, where as in case of traditional levelling it was opposite due to less time required for less earthwork. Until now this technology is successfully used and being in use with several farmers. This is the first step towards the Precision Agriculture in Karnataka with precise land development. Advantges of laser land levelling technology in paddy cultivation are presented in Table 14.3.

Table 14.1: Soil-Moisture Content Readings for Conventional and Laser Levelled Fields during 2010-11

Particulars	Laser Levelled Field			Conventional Levelled Field		
	After 1st Rainfall of 19 mm	After 2nd Rainfall of 70 mm	Avg.	After 1st Rainfall of 19 mm	After 2nd Rainfall of 70 mm	Avg.
Average moisture content, per cent	30.64	38.26	-	30.38	36.60	-
SD (10 x10m grids)	3.73	3.32	3.53	9.11	12.35	10.73
Uniformity coefficient, per cent	92.9	94.36	93.63	75.9	72.93	74.41

Table 14.2: Overall Comparison of Laser Levelling and Conventional Levelling

Particulars	Laser Levelling	Conventional Levelling
Per cent SD reduction	86.6	43.4
Average levelling Index after levelling (cm)	1.3	11
Average Moisture distribution Uniformity(Cu m)	93.94	77.7
Cost/ha (Rs.)	9920	6123
Cost/h, (Rs.)	745	580
Field capacity(ha/h)	0.075	0.095

Table 14.3: Overall Advantages of Laser Levelling Technology (Farmers' feedback)

Particulars	Values, per cent
Increase in cultivable land	3 to 5
Decreased fertiliser usage	25 to 30
Water saving	25 to 30
Labour saving	45 to 50
Increased yield	25 to 35
Energy saving	30 to 40

Conclusion

There is huge opportunity to popularise this technology in Karnataka. The overall benefits of Precise (Laser assisted) levelling to the farmers are increase in cultivation area, yields and water use efficiency, reduction in time and water requirement for irrigation, better and uniform crop establishment, saving of fuel and energy, easy crop management as well as decreased drudgery.

References

Michael, A. M., 2011. Irrigation Theory and practice, second edition, Vikas publishing House Pvt. Ltd., Noida-201301 (UP), India, pp: 398-399.

Rickman, J.F., 2002. Manual for laser land leveling. Rice-Wheat Consortium Technical Bulletin Series 5. New Delhi–110 012, India: Rice-Wheat Consortium for the Indo-Gangetic plains, p. 24.

15

Fertigation in Important Field and Horticulture Crops

M.V. Ravi, H.S. Latha, M.R. Umesh, G.V. Shubha
and K.K. Amrutha

Water and nutrient are the main factors of production in irrigated agriculture and are the major inputs in contributing higher productivity. In intensive agriculture, both fertiliser and irrigation management have contributed immensely in increasing the yield and quality of crops. The method of fertiliser and irrigation application affects the efficiency of these inputs in arid and semi arid regions. Improvement of the use efficiency of these valued inputs is of utmost importance because these are costly and scarce. Under this disadvantaged condition the use efficiency of these inputs is also very low. However, high value crops and green houses may lead to greater efficiency of the two most critical inputs in high production.

In India, since last two decades drip irrigation has received greater attention of both the farmers and the government in view of its well proven advantages in both water scarce and sufficient areas. Application of plant nutrients by dissolving them in irrigation water particularly with the drip system is termed as fertigation. Studies conducted in India and elsewhere have indicated encouraging results of fertigation technology in potato, capsicum, maize, sunflower, banana and pomegranate. Drip irrigation and fertigation go hand in hand in order to improve efficiency of water and nutrients in crop production. Drip irrigation permits application of fertilizers through irrigation water directly at the site of high concentration of root activity and thus help improve the fertilizer use efficiency in crop production. Drip fertigation optimize the use of water and fertilizer enabling to harness high crop yield, simultaneously ensuring a healthy soil and environment. Further advantages occur via fertigation through a subsurface drip irrigation system. These are reduced water evaporation, larger wetted soil volume and a deep rooting pattern. The factors that governs the fertigation are soil types, crops, methods of irrigation used, water quality, types of fertilisers available, economic feasibility *etc*. Fertigation has become an attractive method of fertilisation in modern intensive agriculture systems. This assumes added importance after the introduction of micro-irrigation system like drip in irrigated agriculture.

Benefits of Fertigation

Fertigation had many advantages like higher water and nutrient use efficiency, resource saving; labour, time, higher productivity and product quality, reduced environmental pollution, effective weed management, reduced soil compaction, gives flexibility in farm operations, effective use of undulating soils and manipulating plant growth and development. It increases the yield and economics of most of the high value crops under drip irrigation.

Nutrients which can be Fertigated

The major nutrients like nitrogen and potassium are most commonly fertigated for many field and horticulture crops. Generally all the nitrogen fertilizers are suitable for drip fertigation since they cause little clogging and precipitation problems except $(NH_4)_2SO_4$ which may cause precipitation of $CaSO_4$ in hard calcium rich water. Urea is well suited for injection through drip irrigation since it is highly soluble and dissolves in non-ionic form and does not react with the substances in the water. Nitrate salts are characteristically soluble and are well suited for use in drip irrigation. Application of potassium fertilizers does not cause any precipitation as salts except in the case of K_2SO_4 with irrigation water containing high amount of calcium. Common K sources such as potassium sulphate, potassium chloride and potassium nitrate are readily soluble in water. These fertilizers move freely into the soil and some of the K ions are exchanged on the clay complex and are not readily leached away. Phosphorus has not been generally recommended for application in drip irrigation system because of its tendency to cause clogging and its limited movement in the soil. If irrigation water is high in Ca and Mg, precipitates of insoluble calcium and magnesium phosphates may result from the application of inorganic phosphates. But the addition of H_3PO_4 to the irrigation water maintained a low pH and prevented the precipitation of insoluble salts, thus allowing the introduction of P through drip irrigation systems. Secondary nutrient like Ca, Mg and S should be applied through their conventional sources like gypsum and dolomite directly into soil. Micronutrients such as iron, manganese, zinc and copper can be applied through irrigation water as chelated form (EDTA) without causing any precipitation problem.

Selection and Compatibility of Fertilizers

Liquid fertilizers are best for fertigation as they readily dissolve in irrigation water, but lack of easy availability and high cost restrict their use. Fertigation using granular fertilizers pose several problems like differences in their solubility in water, compatibility among different fertilizers and problem of filtration of undissolved fertilizers. Fertigation can be affected by using single or multiple nutrient fertilizers in their solid or liquid form. Selection and compatibility of fertilizers also depends on the four main factors *viz.*, plant type and stage of growth *e.g.* tomato is very sensitive to high NH_4 concentration; Plants are more sensitive to form of N at fruiting stage; Soil conditions (moisture condition at field capacity level), water quality (pH of the water has to be near to neutral and its EC to be within acceptable limits (< 1 dS/m) and Fertilizer characteristics and price. Some of the desirable characteristics of the fertilizer material for use in fertigation are full solubility, quick dissolution in water, fine grained product, high nutrient content in the saturated solution, compatibility with other fertilizers, absence of chemical interaction with irrigation water and minimum content of conditioning agents, no clogging of filters and emitters, low content of insoluble salts (< 0.02 per cent), no drastic change of water pH (3.5< pH> 9.0) and low corrosives for control and head system.

Fertilizer Management under Fertigation

Generally $1/4^{th}$ of RDF should be applied as pre-plant as it ensures nutrient supply to plants during early stage when irrigation may not be required. In coarse-textured soils it is essential to supply only a part of RDF through fertigation with low rate but high frequency and, rest as pre-plant to reduce leaching losses. Fertilizers can be injected daily, on alternate days or weekly depending on irrigation frequency, soil type, daily nutrient requirement of the crop *etc.* The effectiveness of fertigated nutrients can be enhanced when injected at the end of irrigation run, with only 30-40 minute period of clear water to flush the nutrients from the system. When saline irrigation water is used it is necessary to reduce pH of irrigation to about 5.5. Balance between NH_4/NO_3 supply and high NH_4 leads to decrease in soil solution pH and uptake of other cations and also toxicity to plant roots; NO_3 uptake enhances P and Fe uptake but increase soil solution pH to undesirable levels and optimum $NO_3:NH_4$ should be 80:20.

Filtration System

Filtration is prerequisite for fertigation to avoid clogging of the drip lines and emitters and to maintain the uniformity of water and fertilizer application. The type of filtration system will depend on the source and quality of the water. In fertigation system a second filtration system after fertilizer container is necessary to remove particulate matter or precipitates. Deep well water may contain soluble divalent iron, which on contact with phosphate may produce gel-like precipitate that can block the trucklers and filters.

Fertigation Equipments

Fertiliser can be injected into drip irrigation system by selecting appropriate equipment. Commonly used fertigation equipments are:

i) Venturi pumps; ii) Fertiliser tank and iii) Fertiliser injection pump

i) Venturi Pumps/Injector

This is a very simple and low cost device. A partial vacuum is created in the system which allows suction of the fertilisers into the irrigation system through venturi action. The vacuum is created by diverting a percentage of water flow from the main and pass it through a constriction which increases the velocity of flow thus creating a drop in pressure. When the pressure drops the fertilisers solution is sucked into the venturi through a suction pipe from the tank and from there enters into irrigation stream. Although simple and with greater uniformity of dosing the fertilisers tank the venturi cause a high pressure loss in the system which may results in uneven water and fertiliser distribution in the field. The suction rate of venturi is 30-120 l h^{-1}.

ii) Fertiliser Tank

In this systems part of irrigation water is diverted from the main line to flow through a tank containing the fertiliser in a fluid or soluble solid form, before returning to the main line, the pressure in the tank and the main line is the same but a slight drip in pressure is created between the off take and return pipes for the tank by means of a pressure reducing valve. This causes water from main line to flow through the tank causing dilution and flow of the diluted fertiliser into the irrigation stream. With this system the concentration of the fertiliser entering the irrigation water changes continuously with the time, starting a high concentration. As a result uniformity of fertiliser distribution can be a problem. Fertiliser tanks are available in 90, 120, 160 liters capacity.

iii) Fertigation Pump

These are piston or diaphragm pumps which are driven by the water pressure of the irrigation systems and such as the injection rate is proportional to the flow of water in the system. A high degree of control over the fertiliser injection rate is possible, no serious head losses are incurred and operating cost is low. Another advantage is that if the flow of water stops, fertiliser injection also automatically stops. This is perfect equipment for accurate fertigation. A suction rate of pumps varies from 40 to 160 litres per hour.

Efficacy of Drip Fertigation

The benefit of drip irrigation mainly depends on the practice of fertigation because drip has a special feature, which is absent in other system of irrigations. In drip fertigation only 30-40 per cent of the soil is moistened by the emitters. This is true in case of orchard crop. If the fertilisers and water are applied separately, the fertiliser use efficiency decreases because the fertiliser nutrients do not get dissolved in the dry zones where the soil is not wetted. As a result, the benefits are not fully expressed. Traditional fertilisation is not appropriated, not convenient and efficient as the drip fertigation. Drip fertigation is therefore, the best means of fertilisation to the root zone of the crops. Further, the relative requirements of N, P and K vary with crop growth stages. The uniform and required quantity of nutrients can be supplied through drip fertigation. Hence the nutrient use efficiency will be more under drip fertigation. It has been reported that with fertigation, fertilizer use efficiency can be enhanced up to 95 per cent (Table 15.1).

Table 15.1: Fertilizer Use Efficiency in Fertigation

Nutrient	Soil Application	Drip + Soil Application	Drip + Fertigation
N	30-50	65	95
P_2O_5	20	30	45
K_2O	60	60	80

Source: Satisha, 1997.

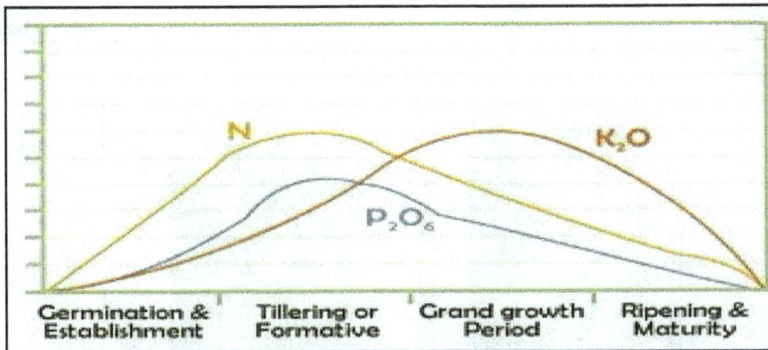

Figure 15.1: Relative Requirement of NPK at different Crop Growth Stages.

Nutrients Availability in soil under fertigation

Anitta and Muthukrishnan (2013) reported that the nitrogen availability steadily increased with increased depth upto 30 cm after that declined in all the distances. The highest available phosphorus in soil was confined to 0-15 cm of soil layer under all fertigation levels. The available phosphorus decreased with increase in distance and soil depth. With regards to potassium, soil K content was significantly higher in the surface soil than in the subsoil, this might be due to majority of applied K was held in the surface soil and the downward movement was slow (Figure 15.2).

Nutrient Uptake at Physiological Stages

Generally the demand for N, P and K will be higher at vegetative period; at fruit ripening stage there will be high demand for N and K and reduced demand for P in lettuce and tomato crops (Figure 15.3).

Fruit trees, flowers and greenhouse crops are always fertigated; while open field vegetables and field crops are either totally fertigated or have some level of fertigation, depending on initial soil fertility and basic fertilization. Drip fertigation is found to be well suited for horticultural crops. In India, the adoption of this twin technology has resulted in enhancement of horticultural production. Row crops are most suited for fertigation. Fertigation is by far the most common, and in some cases the only method of fertilizing the green houses, orchard, vegetables and drip irrigated field crops such as sugarcane cotton, maize *etc.* (Table 15.2).

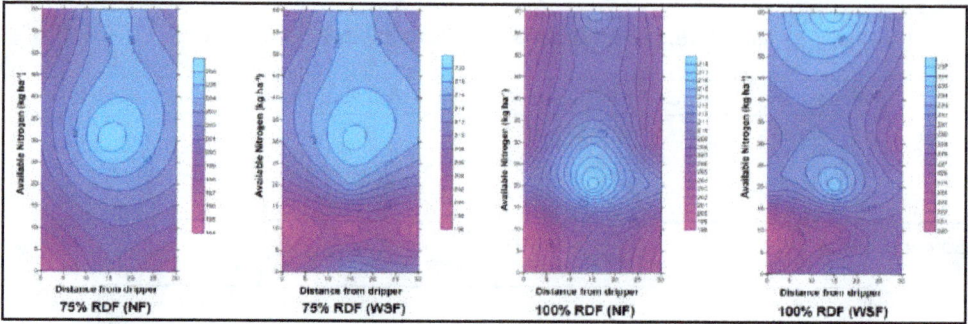

Nitrogen

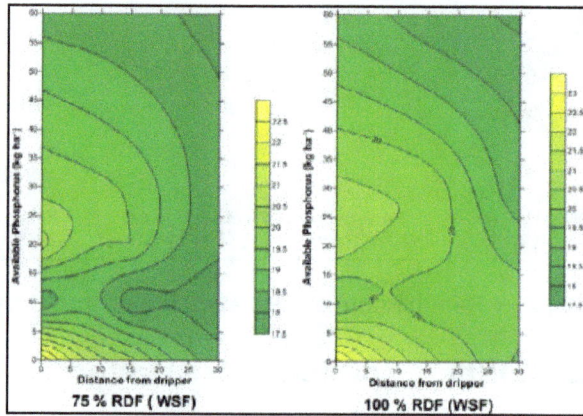

Phosphorous

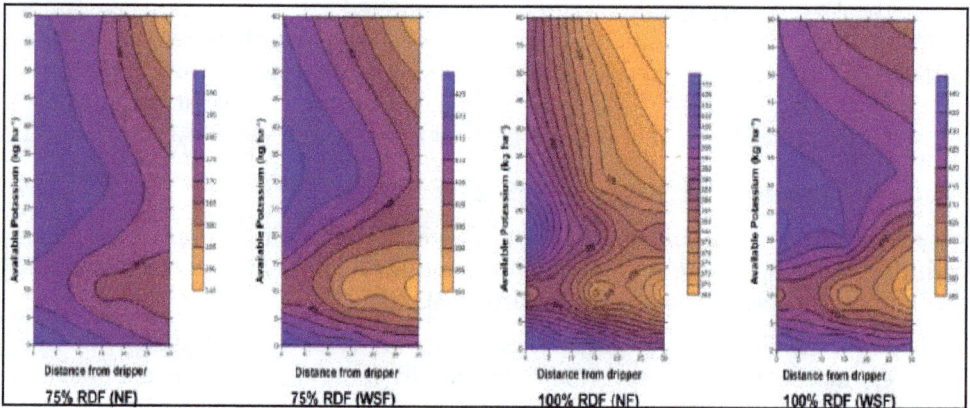

Potassium

Figure 15.2: Nutrient Dynamics under Drip Fertigation.

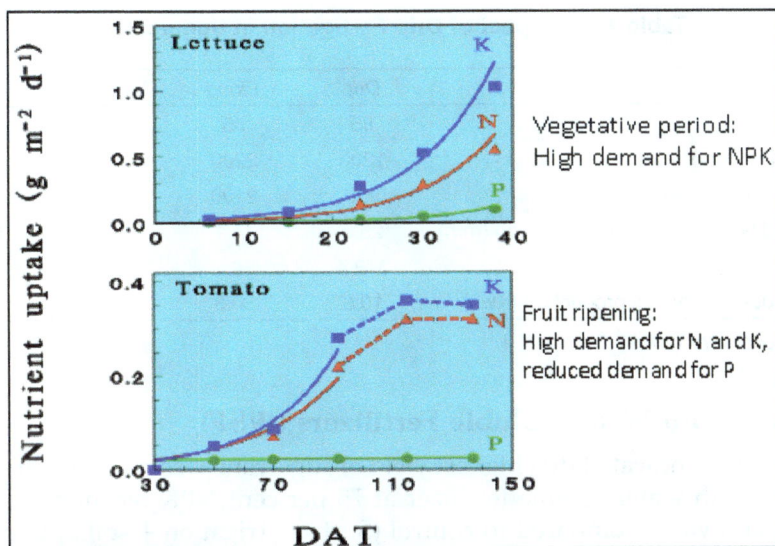

Figure 15.3: Rates of Uptake of N, P and K during different Physiological Growth Stages of Tomato and Lettuce Crops Suitable for Fertigation.

Table 15.2: Crops Suited for the Drip Irrigation

Orchard crops	Grapes, banana, pomegranate, orange, citrus, tamarind, mango, fig, lemon, custard apple, sapota, guava, pineapple, coconut, cashew nut, papaya, aonla, litchi, watermelon, Muskmelon, etc.
Vegetables	Tomato, chilly, capsicum, cabbage, cauliflower, onion, okra, brinjal, bitter gourd, bottle gourd, ridge gourd, cucumber, peas, spinach, pumpkin
Cash crops	Sugarcane, cotton, areca nut, strawberry etc.
Flowers	Rose, carnation, gerbera, anthurium, orchids, jasmine, lily, mogra, tulip, dahilia, marigold etc.
Plantation	Tea, rubber, coffee, coconut etc.
Spices	Turmeric, cloves, mint etc.
Oil seeds	Sunflower, oil palm, groundnut etc.

Source: Soman (2009).

Research Evidences on Fertigation in Field Crops

Impact of Drip Fertigation in Sugarcane

Sugarcane is a very important commercial crop grown in the country. India has the second position in the sugarcane areas and production next to Brazil in the world. The use of fertiliser and irrigation water is also high. Various components of high tech farming are introduced from time to time in this crop to enhance the productivity. Fertigation is one such agro-technique which has proved to be a catalyst to boost the productivity of sugarcane (Table 15.3). Water and plant nutrient are the key components to enhance sugarcane productivity.

Table 15.3: Impact of Drip Fertigation in Sugarcane

Particulars	Drip	Flood	Gains Over	Flood
Yield (t ha⁻¹)	85	55	54.5	30
Water saving (mm)	1200	2200	45.5	1000
Electricity consumption	900	2100	58.5	1260
Water used per tonnes cane production (mm)	25.9	14.1	44	64.75
Cost per tonne (Rs.)	379.4	54.1	29.9	161.6
Electricity use per tonnes production (Kwh)	10.6	39.3	73	28.7

Source: Neena *et al.* (2009)

Sugarcane with Water Soluble Fertilizers (WSF)

The results indicated that highest yield of sugarcane was recorded under drip fertigation with water soluble fertilizer at 75 per cent NPK recommended dose (212.35 t ha⁻¹) when compared to control (Surface irrigation + soil application of NF at 100 per cent NPK does (155.20 tha⁻¹). This was followed by fertigation with WSF at 100 per cent NPK recommended dose (206.65 t ha⁻¹) (Table 15.4). The highest sugar yield of sugarcane was recorded under drip fertigation with water soluble fertilizer at 75 per cent NPK recommended dose (31.80 t ha⁻¹) when compared to control (surface irrigation + soil application of NF at 100 per cent NPK dose, 22.97 t ha⁻¹). This was followed by fertigation with WSF at 125 per cent NPK recommended dose (31.42 t ha⁻¹).

Table 15.4: Drip Fertigation in Sugarcane (Co.853) with Water Soluble Fertilizers

Treatments	Cane Yield (t ha⁻¹)	Sugar Yield (t ha⁻¹)	Per cent Water Saving	Discoun-ted BC Ratio	Net Income (Rs. ha⁻¹)
Fertigation with WSF at 125 per cent NPK	208.9	31.42	28.95	6.39	55501
Fertigation with WSF at 100 per cent NPK	206.6	30.69	28.95	6.69	58811
Fertigation with WSF at 75 per cent NPK r	212.4	31.80	28.95	7.48	67687
Drip irrigation + soil application of NF at 100 per cent NPK	172.2	25.25	28.95	5.90	49995
Surface irrigation + soil application of NF at 100 per cent NPK	155.2	22.97	-	1.92	50415
C.D. (p=0.05)	2.14	-	-	-	-

Source: Asoka Raja (2002a).

Sugarcane with Conventional Fertilizer (NF)

Fertigation at 100 per cent dose has registered a higher cane yield of 173.3 t ha⁻¹, while soil application at same dose has given only 135.3 t ha⁻¹ (thus 28.09 per cent increase over control). Similarly fertigation at 100 per cent dose also recorded higher sugar yield (23.19 tha^{-1}) with higher B:C ratio (1.82) compared to control (Table 15.5).

Table 15.5: Drip Fertigation with Conventional Fertilizers in Sugarcane

Treatment	CCS (per cent)	Cane Yield (tha^{-1})	Sugar Yield (t ha^{-1})	B:C Ratio	Per cent Water Saving
Drip irrigation with N and K in 4 splits	13.38	173.3	23.19	1.82	8.9
Control (Surface irrigation and soil application of N and K at 100 per cent dose)	12.84	135.3	16.37	1.78	-

Fertigation in Cotton Hybrid with Conventional Fertilizers (NF)

Application of fertilizers through drip irrigation (100 per cent N and K in 6 equal Splits) in hybrid cotton (TCHB 213) increased the kapas yield (2367 kg ha^{-1}) which was 43.72 per cent higher compared to surface irrigation and soil application of 100 per cent NPK dose (Table 15.6). Fertigation at 100 per cent recommended dose of N and K as urea and potash applied in 4 equal splits (Basal, 35 DAS, flowering and boll formation) and 6 splits (at 20 days interval from sowing) were found to be superior to conventional fertilization. Increasing the splits (6) had increasing B:C ratio compared to 4 splits (1.72).

Table 15.6: Fertigation with Conventional Fertilizers in Hybrid Cotton

Treatment	Kapas Yield (kg ha^{-1})	B:C Ratio	Per cent Increase
Fertigation with N and K at 100 per cent dose in 4 equal splits	2239	2.49	35.94
Fertigation with N and K at 100 per cent dose in 6 equal splits	2367	2.63	43.72
Fertigation with N and K at 75 per cent dose in 4 equal splits	1892	2.20	1487
Fertigation with N and K at 75 per cent dose in 6 equal splits	2112	2.76	28.23
Control - Surface irrigation + soil application of NPK	1647	1.82	-
CD	224	0.319	

Fertigation in Maize Hybrid

The fertilizer solution was prepared by dissolving the required quantity of fertilizer with water in 1:5 ratios and injected into irrigation through venture assembly. Considering the nutrient uptake pattern at phonological growth phases of maize, the fertigation schedule was worked out (Table 15.7). Fertigation was given once in three days.

Table 15.7: Fertigation Schedule for Maize

Crop Stages	Quantity (per cent)		
	N	P	K
Vegetative stage (6-30 days)	25	25	25
Reproductive stage (30-60 days)	50	50	50
Maturity stage (60-75 days)	25	25	25
Total	100	100	100

Source: Anitta Fanish and Muthukrishnan (2013).

Table 15.8: Effect of Drip Fertigation on Grain Yield of Maize

Treatments	Grain Yield (kg ha^{-1})
Drip fertigation +75 per cent RDF(NF)	5885
Drip fertigation +100 per cent RDF(NF)	6321
Drip fertigation +75 per cent RDF (50 per cent P and K-WSF)	6578
Drip fertigation +100 per cent RDF(50 per cent P and K-WSF)	7309
Drip fertigation +100 per cent RDF	5386
Surface fertigation +100 per cent RDF	4720
CD (p=0.05)	235

NF: Normal fertilizers; WSF: Water soluble fertilizers; RDF: Recommended dose of fertilizer.

Source: Anitta Fanish and Muthukrishnan (2013).

Generally the maize grain yield increased with increase in fertilizer level (Table 15.8). Drip fertigated maize at 100 per cent RDF with 50 per cent P and K through WSF recorded significantly higher grain yield of 7.3 t ha^{-1}. The yield increases over drip irrigation with soil application of fertilizer was 35 per cent. Application of water soluble fertilizer also influenced the grain yield of maize compared to straight fertilizer. Drip fertigation with 100 per cent RDF in which 50 per cent P and K as WSF increased the grain yield to the tune of 15.5 per cent compared to drip fertigation of 100 per cent RDF with normal fertilizer. The increase in yield under 100 per cent RDF with P and K as WSF might be due to the fact that fertigation with more readily available form has resulted in higher availability of all the three (NPK) major nutrients in the soil solution which led to higher uptake and better translocation of assimilates from source to sink thus in turn increased the yield.

Fertigation in Horticulture Crops

As the horticultural crop production is getting momentum in India, where drip fertigation use is more convenient and economically attractive, the potential of increase in the drip fertigation is bright. Cultivation of non traditional vegetable like green capsicum is more profitable than traditional vegetables. Coloured capsicum cultivation through high-tech agriculture provides still higher income.

Fertigation in Coloured Capsicum

A success story of coloured capsicum cultivation through high-tech agriculture using fertigation by the farmer Mr. Minatai Visnu Jagtap, Village Pimpalgao Vasant, P. O. Pimpalgao Vasant, District Nasik, Maharashtra (India) is find a place here.

Techniques Adopted

In the first week of June, 2008 hybrid coloured capsicum seeds were planted in nursery. Three feet wide beds were made up of soil, coco pit, and organic manure. Seedlings were planted in row spacing of 18 inches and plant to plant spacing of 12 inches. The total plant population in 0.4 ha was 11000. A drip lateral was placed in between the two rows. Other practices followed were as follows:

1. After planting, daily irrigation through drip for 15-20 minutes.
2. Daily fertigation after 45 days of planting as mentioned in Table 15.9.
3. In addition, top dressings of fertiliser were also done as indicated in Table 15.10.
4. Disease –pest infestation was less.
5. Powdery mildew, downy mildew and sucking insect attack were suitably taken care of.
6. Harvesting started after two months of planting
7. Harvesting was done thrice in a week in the morning.

Fertigation in Hybrid Tomato with Speciality Liquid and Water Soluble Fertilizers

Drip fertigation in hybrid tomato with WSF at 75 per cent NPK applied drip through at 80 per cent of 2 days CPE was found to be superior technology, registering 26.0 per cent increased fruit yields, 22.99 per cent water saving, 25 per cent saving in fertilizers, 57.67 per cent increased fertilizer use efficiency with higher B:C ratio of 3.47 over soil application of normal fertilizers (Table 15.11).

Table 15.9: Weekly Fertigation Schedule

Day	Name of the Fertilizer	Quantity (kg)
Monday	Calcium nitrate	6
Tuesday	12:61:0	2
Wednesday	13:40:13	3
Thursday	13:0:45	4
Saturday	Zinc sulphate + hexolin	As per recommendation
Sunday	Nicolef + Magnesium sulphate	As per recommendation

Source: Biswas (2010).

Table 15.10: Top Dressing of Plant Nutrients (per acre)

Name of the Fertilizer/Manure	Quantity (kg)	Time of Application
DAP	100	As those were mixed together and applied thrice at an interval of 2 months
SSP	50	
MOP	50	
Borocole	100	
Organic manure	150	

Source: Biswas (2010).

Table 15.11: Comparison of Source of Fertilizers for Fertigation in Hybrid Tomato

Treatments	Fresh Root Yield (t ha⁻¹)	Per cent Increased Yield Over Control	Net Income (Rs. ha⁻¹)	Discounted B:C Ratio
MSF 100 per cent WSF – 2 days	42.31	65.73	75868	9.39
MSF 75 per cent WSF – 2 days	41.93	64.24	76582	9.47
MSF 50 per cent WSF – 2 days	37.52	46.96	67198	8.48
MSF 100 per cent WSF – 4 days	39.53	54.83	68918	8.66
MSF 75 per cent WSF – 4 days	37.24	45.86	64857	8.24
MSF 50 per cent WSF – 4 days	34.58	35.45	59848	7.79
MSI 100 per cent NF – 2 days	35.51	39.09	63917	8.18
MSI 100 per cent NF – 4 days	32.10	25.73	55394	7.24
Control	25.53	-	52538	4.65
S.Ed	0.76	-	-	-
C.D. (p=0.05)	1.60	-	-	-

Source: Asoka Raja (2002c).

Micro Sprinkler Fertigation in Radish with Water Soluble Fertilizer (WSF)

Micro sprinkler fertigation once in 2 days with 75 per cent NPK dose with water soluble fertilizers like Mono Ammonium Phosphate (MAP) and Potassium Nitrate (multi-K) has resulted 41.93 t ha⁻¹ which was 64.24 per cent increase over control (surface irrigation and soil application there is water soluble fertilizers at 100 per cent (Table 15.12). Through micro sprinkler irrigation there is water saving of 62.42 per cent compared to surface irrigation (control). Micro sprinkler fertigation once in 2 days at 75 per cent dose with WSF has also registered higher net income of Rs. 76,582 with a high discounted B:C ratio of 9.47. Thus an additional net income of Rs. 24, 44 was realized compared to surface irrigation and conventional fertilization.

Table 15.12: Fresh Root Yield of Radish under Micro Sprinkler Fertigation

Treatments	Fruit Yield (t ha⁻¹)	Water Use (mm)	Per cent Saving	Per cent increase in FUE	B:C
Liquid fertilizers	62.394	390.6	22.99	57.67	2.48
Water soluble fertilizers	63.916	390.6	22.99	57.67	3.47
Normal fertilizer	50.603	390.6	22.99	-	3.06

Source: Shobana and Asoka Raja (2002).

Fertigation in Banana

Banana with Water Soluble Fertilizers (WSF)

Drip fertigation in banana (Nendran) with water soluble fertilizer at 125 per cent NPK has registered the highest fruit yield of 42.65 tha⁻¹ which was 66.76 per cent increase over surface irrigation and soil application of normal fertilizers (Table 15.13). Fertigation at 100 per cent dose also registered higher fruit yield of 35.1t ha⁻¹ when compared to control (25.6 t ha⁻¹).

Table 15.13: Drip Irrigation with Specially Water Soluble Fertilizers in Banana (Nendran)

Treatments	Yield (t ha⁻¹)	Per cent Water Saving	B:C Ratio	Net Income (Rs.ha⁻¹)
Dip fertigation with WSF at 125 per cent NPK	42.7	35.2	1.96	134768
Dip fertigation with WSF at 100 per cent NPK	37.5	35.2	1.98	117804
Dip fertigation with WSF at 75 per cent NPK	35.1	35.2	2.27	114991
Drip irrigation + soil application of NF at 100 per cent NPK	30.3	35.2	2.01	94741
Surface irrigation + soil application of NF at 100 per cent NPK	25.6	-	1.71	80791

Banana with Conventional Fertilizer (NF)

Fertigation in banana (Robusta) with conventional fertilizer like urea and potash with 25 LPD + 100:30:150 g NPK/plant has registered higher fruit yield of 95.00 t/ha in plant crop (Table 15.14). Thus there an increased fruit yield of 61.07 per cent was achived through fertigation compared to basin irrigation and soil application of conventional fertilizers (200:30:300 g NPK/plant).

Economics of Micro Irrigation and Fertigation

The cost for installing drip irrigation varies from Rs.20,000 to 25,000/ha for wide spaced crops like coconut, mango *etc.* to Rs. 50,000 to 70,000/ha for closely spaced crops like sugarcane, cotton, vegetables *etc.* The cost of the system depends upon the crop, spacing, quantity of water required, distance from water source *etc.* The economics of mico-irrigation has been calculated with and without fertigation and presented in Table 15.15. It is observed that the payback period is about one year for most of the crops and the benefit cost ratio varies from 2 to 5.

Table 15.14: Fertigation with Conventional Fertilizers in Banana (cv. Robusta)

Treatments	Bunch Weight (kg)	Yield (t ha⁻¹)	Per cent Increase over Conventional	Cost Benefit Ratio
Plant crop 25 LPD + 100:30:150 g NPK plant⁻¹	38.00	95.00	61.07	1:1.78
Ratoon crop 25 LPD + 50:30:225 g NPK plant⁻¹	44.42	111.05	88.28	1:3.21
Basin irrigation + 200:30:300 g NPK plant⁻¹soil application	23.59	58.98	-	1:0.93

Source: Kumar (2003).

Table 15.15: Cost of Micro Irrigation with and without Fertigation System for Various Crops

Crops	Spacing (m)	System Type	System Cost (Rs. ha⁻¹)	
			Without Fertigation	With Fertigation
Coconut	8.0x8.0	Dipper	20316	24049
Mango	10x10	Dipper	17784	21517
Citrus	6.0x6.0	Dipper	25012	28745
Sapota	9.0x9.0	Dipper	18998	22731
Pomogranate	4.5x2.7	Dipper	27952	31685
Grapes	2.7x1.8	Dipper	46769	50502
Papaya	1.8x1.8	Dipper	45000	48500
Banana	1.8x1.8	Dipper	45000	48733
Sugarcane, Mulberry, cotton, roses	1.8x0.6	Dipper	70000	63956
Vegetables	0.75x0.60	Dipper		73733

Source: WTC Annual Report, 2003.

Constraints and their Solutions for Successful Adoption of Fertigation

High Initial Cost

Cost can be brought down by cost cutting measures like use of micro tubes, paired row system *etc.*

Clogging of Lines

Due to chemical precipitation: For HCO_3 precipitation use of acid fertilizers and acids like H_3PO_4, HNO_3, and HCl and, check the compability and solubility of solid fertilizers before use.

Due to Microorganisms

Use of acids/chlorine, flush the system after fertigation

Salt Injury

Severe problem with saline irrigation water and in arid climate. When saline irrigation water is used for irrigation apply extra water for leaching of salts beyond crop root zone. Fertigation with NO_3 as it competes with Cl ions. Use of plastic mulch/sub-surface irrigation to reduce evaporation

Nutrient Deficiency

In heavy soils due to low water infiltration denitrification may occur at high soil temperature and hence low concentration of N and regulation of water supply is required. Hydrolysis of urea may lead to NH_3 toxicity or NH_3 volatilization which may pose acidification of irrigation water.

Oxygen Deficiency

Continuous water supply may lead to exclusion of oxygen from saturation zone and hence deliver optimum amount of water is needed.

Conclusion

Fertigation offers an opportunity to optimize field and horticulture crops production system with respect to both irrigation and fertilization. It provides variety of benefits to users like high crop productivity and quality, resource use efficiency, environmental safety, flexibility in operations, effective weed management and successful crop cultivation on fields with undulating topography. It is considered eco-friendly as it avoids leaching of nutrients especially $N-NO_3$. Vegetables have been found responsive to fertigation due to wide spacing nature, continuous need of water and nutrients at optimal rate to give high yield with good quality and high capital turnover to investments. Even though the initial cost of establishing the fertigation system is higher but in long term basis it is economical compared to conventional methods of fertilization as it brings down the cost of cultivation. To get the desired results it requires high management skills at operator level like selection of fertilizers, timing and rate of fertilizer injection, watering schedule as well as the maintenance of the system.

Future Thrust

Need to develop recommendations for the most suitable fertilizer formulations including the basic nutrients (NPK) and microelements according to local soil type, climate, crops and their physiological stages. Need to work on reducing the initial cost of establishment through continuous research and development in technology which suits best to Indian conditions. Therefore to make the agriculture sustainable and economically viable and to ensure food and nutritional security of the burgeoning population there is need to promote the fertigation at large scale by the concerned stakeholders.

References

Anitta Fanish, S., Muthukrishnan, P., 2013, Nutrient Distribution Under Drip Fertigation Systems. World J. Agril. Sci., 9 (3): 277-283.

Asokaraja,N.2002a, Maximising the productivity and quality of banana and sugarcane with water soluble fertilizers through drip fertigation. Annual Report. 2001-2002 WTC, TNAU,Coimbatore.

Asoka Raja, N., 2002b, Maximizing the productivity and quality f banana in sugarcane with water soluble fertilizers through drip fertigation. Annual report, 2001 – 2002, WTC, TNAU, Coimbatore.

Asoka Raja, N., 2002c, Performance evaluation of drip fertigation system with liquid and water soluble fertilizers for increasing the yield and quality of vegetable crop (tomato), Final report of ICAR, Adhoc project 1999-2002, WTC, TNAU, Coimbatore.

Biswas, B.C., 2010, Fertigation in High Tech Agriculture. A Success Story of A Lady Farmer. *Fert. Marketing News*, 41 (10),4-8.

Kumar, N., 2003, Fertigation with conventional fertilizers in Banana (cv. Robusta). *Annual report of Horticultural College and research Institute*,TNAU, Coimbatore.

Neena, Chauhan and Chandel, J S., 2009, *Indian J. Agric Sci.* 78(5) 389-393.

Rajput, T. B. S., 2010, Role of water management in improving agriculture productivity. *Indian J Fertilisers* 6 (4).

Satisha, G.C., 1997, Fertigation – A new concepts in Indian Agriculture, *Kisan World*, P.30.

Shobana, Asokaraja, 2002, Performance evaluation of micro sprinkler fertigation with water soluble fertilizers on water, fertilizer use and yield of radish,. *M.Sc. (Agri.) thesis*, Department of Agronomy, TNAU, Coimbatore.

Soman, P, 2009, Improving Water Use Efficiency to Enhance Crop Productivity, *FAI Annual Seminar*.

WTC *Annual Report*, 2003, Annual report of water technology centre, TNAU, Coimbatore.

16

Micro-Irrigation to Enhance Water Productivity

G.V. Srinivasa Reddy, M.R. Umesh and N. Anand

Micro-irrigation refers to application of small quantity of water per unit time for longer period through efficient devices. So, preferably drip and micro-sprinklers irrigation are referred to micro-irrigation. Drip irrigation is the important technology, which has to be adopted by farmers to enhance water productivity (Alam and Kumar, 2001). In command areas also, the micro-irrigation plays an important role in lift irrigation schemes as well as in gravity irrigation practices, as stored water is used for crop production at many places.

Drip irrigation is the method of frequent and slow application of water to the soil, near to the root zone of the plants through mechanical devices called emitters. Drip irrigation is also known as trickle irrigation which differs from the other conventional methods of water application. With other irrigation methods, we are irrigating the soil whereas, drip irrigation system applied water is directly in the vicinity of the root zone, wetting a limited amount of surface area and depth of soil. Drip irrigation applies water slowly almost matching with the consumptive water use by the plant to keep the sufficient soil moisture for plant growth (Ashwani Kumar, 1996). Water is discharged drop by drop at very low rate of a few litres per hour. This system minimizes the water losses by avoiding deep percolation, runoff and evaporation.

Technical Details of Various Components of Drip Irrigation System

The various components used in drip irrigation system are listed below:

1. Pumping set, 2. Filters, 3. Fertilizer tank and applicators, 4. Main line, 5. Sub main, 6. Lateral, 7. Drippers, 8. Other fittings and accessories (Flow control valves, Non return valve, Vacuum release valve, Air release valve, Pressure regulators, Reducers, Elbows, End caps, Couplers, Tee, Reducing tee, Male threaded adapter, Female threaded adapter, Service saddle, Water meter, Couplers, Flush valve, Pressure gauge, Grommet and take off, *etc.*)

1. Pumping Set

Pump is required to carry water from the source through the main line and laterals up to the sprinkler head or nozzle from where it is sprayed on the field and crops. The choice of the pump set will vary with discharge, pressure and the vertical distance to the water source. The centrifugal pumps are usually employed to lift water from open sources where suction head is less than 8 m. For bore wells, submersible pumps are used.

2. Filters

Filters are called as heart of micro-irrigation system. The filters remove the suspended impurities from the irrigation water and prevent the blockage of drippers (Mane *et al.*, 2008). The type of filtration depends upon the source of water used and the size of the nozzle. Generally two stage filter unit (use of one filter among primary filters and one filter among secondary filters depending on water source) is recommended.

Primary Filters

1. Sand Filter (Gravel Filter)

☆ Filters suspended particles and algae

☆ Must for open source of water

2. Hydro Cyclone Filter

☆ Filters sand material

☆ Used when water source contains sand (Rivers, deep bore wells, canals *etc.*)

Secondary Filters

1. Screen Filter

☆ Filters the inert material depending upon hole size of the screen

2. Disc Filter

☆ Most effective filtration will be done through disc filter

3. Fertilizer Tank and Applicators

The fertilizer may be effectively applied through drip irrigation system. The tank is provided to mix the required fertilizers. The applicators (by-pass flow/pressure by pass flow/venturi/injection pumps) are used to deliver fertilizer solution through the drip system.

4 and 5. Main and Sub Main Lines

The main and sub mains are made up of Ploy Vinyl Chloride (PVC) and the different sizes of PVC pipes available in the present market are presented below:

Inch	0.5	0.75	1	1.25	1.5	2	2.5
mm	20	25	32	40	50	63	75
Inch	3	4	5	6	7	8	9
mm	90	110	140	160	180	200	225

The 6 metre (20 feet) PVC pipes are available in the market and outer diameters of the pipes are considered.

The different grades of PVC pipes are presented below:

Sl.No.	Grade	Pressure (kgf/cm²)	Thickness	Colour code
1	Class I	2.5	Least	Red
2	Class II	4.0	Intermittent	Blue
3	Class III	6.0	Intermittent	Green
4	Class IV	10.0	Highest	Yellow

The maximum pressure that the particular class of pipes can withstand is shown in the above table. The main line carries the water from filtration system to the sub main. They are usually made up of rigid PVC to avoid corrosion and clogging. Usually they are placed 60 to 90 cm below the ground, so as not to interfere with cultivation practices. Their diameter is based on the system flow capacity. The frictional head loss in main pipes should not be more than 5 m per 1000 m running length of main pipeline. The sub main distribute the water from main line to the laterals. They are also buried in the ground below 45 cm and made of rigid PVC material. The diameter of sub main is usually smaller than main line, for economy. There may be number of sub mains from one main line depending upon the plot size and crop type.

6. Laterals

The lateral distributes water to the emitter which delivers water directly to the root zone. Laterals are small diameter flexible pipes or tubing made of low density poly ethylene (LDPE) or linear low density poly ethylene (LLDPE) and having 12, 16 and 20 mm diameter. They are coloured black to avoid algae growth inside and minimize the damaging effect of ultraviolet radiation. They can withstand maximum pressure of 4 kg/cm^2. They are connected to sub main at predetermined distance. The pressure variation between two extreme points of a lateral should not be more than 15 to 20 per cent and discharge variation should not be more than 10 per cent. On sloping ground, the laterals are placed along the contour or across the slope (Keller and Karmeli, 1974).

7. Drippers

The dripper is an emitter for discharging water from lateral to the soil. The drippers are also called as emitters. They discharge water from lateral on to the soil near the plants. There are various types and sizes of drippers, based on different operating principles. They are made up of plastic such as polyethylene or polypropylene. Their discharge ranges between 1 to 16 l/h. Each dripper has its own characteristics, advantages and limitations, which determines its use. The drippers can be classified according to working principle, discharge, type, structure, working pressure, durability, pressure compensating and non-pressure compensating properties. Pressure compensating drippers are used in undulated lands for drip irrigation in order to avoid variations in discharge rates of drippers.

The main principle in dripper selection is to achieve the lower discharge with longer size of water passage. The large water passage is essential to minimize clogging and provide the lower discharge for cheapest set-up. Therefore, an emitter is necessary. Emitters may be punched externally on the lateral or may be fitted in the lateral. Accordingly, they are called as on line or in line emitters, respectively.

8. Other fittings and Accessories

(Flow control valves, Non return valve, Vacuum release valve, Air release valve, Pressure regulators, Reducers, Elbows, End caps, Water meter, Couplers, Pressure gauge, *etc.*):

The main function of the fittings and accessories used in the drip irrigation system are presented below:

- ☆ *Flow control valves:* Valves control the passage of water through the pipe network and helps in diversions.
- ☆ *Non return valve:* The function of this valve is to prevent return flow of water.
- ☆ *Vacuum release valve:* The function of this valve is to ensure that there is no return flow of water floe of water to the potable water system and that is affected by introducing air to the pipeline whenever vacuum is created.
- ☆ *Air release valve:* Air release valve is normally open valve. When the pressure within the system exceeds atmospheric pressure, air is expelled.
- ☆ *Pressure regulators:* Regulates excessive pressures.
- ☆ *Reducers:* The reducers are necessary for coupling pipes of different

diameters.

☆ *Elbows:* They are used at joints for changing the direction of water flow.

☆ *End plug:* These are placed at the end of a line to close the pipe.

☆ *Water meter:* Water meter is used for measuring the total quantity of water delivered through the concerned section of pipe.

☆ *Flange, coupling and nipples:* These are necessary for making proper connection to the pump and suction delivery.

☆ *Pressure gauge:* It is necessary to know whether the sprinkler is working with the desired pressure in order to deliver the water uniformly.

☆ *Flush valve:* It is provided at the end of each sub main to flush out the water and dirt accumulated at the end of sub main.

☆ *Grommet and take –off:* These are used to connect the lateral to sub main. A hole is punched with hand drill of predetermined size in sub main. The grommet is fixed into the hole of on sub main. Take-off is pressed into the grommet with take off punch up to the step provided. Grommet acts as a seal. The sizes are different for 12, 16 and 20 mm lateral diameter.

☆ *End caps:* They are used to close the lateral ends, sub main ends (some times) and main line ends.

Clogging

Clogging of emitters is one of the major problems in drip irrigation (Sivanappan *et al.*, 1987). The various contaminants are physical factors (sand, silt, clay, plastic *etc.*), chemical factors (calcium and magnesium carbonate, calcium sulphate, heavy metals, fertilizers *etc.*) and biological factors (bacteria and algae *etc.*). The various preventive and corrective measures can be taken to avoid clogging (Bucks *et al.*, 1983). The preventive measures include soil and water testing, selection of filters, pump suction location, use of safety valves, selection of proper sizes of pipes, proper design etc,. The corrective measures include regular flushing of filters, flushing of sub-main and laterals, chemical treatments (acid and chlorination) *etc.* The acid treatment may be continuous or intermittent. The commercially available acids like sulphuric acid (H_2SO_4), hydrochloric acid (HCl), nitric acid (HNO_3) and phosphoric acid (H_3PO_4) can be used for acid treatment. While, bleaching powder ($CaOCl_2$) and Sodium hypochloride (NaOcl) can be used for chlorination treatment.

References

Alam, A., Kumar, A. 2001. Micro-irrigation system – Past, Present and Future. Eds. Singh, H.P., Kaushish, S.P., Kumar, A., Murthy, T.S., Samuel, J.C. in "Microirrigation" CBIP, pp.17.

Ashwani Kumar, 1996. Prospects, potential and limitation of drip irrigation system. *J. Water Mangt.*, 4(1-2), pp. 24-27.

Bucks, D.A., Nakagama, F.S., Warrick, A.W. 1983. Principles, practices and potentials of trickle (drip) irrigation. *Adv. Irrin.* 1: 219-299.

Keller, J., Karmeli, D. 1974. Trickle Irrigation Design Parameters. Transaction, ASAE, 17(4), pp: 678.

Mane, M.S., Ayare, B.L., Magar, S.S. 2008. Principles of drip irrigation system. Published by Jain Brothers, New Delhi.

Sivanappan, R.K., Padma Kumari, O., Kumar, V. 1987. Drip Irrigation, Keerthi Publishing House Pvt. Ltd. Coimbatore, Tamil Nadu.

17

Methods of Irrigation, Water Measurement and Estimation of PET

M.R. Umesh, M.Y. Ajaykumar and N. Manjunatha

Water is considered to be precious resource of the world and continued as most wanted commodity. Among various uses of water, agriculture sector needs much higher than domestic, industries and special purposes like gardening, cleaning, *etc*. As quoted by several conservationists, in future global disputes and wars would be only on water crisis. Even at national level major river disputes continued over decades because of water shortage during droughts and summer months. The major concern in major commands is lower irrigation efficiency and shrinking of storage capacity. Inspite of relatively enhanced application and distribution efficiencies irrigation efficiency is much lower than 30 per cent in most of the commands. At large scale most of the crop plans and actions to enhance water available period was not successful. Judicious and precise use of water is most important rather use it unscientifically for single crop.

Crop diversification can reduce quantum of water use and it will spread throughout the year. In commands water availability period depends on success of monsoon and water in the reservoir. During rabi/summer 2015 in TBP command, cultivation of paddy was restricted due to failure of monsoon and low water storage in reservoir. However, it would have been possible to grow remunerative crops over paddy with just increase in irrigation efficiency by 5-10 per cent. The knowledge of different crops is needed to select water efficient and economical crop. In addition to economics water requirement is also vital for selection of specific crop. Maize, *rabi* sorghum, sunflower, gum guar, mustard are found better over paddy in water scarce condition. In TBP and UKP commands mustard was picked up especially in paddy fallows of tail ends. It was found remunerative with least investment and water. The water requirement depth for each irrigation and irrigation interval for different crops, soil types and climate during peak period for water was given in Table 17.1.

Table 17.1: Estimated Irrigation Schedule for Major Field Crops in Peak Periods

Crop	Intervals in Days											
	Sandy Soil				Loamy Soil				Clayey Soil			
	1*	2*	3*	Depth (mm)	1	2	3	Depth (mm)	1	2	3	Depth (mm)
Banana	5	3	2	25	7	5	4	40	10	7	5	55
Cotton	9	6	5	40	11	8	6	55	14	10	7	70
Sorghum	8	6	4	40	11	8	6	55	14	10	7	70
Groundnut	6	4	3	25	7	5	4	35	11	8	6	50
Maize	8	6	4	40	11	8	6	55	17	10	7	70
Peas	6	4	3	30	8	6	4	40	10	7	5	50
Soybean	8	6	4	40	11	8	6	55	14	10	7	70
Sugarcane	8	6	4	40	10	7	5	55	13	9	7	70
Sunflower	8	6	4	40	11	8	6	55	14	10	7	70
Wheat	8	6	4	40	11	8	6	55	14	10	7	70
Tomato	6	4	3	30	8	6	4	40	10	7	5	50

1*– Low temperature (15ºC); 2* medium temperature 15-25º C; 3*- High temperature > 25ºC.

Crop water requirement (WR) depends on crop, climate and soil factors at different growth stages. It can be minimized either by selection of water efficient cultivars, soil and water conservation measures or with precise water application. An account of water requirement will be worked out by accurate measurement of irrigation water in a unit time. With available water measurement techniques choose most appropriate for the situation. Even crop planning also depends on water measurement and water supply period. In this chapter different water measurement techniques and crop planning is discussed in detail.

Units of Water Measurement

There are various units to express the volume of water stored in ponded structures/tanks *viz.*, litres, cubic meter, hectare-centimetre, hectare- meter, acre-inch, acre-foot and so on. Whereas, flowing wateris measured as amount of water flow per unit time and expressed in litre/second (lps), cubic feet/second, cubic meter/sec (cumec), cubic feet per second (cusec), thousand million cubic feet (TMC),

Conversion of Volume of Water

$$
\begin{aligned}
1 \text{ ha-cm} &= 10{,}000 \text{ sqm} \times 1 \text{ cm} \\
&= 10{,}000 \times m \times m \times 1 \text{ cm} \\
&= 10{,}000 \times 100 \text{cm} \times 100 \text{cm} \times \text{cm}
\end{aligned}
$$

$$= 10,000,00,00 \text{ cm}^2 \text{ x cm}$$

$$= 10,000,00,00 \text{ cc}$$

$$= 1,00,000 \text{ Litre of Water } (1000 \text{ cc} = 1\text{litre})$$

1 ha-cm = 100 m^3

Therefore 1 ha-cm = 1,00,000 litre of water

$$= 1 \text{ lakh litre of water}$$

Methods of Water Measurement

i. Volume Method

☆ Collect the flow in a container of known volume for a measured period of time

☆ Ordinary bucket or barrel used as container

$$\text{Discharge rate} = \frac{\text{Volume of container litres}}{\text{Time required fillingin seconds}}$$

ii. Velocity Area Method

Velocity of flow in the channel is measure by some means and the discharge is calculated from the area of cross section

Q= A x V

where,

Q = Discharge rate m^3/s

A = Cross section area of canal (m^2)

V = Velocity of flow (m/s)

iii. Float Method

$$Q = 0.85 \frac{(a+b)}{2} \times H \times V$$

where,

Q = Discharge cm^3/s

a = Channel width at the flow surface (cm)

b = Channel width at the bottom (cm)

H = Flow depth in the channel (cm)

V = Velocity in the channel cm s^{-1}

iv. Water Meters

☆ Utilize multi blade conical propeller made of metal, plastic or rubber, rotating in a vertical plane and geared to a totalizer in counter.

Types

☆ Low pressure line meters

☆ Open flow meters

☆ Vertical flow or hydrant type meters

v. Coordinate Method

Water flowing from wells discharge vertically or from small pumping plants discharge horizontallyis measured by coordinate method. It will measure both horizontal (X) and vertical distance (Y) from some reference point at the end of pipe. Coordinates are measured from the center of the end of the pipe with following formulae.

$$Q = \frac{Ca \, X \, \sqrt{g}}{\sqrt{2Y}}$$

where,

Q = Discharge rate m³/s

C = Coefficient of contraction

A = Cross sectional area of pipe m²

X = Distance at which water falls on soil surface

Y = Height at which water falls

A = Acceleration due to gravity m/sec²

vi. Water Measurement by Flumes, Weirs and Orifice

A weir is a notch or opening of some definite form installed on a channel or a stream through which waterfalls. Many materials like wood, concrete, mild steel, rigid PVC *etc.* can be used for construction of a weir. Mainly a cut on a sheet metal is used for fabrication of weirs to be installed at variations locations in a channel. Weirs can be classified as (i) broad crested and (ii) sharp crested weirs. The sharp crested weir is mostly used for measurement of irrigation water. It is nothing but a weir with thin edge such that the sheet of water flowing over it has the minimum contact area with it. The bottom most portion of the weir in touch with the water is called the weir crest. The sheet of water flowing over the weir is called the nappe. The top surface is the upper nappe and the bottom surface is the lower nappe. The depth of flow over the crest is known as head (H). It is measured at minimum distance of 4 H upstream from the crest. The horizontal distance from the end of the crest to the side of the channel is known as end construction. If both ends of the crest are away from the sides of the channel, then there are two end contractions.

If the length of the crest is same as the width of the channel, then there is no end contraction. The weir is said to have free flow condition if the surface of water downstream is below the crest level so that the nappe is surrounded by air. If the downstream water level is higher than the weir crest level then it is a submerged flow condition. Apart from being classified as sharp crested and broad crested, weirs can also be classified according to the shape of the notch.

VIa. Rectangular Weir

The rectangular weir takes its name from the shape of the notch. They are used to measure high discharges. Its crest is horizontal and the sides are vertical. In case the crest length is same as that of the channel width, it is known as suppressed weir, otherwise contracted weir. The discharge through rectangular weirs may be computed by the Francis' Formula stated below:

$$Q = 0.0184 \, L \, H^{3/2}$$

where,

Q = Discharge, l/second

L = Length of crest, cm

H = Head over the weir, cm

VIb. Trapezoidal Weir

The trapezoidal weir has a horizontal crest and the sides slope outward to give the notch a trapezoidal cross section. Commonly a side slope of 1 horizontal to 4 vertical is used and it is named as Cipoletti weir. It does not require any correction for end contractions and is used for measurement of medium discharges. Normally it is a sharp crested weir. It is named after its inventor Cesare Cipoletti, an Italian engineer. The discharge through Cipoletti weir is computed by $Q = 0.0186 \, LH^3/2$

VIc. Triangular or V-notch Weir

The 90° V-notch weir is commonly used to measure small and medium size streams. The advantage of the V-notch weir is its ability to measure small flows accurately. Thus the sides of the notch make an angle 45° with the vertical which gives a slope of one horizontal to one vertical. The discharge through a 90° V-notch weir may be computed by $Q = 0.0138 \, H^{5/2}$

Vid. Orifice

Orifices in open channels are usually circular or rectangular openings in a vertical bulkhead through which water flows. For measurement of water, orifices are fabricated by making an accurate cut of proper size and shape in a mild steel sheet, aluminium plate *etc.* Proper machining is done to have a sharp edge through which water flows. The cross sectional area of the orifice is small in relation to stream cross-section. These conditions allow complete contraction of the stream flow and the velocity of approach becomes negligible. Orifice may operate under free flow or submerged flow conditions. Under free flow conditions, the flow from the orifice discharges entirely in to air. In submerged flow orifices, the downstream water level is above the top of the opening and the flow discharges through opening into water. Free flow orifice plates can be used to measure comparatively small streams like the flow into border strips, furrows or check basins.

The discharge through an orifice is calculated by

$$Q = 0.61 \times 10^{-3} a \sqrt{2gH}$$

where,

Q = Discharge through orifice, l/s

A = Area of cross section of the orifice, cm²

g = Acceleration due to gravity cm/sec² (981 cm/sec²)

H = Depth of water over the center of the orifice (on the upstream side) in case of free flow orifice, or the difference in elevation between the water surface at the upstream and downstream faces of the orifice plate in case of submerged orifice, cm

VIe. Parshall Flume

There are several disadvantages for measurement of flow by weirs and orifices. They require considerable head loss, get silted up easily and the accuracy of measurement is affected. There are several other limiting conditions for installation, most of these disadvantages of weirs and orifices are largely overcome by use of Parshall flumes. The Parshall flume is an open channel type-measuring device that operates with a small drop in head. The loss of head for free flow limit is only about 25 per cent of that for weir. It is a self-cleaning device, sand and silt in the flowing water does not affect its operation or accuracy. (i) a converging upstream section, (ii) a throat which is a constructed section and (iii) a diverging downstream section. The floor of the upstream converging section is level and thus walls converge towards the throat. The floor of the throat is inclined downwards, but the walls are parallel. The floor of the diverging section slopes upwards and the walls diverge downstream. The size of the flume is determined by the width of the throat of the flume.

Parshall Flumes allow reasonably accurate measurement even when partially submerged. The velocity of the approaching stream has very little influence on its operation. Discharge through the flume can occur under either free flow or submerged flow conditions. To determine the discharge, two scales, H_a and H_b are provided at the upstream and downstream sections of the flume. Only H_a needs to be measured under free flow conditions. Free flow conditions are satisfied if the degree or percentage of submergence as represented by the ratio H_b/H_a is within the following limits. In recent version of Parshall flumes are made with readymade tables indicate Ha and H_b values correspond to total volume flow. There are different capacity Parshall flumes with 0.5 to 2 cusecs.

Crop Planning

Water requirement of crops is quantity of water needed for normal growth and yield may be supplied by precipitation or by irrigation or by both. The quantum of WR significantly influenced by crop, soil, climatic and management factors. The major crop factors are variety, growth stage, duration, plant growth and growing season *etc.* The soil factors are texture, structure, depth, topography, hydraulic conductivity, reflectivity *etc.* Whereas climatic factors includes temp, RH, wind velocity, sunshine hours, advective energy and crop management factors are tillage, fertilization, weeding, *etc.*

Crop water requirement is vital for crop planning. Based on water available period different crops will be selected for the region. An account of season, type of crop, ET, stage of the crop contribution from soil and rainfall is required for crop planning in commands. Water requirement of different crops irrespective of the season is provided in table. It depends on duration of crop, evaporative demand of climate and genetic makeup to water loss through transpiration. Among field crops WR is highest for paddy followed by maize and sorghum.

Low-Energy Precision Application (LEPA)

The low energy precision application (LEPA) **irrigation** concept was developed primarily to allow irrigators in arid and semi-arid areas to maximize the use of their total water resource and significantly increase irrigation efficiencies. It was particularly targeted to those areas experiencing declines in water availability due to dropping water tables, dwindling surface supplies, or supply decline from other socio-economic reasons.

The LEPA is a type of center-pivot irrigation system that was equipped with double-ended drag socks hanging down from a large water carrying pipes that apply water to the alternate rows and has the application efficiency of approximately 90–95 per cent. It distributes water directly to the furrow at very low pressure (6-10 psi) through sprinklers positioned 12-18 inches above ground level. Conventional high pressure impact sprinklers are positioned 5-7 ft. above the ground, so they are very susceptible to spray evaporation and to wind-drift, causing high water loss and uneven water distribution. LESA is same as LEPA except that instead of double-ended drag socks it has small water sprayers with nozzles very close to the ground that gently sprays water onto the crops with the application efficiency of 80–90 per cent. LEPA is designed to apply water more efficiently for center pivot irrigation systems. This reduces water use and water pump energy consumption by 15-30 per cent. It can provide more uniform irrigation application for all of your crops through the conversion of your center pivot irrigation system and lowering sprinkler heads so they are closer to crops. This greatly reduces water evaporation during irrigation, as well as reducing the overall pressure and energy required to efficiently water crops for a true low pressure way to save! LEPA is a great addition for agricultural producers on sandy soils and is currently being tested on additional soil types.

Considerations for Producers

1. **Simple Changes Equal Major Benefits for Crops:** The real value in a new LEPA system lies in its simplicity. Just by moving sprinklers closer to your crops you will save water, reduce energy consumption, and save big on operating costs.

2. **Increase yield with more efficient irrigation:** More efficient irrigation helps you save on water and energy use and can potentially lead to a greater overall yield for your crops.

Benefits LEPA irrigation systems place irrigation water directly into the furrows of the growing crops with nozzles placed very close to the soil surface. Utilizing this type of water placement reduces water losses by minimizing evaporation from leaf surfaces and wind drift. The results of improved placement and lower evaporative losses are conservation of water resources and energy.

The mechanical component of a LEPA irrigation system is a moving truss system with water conveyance tubes extending from the system mainline to near the soil surface, where correctly sized orifices control deposition of water to individual soil furrows.

Irrigation runoff prevention from the furrows as well as rainfall capture is accomplished with soil surface storage enhancement, which is primarily by furrow diking. Other soil manipulation techniques such as deep chiseling for runoff prevention are sometimes applicable for coarse-textured soils.

The capture of rainfall by surface storage in addition to *irrigation* runoff prevention and increased irrigation efficiency combine to enhance the total water resource utilization. Evapotranspiration is also altered since the soil evaporation component is decreased due to alternate furrow application capabilities.

Initial irrigation efficiency tests of the LEPA concept involved every furrow LEPA irrigation, which was compared to both graded furrow and sprinkler methods. The tests showed LEPA application, distribution, and water use efficiencies alongwith energy saving potential were superior to that of furrow and sprinkler delivery systems. Additional tests were initiated to evaluate agronomic yield response to LEPA irrigation under both alternate furrow and every furrow application. In these tests *LEPA* was compared to drip irrigation on cotton, corn, and soybeans, with irrigation initiation based on soil matric potential.

Several years have been devoted to establish optimum management criteria for LEPA irrigation on different crops. The most extensive study has been on cotton. Additional data continues to be assimilated for other crops. Almost without exception (excluding vegetables), favorable yield responses are obtained from alternate furrow irrigation and are possible only with LEPA if a moving irrigation system is involved. Crops also respond to irrigation frequency and in general more frequent irrigation produces higher yields. This enhances the LEPA irrigation method because it decreases runoff potential. There are also interactions between the quantity of water applied and the frequency with which it is applied on crop yield.

Surge Flow

Surge flow irrigation is a type of furrow irrigation that applies surges of water intermittently rather than in a continuous stream. These surges alternate between two sets of furrows for a fixed amount of time. The alternate wetting and "resting" time for each surge slows down the intake rate of the wet furrow and produces a smoother and hydraulically improved surface. By doing so, the next surge travels more rapidly down the wet furrow until it reaches a dry furrow. Surge irrigation provides more uniform water distribution, limits deep percolation, and can reduce tail water runoff. Water infiltration varies substantially based on the type of soil, soil compaction, and soil preparation. Surge flow does not work well on compacted soils, so it is more effective during pre-plant irrigation and the first seasonal irrigation following cultivation. Surge flow can cut water losses by up to 30 per cent in clay soils and can save more than 35 per cent of energy costs compared to simple furrow irrigation. Savings in energy and pumping costs can pay for the cost of surge irrigation valves within two years. Monitoring soil moisture is important for establishing on-off cycles for surge irrigation, and cycle length should be adjusted according to soil type. To accurately determine how much water is being applied, meters should be installed or readings from portable meters should be requested from the local water district. Surge irrigation increases fertilizer application efficiency and lowers salt loading by reducing deep percolation. It may not, however, improve yields when used on short level furrows where irrigation is relatively efficient. Using a computer program, some surge valves allow irrigators to adjust the valve controller for individual farm characteristics such as soil type, moisture content, slope, furrow size, infiltration rate and compaction.

Table 17.2: Water Requirement for different Crops

Crop	Water Requirement (mm)	Crop	Water Requirement (mm)
Paddy	900 – 2500	Tomato	600 – 800
Wheat	450 – 650	Potato	500 – 700
Sorghum	450 – 650	Pea	350 – 500
Maize	600 – 800	Onion	350 – 550
Sugarcane	1500 – 2500	Bean	300 – 500
Sugarbeet	550 – 750	Cabbage	380 – 500
Groundnut	500 – 700	Banana	1200 – 2200
Cotton	700 – 1300	Citrus	900 – 1200
Soybean	450 – 700	Grapes	500 – 1200
Tobacco	400 – 600	Pineapple	700 – 1000

To convert these values into volume basis follow these steps

For one hectare area

Volume = Area x Depth

$V = 10,000 \text{ m}^2 \times 0.001 \text{ m}$

$= 10 \text{ m}^3 \ (1 \text{ m}^3 = 1000 \text{ litre})$

$= 10000 \text{ litre/ha}$

Table 17.3: Estimated Evapotranspiration (ET) and Water Requirement of Crops grown In UKP Command

1	2		Initial Stage			Developmental stage			Mid Season		
			3	4	5	6	7	8	9	10	11
Crops/Season	Crop Duration Month and Year	Duration (Days)	Kc	E (mm/day)	ET	Kc	E (mm/day)	ET	Kc	E (mm/day)	ET
					Kharif						
Rice	July-Nov-15	150	1.125	4.5	253.13#	1.3	4.33	225.2	1.2	4.33	155.9
Maize	July-Oct-15	120	0.4	4.5	54.00	0.78	4.2	98.28	1.13	4.27	144.8
Cotton	July 15-feb 15	180	0.45	4.34	87.89	0.75	4.27	192.2	1.15	4.27	245.5
Groundnut	July-Oct-15	120	0.45	4.5	60.75	0.75	4.19	94.28	1.03	4.27	131.9
Soybean	July-Oct-15	110	0.35	4.5	47.25	0.75	4.2	94.5	1.08	4.27	138.3
Sunflower	July-Oct 15	110	0.35	4.5	47.25	0.75	4.2	94.5	1.13	4.27	144.8
Onion	July-Oct-15	120	0.5	4.5	67.50	0.75	4.2	94.5	1.03	4.27	131.9
Chilli	July 15-feb 15	180	0.5	4.5	67.50	0.78	4.2	196.6	1.13	4.27	193
					Rabi/summer						
Rice	Dec-March	120	1.125	4.58	154.58	1.3	4.83	188.4	1.2	5.97	214.9
Wheat	Nov 15-Feb 15	120	0.35	4.17	43.79	0.75	4.19	94.28	1.13	4.58	155.3
Rabi sorghum (Irrigated)	Oct 15-Jan 15	120	0.35	4.27	44.84	0.79	4.35	103.1	1.08	4.17	135.1
Groundnut	Dec-March-15	120	0.45	4.58	61.83	0.75	4.83	108.7	1.03	5.97	184.5
Onion	Dec-March-15	120	0.5	4.58	68.70	0.75	4.83	108.7	1.03	5.97	184.5
Sunflower	Dec-Feb 15	110	0.35	4.35	45.68	0.75	4.27	96.08	1.13	4.35	147.5

	12	13	14	15	16	17	18	19	20
		Late season		Sub total	Total	Conveyance Losses (30 per cent)	Col 16 +17	Application Losses (30 per cent)	Col 18 +19 Total WR (mm)
	Kc	E (mm/day)	ET						
Kharif									
Rice	1.00	4.62	138.6	772.77	972.77*	389	1361.87	408.56	1770.43
Maize	0.88	4.35	114.8	411.87	411.87	165	576.62	172.99	749.61
Cotton	0.85	4.18	159.9	685.45	685.45	274	959.62	287.89	1247.51
Groundnut	0.80	4.35	104.4	391.37	391.37	157	547.92	164.37	712.29
Soybean	0.75	4.35	97.88	377.97	377.97	151	529.16	158.75	687.91
Sunflower	0.75	4.35	97.88	384.38	384.38	154	538.13	161.44	699.57
Onion	0.88	4.35	114.8	408.78	408.78	164	572.30	171.69	743.99
Chilli	0.88	4.35	153.1	610.18	610.18	244	854.26	256.28	1110.53
Rabi/summer									
Rice	1.00	7.67	230.1	787.97	987.97*	395	1383.15	414.95	1798.10
Wheat	0.71	5.97	127.2	420.48	420.48	168	588.68	176.60	765.28
Rabi sorghum (Irrigated)	0.78	4.19	98.05	381.08	381.08	152	533.52	160.06	693.57
Groundnut	0.75	7.67	172.6	527.55	527.55	211	738.57	221.57	960.15
Onion	0.88	7.67	202.5	564.34	564.34	226	790.07	237.02	1027.09
Sunflower	0.75	4.17	62.55	351.77	351.77	141	492.47	147.74	640.21

ET=Kc *E* duration of the stage

Total= sum of initial, developmental, Mid-season, Later stages includes 200 mm for puddling

(Doddamani *et al.*, 2015).

If a crop need 400 mm of water 10000 x 400= 40, 00, 000 litre/ha/season

The discharge rate of pump, depth of water per irrigation dictates the irrigation interval.

In general per irrigation optimum depth of water is 5 cm

$= 10,000m \times 0.05$ m

$= 500$ m^3 (1m^3= 1000 litre)

$= 5, 00,000$ litre per irrigation

Ex: if crop need 400-500 mm means 40-50 cm of water/season

$= 10,000 \times 0.4$

$= 4,000$ m^3

$= 40,00,000$ litre of total water required per season

So, if divide irrigation requirement per irrigation and total WR

$40,00,000/5,00,000= 8$

So, for complete crop season it is possible to irrigate 8 times with a depth of 5 cm to get optimum yield

Cropping plan for an irrigated commend area of 10,000 ha Alfisols. The storage capacity of reservoir is 10 TMC, ridge level storage 20 per cent, conveyance and seepage loss is 40 per cent, average rainfall during Kharif 400 mm.

Dead storage capacity= 2 TMC

Losses= 4 TMC

Available water= 10 − (2+4)= 4 TMC

Crops	Area Ha	Area Per cent	WR (l)	Water Distributed for Growing Season (TMC)
Paddy	2500	25.0	2×10^7	1.76
Sugarcane	1900	19	1.8×10^7	1.21
Groundnut	1850	18.5	6×10^6	0.39
Maize	1100	11.0	5.5×10^6	0.21
Sunflower	900	9.0	5×10^6	0.15
Sorghum	900	9.0	5×10^6	0.15
Fingermillet	850	8.5	4.5×10^6	0.13
Total	10000	100		4.00 TMC

Situation I

In sub command area of 10,000 ha wherein diversified crops are grown in *Kharif* and *rabi* seasons. Suggest crop planning for available 10 TMC storage capacity of reservoir with dead storage of 20 per cent and irrigation efficiency of 30 per cent. (10 TMC= thousand Million cubic feet)

Situation II

Existing crops and cropping systems of UKP command area during 2015

Sl.No.	Season	Crop	Per cent Area	Area (ha)
1	Kharif	Bajra	10	54087
		Maize	10	54087
		Groundnut	15	81131
		Greengram	5	27044
		Vegetables	7.5	40565
		Sunflower	10	54087
		Total	57.5	311000
2	Bi-seasonal	Tur	2.5	13522
		Chilli	10	54087
		Cotton	10	54087
		Total	22.5	121696
3	Rabi	Rabi jowar	5	27044
		Wheat	10	54087
		Chickpea	5	27044
		Sunflower	10	54087
		Vegetables	5	27044
		Total	35	189305
		Over all Total	115	622000

Reference Evapotranspiration (ET_0)

"A hypothetical reference crop with an assumed crop height of 0.12 m, a fixed surface resistance of 70 s m^{-1} and an albedo of 0.23."

The reference surface closely resembles an extensive surface of green grass of uniform height, actively growing, completely shading the ground and with adequate water. The requirements that the grass surface should be extensive and uniform result from the assumption that all fluxes are one-dimensional upwards.

Estimation of PET

I. Thornthwaite Method

Potential evapotranspiration (PET) for GKVK with 13° 5′ N and 77° 34′ E, 924 m altitude

$$e = 1.6 \, (10t/I)^{a}$$

where,

e = Unadjusted PET cm/month

t = Mean air temperature

I = Annual/seasonal heat index

$I = (t/5)^{1.514}$ Month heat indices

a = Empirical exponent computed by equation

$a = 0.000000675\ I^3 - 0.0000771\ I^2 + 0.01792\ I + 0.49239$

For 2008

Month	Max Temp (°C)	Min Temp (°C)	Mean Temp (°C)	Monthly Heat Index (i) (130 i)	Unadju-stable PET (e)	Correction Factor	Corrected PET
Jan	28.0	13.6	20.80	8.65	6.83	0.982	6.71
Feb	29.4	16.7	23.05	10.11	9.12	0.910	8.30
March	30.2	17	23.60	10.47	9.74	1.030	10.0
April	32.7	19.3	36.0	12.13	12.78	1.036	13.24
May	33.5	20.5	27.0	12.85	14.22	1.098	15.61
June	29.2	19.5	24.35	10.98	10.64	1.072	11.41
July	28.6	19.3	23.95	10.72	10.16	1.104	11.22
Aug	27.1	19.3	23.20	10.21	9.29	1.076	9.99
Sep	28.3	18.7	23.50	10.41	9.63	1.02	9.82
Oct	27.8	17.8	22.80	9.95	8.85	1.014	8.97
Nov	26.8	15.8	21.30	8.97	7.31	0.962	7.03
Dec	26.8	14.7	20.75	8.62	6.78	1.002	6.79

a $= 0.000000675\ I^3 - 0.0000771\ I^2 + 0.01792\ I + 0.49239$

$= 0.000000675\ (124.07)^3 - 0.0000771\ (124.07)^2 + 0.01792\ (124.07) + 0.49239$

$= 1.289 - 0.0118 + 2.22 + 0.49239$

$= 2.82$

1. $e = 1.6\ (10 \times 20.8/124.07)^{2.82} = 6.83$
2. $e = 1.6\ (10 \times 23.05/124.07)^{2.82} = 9.12$
3. $e = 1.6\ (10 \times 23.06/124.07)^{2.82} = 9.74$
4. $e = 1.6\ (10 \times 26/124.07)^{2.82} = 12.78$
5. $e = 1.6\ (10 \times 27/124.07)^{2.82} = 14.22$
6. $e = 1.6\ (10 \times 24.35/124.07)^{2.82} = 10.64$
7. $e = 1.6\ (10 \times 23.95/124.07)^{2.82} = 10.16$
8. $e = 1.6\ (10 \times 23.2/124.07)^{2.82} = 9.29$
9. $e = 1.6\ (10 \times 23.5/124.07)^{2.82} = 6.83$

10. $e = 1.6 (10 \times 22.8/124.07)^{2.82} = 8.85$

11. $e = 1.6 (10 \times 21.3/124.07)^{2.82} = 7.21$

12. $e = 1.6 (10 \times 20.75/124.07)^{2.82} = 6.78$

For 2009

Month	Max Temp (ºC)	Min Temp (ºC)	Mean Temp (ºC)	Heat index (i)	Unadju- stable PET (e)	Correction factor 'e'	Actual 'e' mm/day
Jan	27.6	12.2	19.9	8.10	5.99	0.98	5.88
Feb	30.8	14.2	22.5	9.75	8.53	0.91	7.76
March	32.7	17.4	25.1	11.50	11.68	1.03	12.03
April	34.2	20	27.1	12.92	14.57	1.036	15.09
May	32.1	19.1	25.6	11.85	12.36	1.098	13.57
June	29.4	19.4	24.4	11.02	10.77	1.072	11.55
July	28.7	19.5	24.1	10.82	10.39	1.104	11.47
Aug	28.3	19.2	23.8	10.62	10.02	1.076	10.78
Sep	28	19.2	23.6	10.48	9.78	1.020	9.97
Oct	28.1	17.5	22.8	9.95	8.86	1.014	8.98
Nov	27	17.7	22.4	9.68	8.42	0.962	8.10
Dec	26.8	16.3	21.6	9.17	7.58	1.002	7.59

$a = 0.000000675 \, I^3 - 0.0000771 \, I^2 + 0.01792 \, I + 0.49239$

$= 0.000000675 \, (125.86)^3 - 0.0000771 \, (125.86)^2 + 0.01792 \, (125.86) + 0.49239$

$= 1.35 - 01.22 + 2.26 + 0.49239$

$= 2.88$

1. $e = 1.6 (10 \times 19.9/125.86)^{2.88} = 5.99$

2. $e = 1.6 (10 \times 22.5/125.86)^{2.88} = 8.53$

3. $e = 1.6 (10 \times 25.1/125.86)^{2.88} = 11.68$

4. $e = 1.6 (10 \times 27.1/125.86)^{2.88} = 14.57$

5. $e = 1.6 (10 \times 25.6/125.86)^{2.88} = 12.36$

6. $e = 1.6 (10 \times 24.4/125.86)^{2.88} = 10.77$

7. $e = 1.6 (10 \times 24.1/125.86)^{2.88} = 10.39$

8. $e = 1.6 (10 \times 23.8/125.86)^{2.88} = 10.02$

9. $e = 1.6 (10 \times 23.6/125.86)^{2.88} = 9.78$

10. $e = 1.6 (10 \times 22.8/125.86)^{2.88} = 8.86$
11. $e = 1.6 (10 \times 22.4/125.86)^{2.88} = 8.42$
12. $e = 1.6 (10 \times 21.6/125.86)^{2.88} = 7.58$

For 2010

Month	Max Temp (ºC)	Min Temp (ºC)	Mean Temp (ºC)	Heat Index (i)	Unadju- stable PET (e)	Correction Factor (e)	Actual 'e' mm/day
Jan	27.5	15	21.3	8.97	7.12	0.982	6.99
Feb	31.10	15.7	23.4	10.35	9.38	0.910	8.54
March	34.2	18.7	26.5	12.49	13.51	1.030	13.91
April	34.1	20.5	27.3	13.07	14.74	1.036	15.27
May	32.7	20.6	26.7	12.63	13.10	1.098	14.38
June	29.7	19.7	24.7	11.23	10.99	1.072	11.78
July	27.8	19.2	23.5	10.41	9.49	1.104	10.48
Aug	27.1	19.2	23.2	10.25	9.15	1.076	9.85
Sep	27.2	19.1	23.2	10.25	9.15	1.02	9.85
Oct	28	19	23.5	10.41	9.49	1.014	9.62
Nov	26.3	17.7	22	9.42	7.83	0.962	7.53
Dec	25.7	15.2	20.50	8.47	6.37	1.002	6.38

a $= 0.000000675 \, I^3 - 0.0000771 \, I^2 + 0.01792 \, I + 0.49239$

$= 0.000000675 (127.95)^3 - 0.0000771 (127.95)^2 + 0.01792 (127.95) + 0.49239$

$= 1.35 - 01.22 + 2.26 + 0.49239$

$= 2.88$

1. $e = 1.6 (10 \times 21.3/127.95)^{2.93} = 7.12$
2. $e = 1.6 (10 \times 23.4/127.95)^{2.93} = 9.38$
3. $e = 1.6 (10 \times 26.5/127.95)^{2.93} = 13.51$
4. $e = 1.6 (10 \times 27.3/127.95)^{2.93} = 14.74$
5. $e = 1.6 (10 \times 26.7/127.95)^{2.93} = 13.10$
6. $e = 1.6 (10 \times 24.7/127.95)^{2.93} = 10.99$
7. $e = 1.6 (10 \times 23.5/127.95)^{2.93} = 9.49$
8. $e = 1.6 (10 \times 23.2/127.95)^{2.93} = 9.15$
9. $e = 1.6 (10 \times 23.2/127.95)^{2.93} = 9.15$
10. $e = 1.6 (10 \times 23.5/127.95)^{2.93} = 9.49$
11. $e = 1.6 (10 \times 22/127.95)^{2.93} = 7.83$
12. $e = 1.6 (10 \times 20.50/127.95)^{2.93} = 6.37$

Calculated PET and Rainfall Comparison

Month	PET (mm/month)			Rainfall (mm)		
	2008	2009	2010	2008	2009	2010
Jan	6.71	5.88	6.99	0.0	0.0	0.0
Feb	8.30	7.76	8.54	13.2	0.0	0.0
March	10.03	12.03	13.91	137.4	10.2	34.2
April	13.24	15.09	15.27	1.60	106	112.8
May	15.61	13.57	14.38	98.8	153	137.5
June	11.41	11.55	11.78	32.4	58.8	95.4
July	11.22	11.47	10.48	182.8	55.8	142.2
Aug	9.99	10.78	9.85	249.8	106.8	158.2
Sep	9.82	9.97	9.85	126	231.7	89.4
Oct	8.97	8.98	9.62	208	29.6	19.2
Nov	7.03	8.10	7.53	20.6	49.4	128.6
Dec	6.79	7.59	6.38	1.80	11	9.80
Total	119.12	122.77	124.38	1072.4	812.3	927.3

II. Radiation Method

PET for GKVK is 13ºC 5' N and 77ºC 34' and 924 m

Month	Mean Temp. (ºC)	RH (Per cent)	Wind Velo-city (m/s)	Sun-shine Hours (hr)	N	n/N	RA	RS	W	WX RS	Actual ET_o
Jan.	21.3	70	2.00	8.3	11.42	0.73	12.60	7.75	0.72	5.68	4.7
Feb.	23.4	64	1.83	9.8	11.74	0.83	13.75	9.14	0.73	6.67	5.5
March	26.5	60	1.81	8.3	12.00	0.69	15.00	8.91	0.78	6.97	5.8
April	27.3	40	1.58	8.5	12.42	0.68	15.7	9.26	0.78	7.22	6.0
May	26.7	62	1.92	8.0	12.72	0.63	15.75	8.89	0.78	6.93	5.7
June	24.7	68	2.44	5.9	12.88	0.46	15.60	7.49	0.76	5.69	5.2
July	23.5	74	2.36	3.3	12.78	0.26	15.60	5.93	0.75	4.45	3.7
Aug.	23.2	76	2.00	3.2	12.52	0.26	15.65	5.95	0.74	4.40	3.1
Sep.	23.2	76	1.64	3.8	12.16	0.31	15.15	6.14	0.74	4.54	3.2
Oct.	23.5	74	1.28	5.0	11.80	0.42	14.25	6.56	0.75	4.92	3.5
Nov.	22	77	1.17	4.2	11.48	0.37	13.05	5.68	0.73	4.15	2.9
Dec.	20.5	73	1.36	5.1	11.32	0.45	12.25	6.82	0.72	4.91	3.2

$$Rs = (0.25 + 0.5 \, n/N)Ra$$

$$Jan = (0.25 + 0.5 \times 0.73)^{12.6} = 7.75$$

Feb $= (0.25 + 0.5 \times 0.83)^{12.6} = 9.14$

Mar $= (0.25 + 0.5 \times 0.69)^{12.6} = 8.93$

Apr $= (0.25 + 0.5 \times 0.68)^{12.6} = 9.26$

May $= (0.25 + 0.5 \times 0.63)^{12.6} = 8.89$

June $= (0.25 + 0.5 \times 0.46)^{12.6} = 7.49$

July $= (0.25 + 0.5 \times 0.26)^{12.6} = 5.93$

Aug $= (0.25 + 0.5 \times 0.26)^{12.6} = 5.93$

Sep $= (0.25 + 0.5 \times 0.31)^{12.6} = 6.14$

Oct $= (0.25 + 0.5 \times 0.42)^{12.6} = 6.56$

Nov $= (0.25 + 0.5 \times 0.37)^{12.6} = 5.68$

Dec $= (0.25 + 0.5 \times 0.45)^{12.6} = 5.82$

ET_O for MRS Hebbal 12° 58′ N and 77°C 34′ and 924 m

$n = 9.43$ hr; $N = 11.42$ hr/day; $Ra = 13.75$; $W = 0.71$

$Rs = [0.25 + 0.5 (9.43/11.72)]\ 13.75$

$\quad = (0.25 + 0.5 \times 0.805)\ 13.75$

$\quad = 0.653 \times 13.75$

$\quad = 8.97$

Therefore, $W \times Rs = 0.713 \times 8.97$

$\qquad\qquad\qquad = 6.4$ mm day^{-1}

$ET_O = 6$ mm day^{-1}

III. Blaney – Criddle Method

$ET_0 = C\ [P\ (0.46 + 8)]$ mm/day

where,

$ET_0 =$ Reference crop ET mm/day for month under consideration

$T =$ Mean daily temperature in °C

$P =$ Mean daily per cent of total annual day time hours obtained from table for any given month and latitiude

$e =$ Adjustment factor depends on minimum RH and sunshine hrs and day time wind estimate

Example: PET for GKVK is 13ºC 5' N latitude and 77ºC 34' longitude 924 MSL altitude

Month (2010)	Max Temp (ºC)	Min Temp (ºC)	Mean Temp (ºC)	RH (per cent)	Sun-shine Hours	P	P (0.46 T +8)	N	n/N	ET_0
Jan	27.5	15.0	21.3	70	8.3	0.26	4.63	11.42	0.73	5.1
Feb	31.10	15.7	23.4	64	9.8	0.26	4.88	11.74	0.83	5.3

Month (2010)	Max Temp (°C)	Min Temp (°C)	Mean Temp (°C)	RH (per cent)	Sun-shine Hours	P	P (0.46 T +8)	N	n/N	ET_0
March	34.2	18.7	26.5	59.5	8.3	0.27	5.45	12.0	0.69	5.9
April	34.1	20.5	27.3	60	8.5	0.28	5.76	12.42	0.68	6.2
May	32.7	20.6	26.7	61.5	8.0	0.28	5.80	12.72	0.63	6.3
June	29.7	19.7	24.7	67.5	5.9	0.28	5.54	12.88	0.46	6.0
July	27.8	19.2	23.5	73.5	3.3	0.29	5.45	12.78	0.26	6.7
Aug	27.1	19.2	23.2	76	3.2	0.28	5.23	12.52	0.26	6.5
Sep	27.2	19.1	23.2	75.5	3.8	0.28	5.23	12.16	0.31	6.5
Oct	28	19.0	23.5	73.5	5.0	0.27	5.15	11.80	0.42	6,3
Nov	26.3	17.7	22.0	76.5	4.2	0.26	4.78	11.48	0.37	5.9
Dec	25.7	15.2	20.5	72.5	5.1	0.25	4.43	11.32	0.45	5.8

Jan ET_0 = 0.26(0.46 × 21.3) + 8 = 4.63

Feb ET_0 = 0.26 (0.46 × 23.4) + 8 = 4.88

Mar ET_0= 0.26 (0.46 × 26.5) + 8 = 5.45

Apr ET_0 = 0.26 (0.46 × 27.3) + 8 = 5.76

May ET_0 = 0.26 (0.46 × 26.7) + 8 = 5.80

June ET_0= 0.26 (0.46 × 24.7) + 8 = 5.54

July ET_0 = 0.26 (0.46 × 29.5) + 8 = 5.45

Aug ET_0= 0.26 (0.46 × 23.2) + 8 = 5.23

Sep ET_0 = 0.26 (0.46 × 23.2) + 8 = 5.23

Oct ET_0 = 0.26 (0.46 × 23.5) + 8 = 5.15

Nov ET_0 = 0.26 (0.46 × 22.0) + 8 = 4.78

Dec ET_0 = 0.26 (0.46 × 20.5) + 8 = 4.43

IV. Pan Evaporation Method

Recommended relationship to estimate ET_0 using pan evaporation data is

ET_0 = Epan × Kp

Therefore ETcrop × Kc

1) The mean climatic data for Hiriyur and Chitradurga district are as follows

	January	February	March	April	May
RH Mean	58.1	52.3	46.5	50.3	53.3
Wind velocity (km hr⁻¹)	2.6	2.7	2.8	3.3	53.3
Epan (mm day⁻¹)	5.0	6.1	7.6	8.6	4.3

Pan is situated in dry fallow area which area 100 m windward side of dry fallow and Pan is screened find out ET_O

January	February	March	April	**May**
Kp = 0.65	Kp = 0. 45	Kp= 0.65	Kp = 0.65	Kp = 0.65
ET_0= Epan × Kp	ET_0 = Epan × Kp	ET_0 = Epan × Kp	ET_0= Epan × Kp	ET_0= Epan × Kp
= (5.0/0.9) × 0.65	ET_0= (6.1/0.9) × 0.65	ET_0= (7.6/0.9) × 0.65	= (8.6/0.9) × 0.65	= (7.9/0.9) × 0.65
= 3.6 mm/day	= 4.41 mm/day	= 5.49 mm/day	= 6.21 mm/day	= 5.70 mm/day

2. Pan is situated in dry fallow area with 100 m windward side of dry fallow and pan is screened. Find out the ET of crop using the climate data of year 2010

Month	RH per cent (Mean)	Epan	Wind Velocity km/hr	Wind Velocity km/day	Kp	Screened Pan Epan/0.9	$ET_0 =$ Epan/ 0.9×Kp
Jan	70	6.99	7.4	178	0.65	7.77	5.10
Feb	64	8.54	6.6	158	0.60	9.49	5.70
March	59	13.90	6.5	156	0.60	15.40	9.24
April	60	15.30	5.7	137	0.60	17	10.20
May	62	14.40	6.9	166	0.60	16	9.60
June	68	11.80	8.8	211	0.60	13.10	7.86
July	74	10.5	8.5	204	0.65	11.67	7.59
Aug	76	9.90	7.2	173	0.75	11	8.25
Sep	75	9.90	5.9	142	0.75	11	8.25
Oct	74	9.60	4.6	110	0.75	10.70	8.03
Nov	76	7.50	4.2	101	0.75	8.33	6.25
Dec	73	6.40	4.9	118	0.75	7.11	5.33

V. Penman method of ET estimation

$$ET_0 = C [W. Rn + (1- W) f(u) (ea - ed)]$$

where,

ET_0 = Reference crop evapotranspiration mm/day

C = Adjustment factor to comonsate for the day and night weather conditions

W = Temperature and elevation related weighing factor

Rn = Net radiation is equivalent evaporation mm/day Rn = Rns - RnL

Rns = Net incoming short wave solar radiation

= Ra (1 − L) (0.25 + 0.50 n/N)

where,

Ra = Extra terrestrial radiation

L = Reflection co-efficient = 0.25 for most of crops

n/N = Ratio of actual sunshine hour/maximum possible sunshine hours

RnL = Net long wave radiation

RnL = f (t) f(ed) f (n/N)

1 – W = Temperature and elevation related weighing factor for effect of wind and humidity onET_0

ea = Mean saturation vapour pressure in milibar at mean air temperature

ed = Mean actual vapour pressure of air in milibar

F (n) = Wind related function

This equation consists of two terms

1) The energy (Radiation form)
2) Aerodynamic term (wind and humidity)

The relative importance of each term varies with climatic condition under calm weather condition aerodynamic term is less important than energy term but under windy condition particularly in arid region aerodynamic term become more important

In this equation the suggested wind function applied to condition which are found during summer with moderate wind with maximum RH of 70 per cent and day to night wind ratio of 1.5 to 2.0 for these condition no adjustment is required. If 24 hour wind total are used these will be under prediction of E_{TO} by 15 to 30 per cent in areas where:

☆ Day time wind greatly exceed high wind

☆ Relative humidity approaches 100 per cent

☆ Radiation is high where as in areas with,

☐ Moderate strong wind prevails

☐ Higher humidity is low

☐ Radiation is low

Then equation will over predict ETO and this over prediction increases with decreasing ratio of day time and night time wind velocity. Under these conditions adjustment factor 'C' should be applied.

Example I: Location- 14° 28' 6" N, 75° 55' 3" E and 586.94 m

January

Air temperature: Maximum – 30.1 °C   RH maximum – 65 per cent

Minimum - 16.9 °C   Minimum – 29.7 per cent

Mean - 23.5 °C   Mean - 47.35 per cent

Bright sunshine hours = 10.5 hr; Wind speed at 1.5 m ht = 2.5 km hr⁻¹Day/

night wind speed = 1

Calculation of ET_0

A. Solving Aerodynamic Term

(1-W) f (u) (ea – ed)

1) ea = from table at mean air temperature 23.5 °C = 28.55 m bar

2) ed = (ea × RH mean/100) = (28.95 × 47.35)/100 = 13.71 m bar

3) ea – ed = 28.95 – 13.71 = 15.24 m bar

4) f(u) = wind speed   = 2.5 × 24 = 60 km/day at 1.5 m ht

                     = 60 × 1.06 = 63.6

From the table calculate f(u) at 63.6 km/day

Therefore f(u) = 0.44

5) 1 – W at 23.5 °C at 589 m = 0.263

Therefore aerodynamic term   = 0.263 × 0.44 (15.24)

                                   = 1.76

B. Radiation Term: WRn = W (Rns – Rnl)

6) n = 10.5 hours

N = 11.36 hours for January at 14° 28′ N

n/N = 10.5/11.36 = 0.92

7) Ra from table at 14° 28′ N = 12.30

8) Correction factor extra terrestrial radiation Ra to net solar radiation Rns for given reflection of x = 0.25 and n/N OF 0.92 from table

i.e. (1 – x) (0.25 + 0.5 n/N)

0.25) [0.25 + 0.5(0.92)] = 0.528

9) Rns   = Ra (1 – x) (0.25 + 0.5 n/N)

          = 12.30 × 0.528 = 6.49

10) Rnl   = f(t) f(ed) f (n/N)

f(t)   = 15.3 f(ed) = 0.1815 at 13.15 f (n/N) = 0.93

Rnl   = 15.3 × 0.1815 × 0.93 = 2.58

11) Value of W at 23.5 and at altitude 587 m = 0.737

12) WRn = W(Rns – Rnl)

             = 0.737 (6.49 – 2.58) = 2.88

13) Aerodynamic term and radiation term = 1.76 + 2.88 = 4.64

C = adjustment factor for Rs = (0.25 + 0.5 n/N) Ra

For 4 day/night =1 RH of 65 per cent for Rs 8.73 = 9 =0.77

ET_0 for January = 0.77 × 4.64 = 3.57

February

Latitude: 14° 28' 6" N

Longitude: 75° 53' 3" E

Air temperature   Maximum – 33.5 °C   RH maximum – 61.9 per cent

                           Minimum - 19.5 °C   Minimum – 23.6 per cent

                           Mean - 26.5 °C   Mean - 42.75 per cent

Sunshine hours = 10.6 hr; Wind speed at 1.5 m ht = 2.2 km/hr

Wind speed measures at 1.5 m ht

$ET_0 = \{WRn + (1- W) f (u) (ea – ed)\}$

A. Solving aerodynamic term:$(1-W) f (u) (ea – ed)$

    1) ea = from table at mean air temperature 26.5 °C = 34.65 m bar

    2) ed = (ea × RH mean/100) = (34.65 × 42.75)/100 = 14.81 m bar

    3) ea-ed = 34.65 – 14.81= 19.84 m bar

    4) f(u)   = wind speed = 2.2 × 24 = 52.8 km/day

              = 52.8 × 1.06 = 55.96 = 56

    From the table = 0.4222

    5) 1 – W at 26.5 °C and altitude of 587 m = 0.229

        Therefore aerodynamic term   = 0.229 × 0.422 (19.84)

                                      = 1.92

B. Radiation Term

       WRn = W (Rns – Rnl)

       n = 10.6 hours

    6) N = 11.64 hours for January at 14°

       n/N = 10.6/11.64 = 0.91

    7) Ra from table at 14° 28' N = 13.6

    8) Correction factor extra terrestrial radiation Ra to net solar radiation Rns for given reflection of x = 0.254 and n/N OF 0.91 from table

       i.e. (1-0.25) [0.25 + 0.5(0.91)] = 0.528

    9) Rns   = Ra (1 – x) (0.25 + 0.5 n/N)

             = 13.6 × 0.528 = 7.18

    10) Rnl   = f(t) f(ed) f (n/N)

        f(t)   = 16 f(ed) = 0.1792 f (n/N) = 0.91

        Rnl   = 16 × 0.1792 × 0.91 = 2.63

    11) Value of W at 26.5 and at altitude 587 m = 0.767

    12) WRn = W(Rns – Rnl)

$$= 0.767 \ (7.18 - 2.63)$$
$$= 3.48$$

13) Aerodynamic term and radiation term = $1.92 + 3.48 = 5.4$

C = adjustment factor for Rs = $(0.25 + 0.5 \ n/N)$ Ra

For 4 day/unit =1 RH of 61.9 per cent for Rs 9.58

C = 0.75

ET$_0$ for February = 0.75 × 5.4 = 4.05

March

Latitude: 14° 28' 6" N Longitude: 75° 53' 3" EAltitude = 587 m

Air temperature   Maximum – 36.2 °C   RH maximum – 61.3 per cent

                       Minimum - 21.3 °C   Minimum – 24.1 per cent

                       Mean - 28.75 °C   Mean - 44.7 per cent

Sunshine hours = 10.6 hr Wind speed measures at 1.5 m height is 2.8 km/hr

$ET_0 = \{ WRn + (1- W) \ f \ (u) \ (ea - ed) \}$

A. Solving Aerodynamic Term: (1-W) f (u) (ea – ed)

1) ea = from table at mean air temperature 28.75 °C = 38.66 m bar

2) ed = (ea × RH mean/100) = (38.66 × 44.7)/100 = 17.28 m bar

3) ea – ed = 38.66 – 17.28 = 21.38 m bar

4) f(u)   = wind speed = 2.8 × 24 = 67.2 km/day

            = 67.2 × 1.06 = 71.23 = 56

From the table f(u) = 0.463

5) 1 – W at 28.75 °C and altitude of 587 m = 0.2076

Therefore aerodynamic term = 0.2076 × 0.463 (21.38) = 2.06

B. Radiation Term: WRn = W (Rns – Rnl)

6) N = 12.0 at an altitude of hours for January at 14°28'n/N = 10.6/12 = 0.88

7) Ra from table at 14° 28' N = 14.9

8) Correction factor extra terrestrial radiation Ra to net solar radiation Rns for given relation of x = 0.25 and n/N OF 0.88 from table *i.e.* (1-0.25) [0.25 + 0.5(0.88)] = 0.5175

9) Rns   = Ra (1 – x) (0.25 + 0.5 n/N)

           = 14.9 × 0.5175 = 7.71

10) Rnl   = f(t) f(ed) f (n/N)

    f(t)   = 16.45 f(ed) = 0.163 f (n/N) = 0.894

    Rnl   = 16.45 × 0.613 × 0.894 = 2.40

11) Value of W at 28.75 and at altitude 587 m = 0.7949
12) WRn = W(Rns – Rnl)
    = 0.7949 (7.71 – 2.40) = 4.22
13) Aerodynamic term and radiation term = 2.06 + 4.22 = 6.28
    C = adjustment factor for Rs = (0.25 + 0.5 n/N) Ra
    For 4 day/unit RH of 65.3 per cent for Rs 10.28
    C = 0.78
    ET_0 for March is 0.75 × 6.24 = 4.898

April

Latitude: 14° 28′ 6″ N Altitude = 587 m

Longitude: 75° 53′ 3″ E

Air temperature   Maximum – 36 °C   RH maximum – 72.2 per cent
                    Minimum - 23.4 °C   Minimum – 36.2 per cent
                    Average - 29.7 °C   Average- 54.2 per cent

Sunshine hours = 9.6 hr; Wind speed measures at 1.5 m ht= 3.4 km/hr

ET_0 = {WRn + (1- W) f (u) (ea –ed)}

A. *Solving Aerodynamic Term (1-W) f (u) (ea – ed)*

1) ea = from table at mean air temperature 29.7 °C = 41.83 m bar
2) ed = (ea × RH mean/100) = (41.83 × 54.2)/100 = 22.67 m bar
3) ea – ed = 41.83 – 22.67 = 19.16 m bar
4) f(u)   = wind speed = 3.4 × 24 = 81.6 km/day
          = 81.6 × 1.06 = 86.49 = 86.5
   From the table = 0.502
5) (1 – W) at 29.7 °C and altitude of 587 m = 0.2025
   Therefore aerodynamic term = 0.2025 × 0.502 (19.16) = 1.95

B. *Radiation term:WRn = W (Rns – Rnl)*

N = 12.46 at an altitude of hours for January at 14°28′
6) n/N = 9.6/12.46 = 0.77
7) Ra from table at 14° 28′ N = 15.7
8) Correction factor extra terrestrial radiation Ra to net solar radiation
   Rns for given relation of x = 0.25 and n/N OF 0.77 from table
   i.e. (1-0.25) [0.25 + 0.5(0.77)] = 0.4762
9) Rns   = Ra (1 – x) (0.25 + 0.5 n/N)
         = 15.7 × 0.4762 = 7.48

10) Rnl = f(t) f(ed) f (n/N)

    f(t) = 16.6 f(ed) = 0.1225 f (n/N) = 0.796

    Rnl = 16.6 × 0.1225 × 0.796 = 1.61

11) Value of W at 29.7 and at altitude 587 m = 0.7975

12) WRn = W(Rns – Rnl)

    = 0.7975 (7.48 – 1.61) = 4.68

13) Aerodynamic term and radiation term = 1.95 + 4.68 = 6.63

C = adjustment factor for Rs = (0.25 + 0.5 n/N) Ra

For 4 day/unit RH of 72.2 per cent for Rs 9.97

C = 0.759

ET_0 for April 0.759 × 6.63 = 5.03

May

Latitude: 14° 28′ 6″ N Altitude = 587 mLongitude: 75° 53′ 3″ E

Air temperature   Maximum – 35.1 °C   RH Maximum – 74.5 per cent

    Minimum - 24.3 °C   Minimum – 44.2 per cent

    Mean - 29.7 °C   Mean - 59.35 per cent

Sunshine hours = 9.0 hr; Wind speed measures at 1.5 m ht= 4.9 km/hr

ET_0 = {WRn + (1- W) f (u) (ea – ed)}

A. *Solving Aerodynamic Term (1-W) f (u) (ea – ed)*

1) ea = from table at mean air temperature 29.7 °C = 41.83 m bar

2) ed = (ea × RH mean/100) = (41.83 × 54.4)/100 = 22.77 m bar

3) ea – ed = 41.83 – 22.77 = 19.06 m bar

4) f(u) = wind speed = 4.9 × 24 = 117.6 km/day

    = 117.6 × 1.06 = 124.65

From the table = 0.602

5) (1 – W) at 29.7 °C and altitude of 587 m = 0.2025

Therefore aerodynamic term = 0.2025 × 0.602 (19.06) = 2.32

B. *Radiation Term: WRn = W (Rns – Rnl)*

N = 12.76 at an altitude of hours for January at 14° 28′

6) n/N = 9.6/12.76 = 0.71

7) Ra from table at 14° 28′ N = 15.8

8) Correction factor extra terrestrial radiation Ra to net solar radiation Rns for given relation of x = 0.25 and n/N OF 0.77 from table *i.e.* (1-0.25) [0.25 + 0.5(0.71)] = 0.453

9) Rns $= Ra(1 - x) (0.25 + 0.5 n/N)$

$= 15.8 \times 0.453 = 7.16$

10) Rnl $= f(t) f(ed) f (n/N)$

f(t) $= 16.6 f(ed) = 0.1225 f (n/N) = 0.731$

Rnl $= 16.6 \times 0.12 \times 0.731 = 1.456$

11) Value of W at 29.7 and at altitude 587 m = 0.7975

12) WRn $= W(Rns - Rnl)$

$= 0.7975 (7.16 - 1.456) = 4.55$

13) Aerodynamic term and radiation term = 2.07 + 4.55 = 6.62

C = adjustment factor for Rs = (0.25 + 0.5 n/N) Ra

For 4 day/unit RH of 74.5 per cent for Rs 9.55

C = 0.752

ET_0 for May = 0.752 × 6.62 = 4.98

Example II: Location: G.K.V.K., Bangalore (13° 5' N,77° 34' E,924m)

Year 2010

January 2010

Air temperature	Maximum – 27.5°C	RH Maximum – 92 per cent
	Minimum - 15 °C	Minimum – 48 per cent
	Average - 21.25°C	Average - 70 per cent

Sunshine hours = 8.3 hr Wind speed = 7.4 km/hr

$ET_0 = \{WRn + (1- W) f (u) (ea - ed)\}$

A. *Aerodynamic Term: (1-W) f (u) (ea - ed)*

1. ea = from table at mean air temperature 21.3°C = 25.23 m bar
2. ed = (ea × RH mean/100) = (21.3 × 70)/100 =17.69 m bar
3. ea – ed = 25.27 – 17.69 = 7.58 m bar
4. f(u)   = wind speed = 7.4 × 24 = 177.6 km/day

$= 177.6 \times 1.06 = 188.3$ km/day

From the table calculate f(u) at 188.3 km/day, therefore f (u) = 0.776

5. 1 – W at temp 21.25 at and altitude of 924 m = 0.28

Therefore aerodynamic term = 0.28 × 0.776 (7.58) = 1.65

B. *Radiation term: WRn = W (Rns – Rnl)*

6. n = 8.3, N = 11.42 at an altitude of hours for January at 13.5′

n/N = 8.3/11.42 = 0.73

7. Ra from table at 13.5′ N = 12.6

8. Correction factor extra terrestrial radiation Ra to net solar radiation Rns for given reflection of x = 0.25 and n/N of 0.77 from table *i.e.* (1-0.25) [0.25 + 0.5(0.73)] = 0.46

9. Rns $\quad$ = Ra (1 – x) (0.25 + 0.5 n/N)

$\quad\quad\quad$ = 12.6 × 0.46 = 5.79

10. Rnl $\quad$ = f(t) f(ed) f (n/N)

$\quad$ f(t) $\quad$ = 14.8 f(ed) = 0.155 at 17.69 f (n/N) = 0.76

$\quad$ Rnl $\quad$ = 14.8 × 0.155 × 0.76 = 1.74

11. Value of W at 21.3 and at altitude 924 m = 0.72

12. WRn $\quad$ = W(Rns – Rnl)

$\quad\quad\quad$ = 0.72 (5.79 – 1.74) = 2.916

13. Aerodynamic term and radiation term = 16.5 + 2.916 = 4.56

$\quad$ C = adjustment factor for Rs = (0.25 + 0.5 n/N) Ra

$\quad$ For 4 day/unit RH of 92 per cent for Rs 8

$\quad$ C = 0.82

ET_0 for January = 0.82 × 4.56 = 3.739

February 2010

Air temperature $\quad$ Maximum – 31.1 °C $\quad\quad$ RH Maximum – 90 per cent

$\quad\quad\quad\quad\quad\quad\quad$ Minimum - 15.7 °C $\quad\quad\quad\quad$ Minimum –38 per cent

$\quad\quad\quad\quad\quad\quad\quad$ Average - 23.4 °C $\quad\quad\quad$ Average - 64 per cent

Sunshine hours = 9.8 hr; Wind speed = 6.5 km/hr

ET_0 = {WRn + (1- W) f (u) (ea – ed)}

A. Aerodynamic term: (1-W) f (u) (ea – ed)

1. ea = from table at mean air temperature 23.4 °C = 28.95 m bar

2. ed = (ea × RH mean/100) = (28.95 × 64)/100 =18.53 m bar

3. ea – ed = 28.95 – 18.53 = 10.42 m bar

4. f(u) $\quad$ = wind speed = 6.5× 24 = 156 km/day

$\quad\quad\quad$ = 156× 1.06 = 165.36 km/day

$\quad$ From the table calculate f(u) at 165.36 km/day, therefore f (u) = 0.685

5. (1 – W) at temp 23.4 at and altitude of 924 m = 0.255

$\quad$ Therefore aerodynamic term = 0.255 × 0.685 (10.42) = 1.82

B. Radiation term:WRn = W (Rns – Rnl)

6. n = 9.8, N = 11.68 at an altitude of hours for February at 13.5′

n/N = 9.8/11.68 = 0.83

7. Ra from table at 13.5′ N = 13.75
8. Correction factor extra terrestrial radiation Ra to net solar radiation Rns for given reflection of x = 0.25 and n/N of 0.83 from table *i.e.* (1-0.25) [0.25 + 0.5(0.83)] = 0.498
9. Rns   = Ra (1 − x) (0.25 + 0.5 n/N)

           = 13.75 × 0.498 = 6.8475

10. Rnl   = f(t) f(ed) f (n/N)

    f(t)   = 15.3 f(ed) = 0.1475 at 18.53 f (n/N) = 0.85

    Rnl   = 15.3 × 0.1475 × 0.85 = 1.92

11. Value of W at 23.4 °C and at altitude 924 m = 0.7315
12. WRn = W(Rns − Rnl)

           = 0.7315 (6.84 − 1.92) = 3.60

13. Aerodynamic term and radiation term = 1.82 + 3.60 = 5.42

    C = adjustment factor for Rs = (0.25 + 0.5 n/N) Ra

    For 4 day/unit RH of 90 per cent for Rs 9

    C = 0.87

    = 0.87 × 5.42

    ET_0 for Februaryis 4.72

March 2010

Air temperature   Maximum – 34.2 °C   RH Maximum – 84 per cent

                        Minimum - 18.7 °C   Minimum –38 per cent

                        Average- 26.45 °C   Average - 59.5 per cent

Sunshine hours = 8.3 hr; Wind speed = 6.5 km/hr

$ET_0 = \{WRn + (1- W) f (u) (ea − ed)\}$

A. *Aerodynamic Term: (1-W) ƒ (u) (ea − ed)*

1. ea = from table at mean air temperature 26.5 °C = 34.65 m bar
2. ed = (ea × RH mean/100) = (34.65 × 59.5)/100 =20.61 m bar
3. ea − ed = 34.65 − 20.61 = 14.04 m bar
4. f(u)   = wind speed = 6.5× 24 = 156 km/day

              = 156× 1.06 = 165.4 km/day

    From the table calculate f(u) at 165.4 km/day, therefore f (u) = 0.7015

5. 1 − W at temp 23.4 at and altitude of 924 m = 0.775

    Aerodynamic term = 0.775 × 0.7015 (14.04) = 7.63

B. Radiation Term: WRn = W (Rns – Rnl)

n = 8.3, N = 12.0 at an altitude of hours for March at 13.5′

6. n/N = 8.3/12.0 = 0.692
7. Ra from table at 13.5′ N = 15.0
8. Correction factor extra terrestrial radiation Ra to net solar radiation Rns for given reflection of x = 0.25 and n/N of 0.69 from table *i.e.* (1-0.25) [0.25 + 0.5(0.69)] = 0.45
9. Rns $\quad$ = Ra (1 – x) (0.25 + 0.5 n/N)
    $\qquad$ = 15 × 0.45 = 6.75
10. Rnl $\quad$ = f(t) f(ed) f (n/N)
    f(t) $\quad$ = 16 f(ed) = 0.1375 at 20.61 f (n/N) = 0.722
    Rnl $\quad$ = 16 × 0.1375 × 0.722 = 1.5884
11. Value of W at 26.5 °C and at altitude 924 m = 0.775
12. WRn = W(Rns – Rnl)
    $\qquad$ = 0.775 (6.75 – 1.5884) = 4.0

Aerodynamic and radiation terms = 7.63 + 4 = 11.63

C = adjustment factor for Rs = (0.25 + 0.5 n/N) Ra

For 4 day/unit RH of 85 per cent for Rs 9

C = 0.87

$\qquad$ = 0.87 × 11.63 = 10.1

ET_0 for March is 10.1

April 2010

Air temperature	Maximum – 34.1 °C	RH Maximum – 86 per cent
	Minimum - 20.5 °C	Minimum –34 per cent
	Average - 27.3 °C	Average - 60 per cent

Sunshine hours = 8.5 hr; Wind speed = 5.7 km/hr

ET_0 = {WRn + (1- W) f (u) (ea – ed)}

A. Aerodynamic Term: (1-W) ƒ (u) (ea – ed)

1. ea = from table at mean air temperature 27.3 °C = 36.75 m bar
2. ed = (ea × RH mean/100) = (36.75 × 60)/100 =22.05 m bar
3. ea – ed = 36.75 – 22.05 = 14.7 m bar
4. fu) $\quad$ = wind speed = 8.5× 24 = 204 km/day
    $\qquad$ = 204 × 1.06 = 216 km/day

From the table calculate f(u) at 216 km/day, therefore f(u) = 0.852

5. $1 - W$ at temp 23.4 at and altitude of 925 m = 0. 785

Aerodynamic term = $0.785 \times 0.852 \, (14.7) = 9.83$

B. Radiation term: WRn = W (Rns – Rnl)

n = 8.5, N = 12.42 at an altitude of hours for April at 13.5′

6. n/N = 8.5/12.42 = 0.684

7. Ra from table at 13.5′ N = 15.7

8. Correction factor extra terrestrial radiation Ra to net solar radiation Rns for given reflection of x = 0.25 and n/N of 0.68 from table

i.e. (1-0.25) [0.25 + 0.5(0.68)] = 0.4425

9. Rns = Ra (1 – x) (0.25 + 0.5 n/N)

= 15.7 × 0.4425 = 6.95

10. Rnl = f(t) f(ed) f (n/N)

f(t) = 16.2 f(ed) = 0.13 at 22 f (n/N) = 0.714

Rnl = 16.2 × 0.714 × 0.722 = 1.50

= 5.45

11. Value of W at 27.3 °C and at altitude 924 m = 0.782

12. WRn = W(Rns – Rnl)

= 0.782 (5,45) = 4.26

13. Aerodynamic and radiation term = 9.83 + 4.26 = 14.09

C = adjustment factor for Rs = (0.25 + 0.5 n/N) Ra For 4 day/night RH of 86 per cent for Rs 9.2

C = 0.87

= 0.87 × 14.09 = 12.25

ET$_0$ for April Month is 12.25

May 2010

Air temperature   Maximum – 32.7 °C   RH Maximum – 84 per cent

Minimum - 20.6 °C   Minimum –39 per cent

Average- 26.7 °C   Average - 61.5 per cent

Sunshine hours = 8.0 hr; Wind speed = 6.9 km/hr

$ET_0 = \{WRn + (1- W) f (u) (ea - ed)\}$

A. Aerodynamic Term (1-W) ƒ (u) (ea – ed)

1. ea = from table at mean air temperature 26.7 °C = 35.17 m bar

2. ed = (ea × RH mean/100) = (35.17 × 61.5)/100 =21.62 m bar

3. ea – ed = 35.17 – 21.62 = 13.55 m bar

4. f(u) = wind speed = 6.9× 24 = 165.6 km/day

$$= 204 \times 1.06 = 175.5 \text{ km/day}$$

From the table calculate f(u) at 216 km/day, therefore f (u) = 0.745

5. 1 – W at temp 23.4 at and altitude of 925 m = 0. 767

Aerodynamic term = 0.767 × 0.745 (13.55) = 7.74

B. Radiation Term: WRn = W (Rns – Rnl)

n = 8, N = 12.72 at an altitude of hours for May at 13.5′

6. n/N = 8/12.72 = 0.63

7. Ra from table at 13.5′ N = 15.75

8. Correction factor extra terrestrial radiation Ra to net solar radiation Rns for given reflection of x = 0.25 and n/N of 0.63 from table *i.e.,* (1-0.25) [0.25 + 0.5(0.63)] = 0.42

9. Rns   = Ra (1 – x) (0.25 + 0.5 n/N)

$$= 15.75 \times 0.42 = 6.61$$

10. Rnl   = f(t) f(ed) f (n/N)

f(t)   = 16.04 f(ed) = 0.136 at 21.62 f (n/N) = 0.67

Rnl   = 16.04 × 0.136 × 0.67 = 1.46

11. Value of W at 26.7 °C and at altitude 924 m = 0.777

12. WRn = W(Rns – Rnl)

$$= 0.777 (6.61 – 1.46) = 4.0$$

13. Aerodynamic and radiation term = 7.74 + 4.0 = 11.74

C = adjustment factor for Rs = (0.25 + 0.5 n/N) Ra; For 4 day/night ; RH of 84 per cent for Rs 9

C = 0.87

$$= 0.87 \times 11.74 = 10.21$$

ET_0 for May is 10.21

June 2010

Air temperature	Maximum – 29.7 °C	RH Maximum – 89 per cent
	Minimum - 19.7 °C	Minimum –46 per cent
	Average - 24.7 °C	Average - 67.5 per cent

Sunshine hours = 5.9 hr; Wind speed = 8.8 km/hr

$ET_0 = \{WRn + (1- W) f (u) (ea – ed)\}$

A. Aerodynamic Term (1-W) f (u) (ea – ed)

1. ea = from table at mean air temperature 24.7 °C = 31.2 m bar

2. ed = (ea × RH mean/100) = (31.2 × 21.06)/100 =21.06 m bar

3. ea – ed = 31.2 – 21.06 = 10.14 m bar

4. f(u) = wind speed = 8.8× 24 = 211.2 km/day

$\qquad$ = 211.2 × 1.06 = 224 km/day

From the table calculate f(u) at 216 km/day, therefore f (u) = 0.872

5. $1 - W$ at temp 23.4 at and altitude of 925 m = 0. 243

Aerodynamic term = 0.243 × 0.872 (10.14) = 2.14

B. Radiation Term WRn = W (Rns – Rnl)

6. n = 5.9, N = 12.88 at an altitude of hours for June at 13.5′

n/N = 8/12.72 = 0.63

7. Ra from table at 13.5′ N = 15.6

8. Correction factor extra terrestrial radiation Ra to net solar radiation Rns for given reflection of x = 0.25 and n/N of 0.46 from table

i.e. (1-0.25) [0.25 + 0.5(0.46)] = 0.36

9. Rns = Ra $(1 - x)$ $(0.25 + 0.5 \, n/N)$

$\qquad$ = 15.6 × 0.36 = 5.62

10. Rnl = f(t) f(ed) f (n/N)

$\quad$ f(t) = 15.575 f(ed) = 0.135 at 21.06 f (n/N) = 0.52

$\quad$ Rnl = 0.52 × 0.135 × 15.575 = 1.09

11. Value of W at 24.7 °C and at altitude 924 m = 0.757

12. WRn = W(Rns – Rnl)

$\qquad$ = 0.757 (4.53) = 3.43

13. Aerodynamic and radiation terms = 2.14 + 3.43 = 5.57

C = adjustment factor for Rs = (0.25 + 0.5 n/N) Ra; For 4 day/night; RH of 89 per cent for Rs 7.5

C = 0.88

= 0.87 × 5.57 = 4.90

ET$_0$ for June is 4.90

July 2010

Air temperature $\quad$ Maximum – 27.8 °C $\qquad$ RH Maximum – 94 per cent

$\qquad\qquad\qquad$ Minimum - 19.2 °C $\qquad\qquad$ Minimum –53 per cent

$\qquad\qquad\qquad$ Mean - 23.5 °C $\qquad\qquad$ Mean - 73.5 per cent

Sunshine hours = 3.3 hr; Wind speed = 8.5 km/hr

$ET_0 = \{WRn + (1- W) \, f(u) \, (ea - ed)\}$

A. Aerodynamic term: (1-W) f (u) (ea – ed)

1. ea = from table at mean air temperature 23.5 °C = 28.95 m bar

2. ed = (ea × RH mean/100) = (28.95 × 73.5)/100 =21.3 m bar

3. ea – ed = 28.95 – 21.3 = 7.65 m bar

4. f(u)   = wind speed = 8.5× 24 = 204 km/day

   = 211.2 × 1.06 = 216.2 km/day

   From the table calculate f(u) at 216.2 km/day, therefore f (u) = 0.852

5. (1 – W) at temp 23.5 at and altitude of 924 m = 0. 255

   Therefore aerodynamic term = 0.255 × 0.852 (7.65) = 1.66

B.   Radiation Term: WRn = W (Rns – Rnl)

6. n = 3.3, N = 12.78 at an altitude of hours for July at 13.5′

   n/N = 3.3/12.78 = 0.258

7. Ra from table at 13.5′ N = 15.6

8. Correcion factor extra terrestrial radiation Ra to net solar radiation
   Rns for given reflection of x = 0.25 and n/N of 0.258 from table
   i.e. (1-0.25) [0.25 + 0.5(0.258)] = 0.284

9. Rns   = Ra (1 – x) (0.25 + 0.5 n/N)

   = 15.6 × 0.284 = 4.430

10. Rnl   = f(t) f(ed) f (n/N)

    f(t)   = 15.3 f(ed) = 0.1315 f (n/N) = 0.354

    Rnl   = 15.3 × 0.1315 × 0.354= 0.712

11. Value of W at 23.5 °C and at altitude 924 m = 0.745

12. WRn   = W(Rns – Rnl)

    = 0.745 (3.718) = 2.75

13. Aerodynamic and radiation term = 1.66 + 2.75 = 4.41

    C = adjustment factor for Rs = (0.25 + 0.5 n/N) Ra; for 4 day/night; RH
    of 94 per cent for Rs 6

    C = 0.82

    = 0.82 × 4.41 = 3.62

    ET_0 for July is 3.62

August 2010

Air temperature   Maximum – 27.1 °C   RH Maximum – 94 per cent

                   Minimum - 19.2 °C   Minimum –58 per cent

                   Average - 23.2 °C   Average - 76 per cent

Sunshine hours = 3.2 hr; Wind speed = 7.2 km/hr

ET_0 = {WRn + (1- W) f (u) (ea – ed)}

A. Aerodynamic Term: (1-W) ƒ (u) (ea – ed)

1. ea = from table at mean air temperature 23.2 °C = 28.525 m bar
2. ed = (ea × RH mean/100) = (28.52 × 76)/100 = 21.68 m bar
3. ea – ed = 28.95 – 21.68 = 6.85 m bar
4. f(u) = wind speed = 7.2× 24 = 172.8 km/day
          = 172.8 × 1.06 = 183 km/day

   From the table calculate f(u) at 183km/day, therefore f (u) = 0.766
5. (1 – W) at temp 23.5 at and altitude of 924 m = 0. 258
   = 0.258 × 0.766 (6.85) = 1.35

B. Radiation Term: WRn = W (Rns – Rnl)

6. n = 3.2, N = 12.52 at an altitude of hours for August at 13.5′
   n/N = 3.2/12.52 = 0.255
7. Ra from table at 13.5′ N = 15.65
8. Correction factor extra terrestrial radiation Ra to net solar radiation
   Rns for given reflection of x = 0.25 and n/N of 0.255 from table
   i.e. (1-0.25) [0.25 + 0.5(0.255)] = 0.283
9. Rns = Ra (1 – x) (0.25 + 0.5 n/N)
          = 15.65 × 0.283 = 4.42
10. Rnl = f(t) f(ed) f (n/N)
    f(t) = 15.24 f(ed) = 0.133 f (n/N) = 0.35
    Rnl = 15.24 × 0.133 × 0.35= 0.709
11. Value of W at 23.2 °C and at altitude 924 m = 0.742
12. WRn = W(Rns – Rnl)
          = 0.745 (3.711) = 2.75
13. Aerodynamic and radiation term = 1.35 + 2.75 = 4.1

    C = adjustment factor for Rs = (0.25 + 0.5 n/N) Ra; For 4 day/night RH
    of 94 per cent for Rs 6

    C = 0.82
    = 0.82 × 4.1 = 3.36

    ET$_0$ for August is 3.36

September 2010

Air temperature   Maximum – 27.2 °C   RH Maximum – 92 per cent

Minimum - 19.1 °C   Minimum –59 per cent

Average - 23.2 °C   Average- 75.5 per cent

Sunshine hours = 3.8 hr; Wind speed = 5.9 km/hr

$ET_0 = \{WRn + (1- W) f (u) (ea – ed)\}$

A. *Aerodynamic term: (1-W) f (u) (ea – ed)*

1. ea = from table at mean air temperature 23.2 °C = 28.44 m bar

2. ed = (ea × RH mean/100) = (28.44 × 75.5)/100 =21.47 m bar

3. ea – ed = 28.44 – 21.47 = 6.97 m bar

4. f(u)   = wind speed = 5.9× 24 = 141.6 km/day

         = 141.6 × 1.06 = 150 km/day

   From the table calculate f(u) at 150 km/day, therefore f (u) = 0.67

5. (1-W) at temp 23.2 at and altitude of 924 m = 0. 262

   Aerodynamic term = 0.262 × 0.67 (6.97) = 1.22

B. *Radiation Term: WRn = W (Rns – Rnl)*

6. n = 3.8, N = 12.16 at an altitude of hours for September at 13.5′

   n/N = 3.8/12.16 = 0.312

7. Ra from table at 13.5′ N = 15.15

8. Conversion factor extra terrestrial radiation Ra to net solar radiation Rns for given reflection of x = 0.25 and n/N of 0.312 from table

   i.e. (1-0.25) [0.25 + 0.5(0.312)] = 0.3045

9. Rns   = Ra(1 – x) (0.25 + 0.5 n/N)

        = 15.15 × 0.3045 = 4.61

10. Rnl   = f(t) f(ed) f (n/N)

    f(t)   = 15.24 f(ed) = 0.1375 f (n/N) = 0.38

    Rnl   = 15.24 × 0.1375 × 0.38= 0.796

11. Value of W at 23.2 °C and at altitude 924 m = 0.742

12. WRn = W(Rns – Rnl)

          = 0.742 (4.61 – 0.76) = 2.83

13. Aerodynamic and radiation terms = 1.22 + 2.83 = 4.05

   C = adjustment factor for Rs = (0.25 + 0.5 n/N) Ra; for 4 day/unit RH of 92 per cent for Rs 6

   C = 0.82

   = 0.82 × 4.05 = 3.321

   ET_0 for September is 3.321 mm

October 2010

Air temperature   Maximum – 28 °C   RH Maximum – 93 per cent

                          Minimum - 19.6 °C   Minimum –54 per cent

                Averaghe - 23.8 °C   Average - 73.5 per cent

Sunshine hours = 5.0 hr; Wind speed = 4.6 km/hr

$ET_0 = \{WRn + (1- W) f (u) (ea – ed)\}$

A. Aerodynamic Term: $(1-W) f (u) (ea – ed)$

1. ea = from table at mean air temperature 23.8 °C = 29.46 m bar
2. ed = (ea × RH mean/100) = (29.46 × 73.5)/100 =21.65 m bar
3. ea – ed = 29.46 – 21.65 = 7.81 m bar
4. f(u)   = wind speed = 5.0× 24 = 120 km/day

             = 120 × 1.06 = 127 km/day

From the table calculate f(u) at 127 km/day, therefore f (u) = 0.611

5. (1-W) at temp 23.8 at and altitude of 924 m = 0. 268

Aerodynamic term = 0.268 × 0.611 (7.81) = 1.29

B. Radiation Term $WRn = W (Rns – Rnl)$

6. n = 5.0, N = 11.8 at an altitude of hours for October at 13.5′

   n/N = 5.0/11.8 = 0.42

7. Ra from table at 13.5′ N = 14.25
8. Correction factor extra terrestrial radiation Ra to net solar radiation Rns for given reflection of x = 0.25 and n/N of 0.42 from table

i.e. (1-0.25) [0.25 + 0.5(0.42)] = 0.345

9. Rns   = Ra (1-x) (0.25 + 0.5 n/N)

          = 14.25 × 0.345 = 4.91

10. Rnl   = f(t) f(ed) f (n/N)

    f(t)   = 15.24 f(ed) = 0.1249 f (n/N) = 0.48

    Rnl   = 15.24 × 0.1249 × 0.48= 0.91

11. Value of W at 23.8 °C and at altitude 924 m = 0.748
12. WRn = W(Rns – Rnl)

          = 0.748 (4.0) = 2.99

13. Aerodynamic term and radiation term = 1.29 + 2.99 = 4.28

C = adjustment factor for Rs = (0.25 + 0.5 n/N) Ra; For 4 day/unit RH of 93 per cent for Rs 6.5

C = 0.842

= 0.842 × 4.28 = 3.6

ET_0 for October is 3.6 mm/day

November 2010

    Air temperature   Maximum – 26.3 °C   RH Maximum – 95 per cent

                           Minimum - 17.7 °C   Minimum –58 per cent

                           Average - 22 °C   Average - 76.5 per cent

Sunshine hours = 4.2 hr; Wind speed = 4.2 km/hr

$ET_0 = \{WRn + (1-W)\,f(u)\,(ea - ed)\}$

A. Aerodynamic Term: (1-W) f (u) (ea – ed)

1. ea = from table at mean air temperature 22 °C = 26.4 m bar
2. ed = (ea × RH mean/100) = (26.4 × 76.5)/100 =20.2 m bar
3. ea – ed = 26.4 – 20.2 = 6.2 m bar
4. f(u)   = wind speed = 4.2× 24 = 100.8 km/day

            = 100.8 × 1.06 = 106.8 km/day

    From the table calculate f(u) at 106.8 km/day, therefore f (u) = 0.564

5. (1 – W) at temp 22 at and altitude of 924 m = 0. 27

    Aerodynamic term = 0.27 × 0.564 (6.2) = 0.94

B. Radiation Term WRn = W (Rns – Rnl)

6. n = 4.2, N = 11.48 at an altitude of hours for November at 13.5′

    n/N = 4.2/11.48 = 0.37
7. Ra from table at 13.5′ N = 13.05
8. Correction factor extra terrestrial radiation Ra to net solar radiation Rns for given reflection of x = 0.25 and n/N of 0.37 from table

    i.e. (1-0.25) [0.25 + 0.5(0.37)] = 0.326
9. Rns   = Ra (1 – x) (0.25 + 0.5 n/N)

            = 13.05 × 0.326 = 4.25
10. Rnl   = f(t) f(ed) f (n/N)

    f(t)   = 15. f(ed) = 0.139 f (n/N) = 0.39

    Rnl   = 15 × 0.139 × 0.39= 0.813

    Rn   = Rns – Rnl

           = 4.25 – 0.813= 3.437
11. Value of W at 22 °C and at altitude 924 m = 0.73
12. WRn = W(Rns – Rnl)

           = 0.73 (3.437) = 2.50
13. Aerodynamic and radiation term = 0.94 + 2.99 = 3.44

    C = adjustment factor for Rs = (0.25 + 0.5 n/N) Ra; For 4 day/unit RH of 93 per cent for Rs 5.67

$C = 0.786$

$= 0.786 \times 3.44 = 2.70$

ET_0 for November is 2.7 mm/day

December 2010

Air temperature   Maximum – 25.7°C   RH Maximum – 92 per cent

                          Minimum - 15.2 °C   Minimum –53 per cent

                          Mean - 20.45°C   Mean - 72.5 per cent

Sunshine hours = 5.1 hr; Wind speed = 4.8 km/hr

$ET_0 = \{WRn + (1- W) f (u) (ea – ed)\}$

A. *Aerodynamic Term (1-W) f (u) (ea – ed)*

1. ea = from table at mean air temperature 20.45°C = 24.15 m bar
2. ed = (ea × RH mean/100) = (24.15 × 72.5)/100 =18.23 m bar
3. ea – ed = 24.15 – 18.23 = 5.92 m bar
4. f(u)   = wind speed = 4.9× 24 = 117.6 km/day

              = 117.6 × 1.06 = 124.6 km/day

     From the table calculate f(u) at 124.6 km/day, therefore f (u) = 0.608
5. 1 – W at temp. 20.45 at and altitude of 924 m = 0. 285

     Therefore aerodynamic term = 0.285 × 0.608 (65.92) = 1.03

B. *Radiation Term WRn = W (Rns – Rnl)*

6. n = 5.1, N = 11.32 at an altitude of hours for December at 13.5′

     n/N = 5.1/11.32 = 0.45
7. Ra from table at 13.5′ N = 12.25
8. Correction factor extra terrestrial radiation Ra to net solar radiation Rns for given reflection of x = 0.25 and n/N of 0.45 from table

     i.e. (1-0.25) [0.25 + 0.5 (0.45)] = 0.356
9. Rns   = Ra (1 – x) (0.25 + 0.5 n/N)

            = 12.25 × 0.356 = 4.361
10. Rnl   = f(t) f(ed) f (n/N)

     f(t)   = 14.62. f(ed) = 0.1495 f (n/N) = 0.51

     Rnl   = 14.62 × 0.1495 × 0.51= 1.11

     Rn   = Rns – Rnl

          = 4.361 – 1.11= 3.251
11. Value of W at temp 22 °C and at altitude 924 m = 0.714
12. WRn = W(Rns – Rnl)

$= 0.714 (3.251) = 2.32$

13. Aerodynamic and radiation term $= 1.03 + 2.32 = 3.35$

C = adjustment factor for Rs = (0.25 + 0.5 n/N) Ra; For 4 day/night; RH of 92 per cent for Rs 6

C = 0.82

$= 0.82 \times 3.35 = 2.747$

ET_0 for December month is 2.747 mm/day

Summary

Month	Temperature		Relative Humidity		Sunshine Hours	Wind Speed km/h	Calculated ET0
	Maximum	Minimum	maximum	Minimum			
January	27.5	15	92	48	8.3	7.4	3.739
February	31.1	15.7	90	38	9.8	6.5	4.72
March	34.2	18.7	84	38	8.3	6.54	10.1
April	34.1	20.5	86	34	8.5	5.7	12.25
May	32.7	20.6	84	39	8	6.9	10.21
June	29.7	19.7	89	46	5.9	8.8	4.9
July	27.8	19.2	94	53	3.3	8.5	3.62
August	27.1	19.2	94	58	3.2	7.2	3.36
September	27.2	19.1	92	59	3.8	5.9	3.32
October	28	19.6	93	54	5	4.6	3.6
November	26.3	17.7	95	58	4.2	4.2	2.7
December	25.7	15.2	92	53	5.1	4.8	2.747

18

Scheduling of Irrigation in Major Field Crops

M.R. Umesh, N. Jagadeesh and Shanthappa Duttangarvi

Water is the most important and critical input in man's life especially in agriculture. The pressure for the most efficient use of water for agriculture is intensifying with the increased competition for water resources among various sectors with mushrooming population. In spite of having the largest irrigated area in the world, India too has started facing severe water scarcity in different regions. Efficient utilization of available water resources is crucial for India, which shares 17 per cent of the global population with only 2.4 per cent of land and 4 per cent of the water resources. Improper management of water and nutrient has contributed extensively to the current water scarcity and pollution problems in many parts of the world, and is also a serious challenge to future food security and environmental sustainability. Addressing these issues requires an integrated approach to soil-water-plant-nutrient management at the plant-rooting zone. Proper irrigation management requires that growers assess their irrigation needs by taking measurements of various physical parameters. Some use sophisticated equipment while others use the tried and true common sense approaches. Whichever method used, each has its merits and limitations. Scheduling of irrigation water varies with textural classes; more frequent irrigation interval in sandy and shallow soils as compared to loamy and clayey soils. It also varies with evapo transpiration of a location. The quantity of water to be applied at each irrigation depends on root zone depth, soil profile moisture status before irrigation and field capacity of the soil.

Table 18.1: Irrigation Interval for different Crops Grown in Textural Classes and Levels of ET

	Shallow Soil				Loamy Soil				Clayey Soil			
	ET_0 mm/day				ET_0 mm/day				ET_0 mm/day			
	4 to 5	6 to 7	8 to 9	Net Depth (mm)	4 to 5	6 to 7	8 to 9	Net depth (mm)	4 to 5	6 to 7	8 to 9	Net depth (mm)
Maize	8	6	4	40	11	8	6	55	14	10	7	70
Sorghum	8	6	4	40	11	8	6	55	14	10	7	70
Soybean	8	6	4	40	11	8	6	55	14	10	7	70
Wheat	8	6	4	40	11	8	6	55	14	10	7	70
Cabbage	3	2	2	15	4	3	2	20	7	5	4	30
Tomato	6	4	3	30	8	6	4	40	10	6	5	50
Banana	5	3	2	25	7	5	4	40	10	7	5	55
Onion	3	2	2	15	4	3	2	20	7	5	4	30

In developing any irrigation management strategy, two questions are common: "when do I irrigate?" and "how much do I apply?" To attain maximum yield and quality produce the crops vary in requirement of soil moisture level that has to be maintained during the crop season. Near field capacity most of the plants are efficient to use soil moisture as well as nutrient whose availability are governed by soil moisture, during crop season the continuous loss of water through ET will lead to decrease in both soil and plant water status which is well depicted through the decrease in soil and leaf water potential. So plants fail to extract enough water from soil to maintain their optimal growth rate. The major problem in crop production is to maintain optimum soil moisture (field capacity) throughout the cropping season. In order to provide irrigation water precisely irrigation scientists are trying to estimate crop water requirement through various soil, plant, weather and sensor based parameters. The major losses of water in plant-soil-atmosphere continuum are ET and plant uptake. The scientific scheduling of irrigation is to replenish the evapo-transpirational requirements. The various techniques adopted at large scale are briefed below.

Determining when to Irrigate based on following techniques

1. Soil Driven Parameters as a Basis for Irrigation Scheduling

Estimation of available soil moisture by Feel and appearance, gravimetric, volumetric, tensiometric, gypsum block, neutron probe, pressure plate apparatus, water budget technique, depletion of available water *etc.*

2. Crop Status as a Basis for Scheduling Irrigation

Visual plant appearance and growth, canopy temperature, stomatal resistance, relative water content, leaf water potential, pressure bomb technique, critical stage approach *etc.*

The concept of stress day index based on plant parameters to schedule irrigation was developed; it is a product of plant stress factor and crop susceptibility factor. The former is an indicator of crop water deficiency and degree and duration of deficit in growth stages and the latter is a function of crop, species and its growth stage. Plant water potential as stress indicator further leaf temperature had been correlated with crop water stress. But its measurement is not easy and many measurements have to be made to characterise a field. Stress degree day (SDD) is a difference between canopy temperature and atmospheric temperature when the difference accumulation to certain value irrigation is necessary. It is to be specified for each crop because canopy and air temperature differences are soil, crop and climate specific. Temperature stress day is a difference in temperature between a stressed plot and well-watered plot as an index for scheduling irrigation.

3. Climatic Parameters based Scheduling of Irrigation

Pan evaporimeter, cumulative pan evaporation, IW/CPE

4. Water Budget Technique

The water balance is based on the gain and loss of water is a system. It is worked out using the relationship of evaporative demand that will be met out by deducting losses from precipitation.

$$E = W - (Q_r + Q_l + Q_i + \Delta Q_w + \Delta Q_s)$$

where,

E = Evaporation (mm)

W = Precipitation in the form of rainfall, snowfall and irrigation

Q_r = Drainage, runoff

Q_l = Deep losses to underground layers

Q_i = Water intercepted by plants, leaves *etc.*

ΔQ_w = Change in ground water table

ΔQ_s = Change in available soil moisture

The equation seems to be very easy but practically difficult to workout in large scale. Because, accurate measurement of deep underground losses and runoff was found to be difficult. However, under controlled condition using lysimeter comparatively different losses can be measured.

Reference evapotranspiration (ET_0) is the rate of evapotranspiration from an extended surface of 8 to 15 cm tall green grass cover of uniform height, actively growing completely shading the ground and not short of water. It can be computed by Blaney-Criddle, modified penman, radiation and pan evaporation methods. Each method is calibrated by measured ET_0 for particular season and location. From ET_0 actual crop ET will be measured by ET_0 and ET crop relationship $ET_{crop} = K_c \cdot ET_0$; wherein Kc is crop coefficient.

Scheduling of Irrigation on IW/CPE Ratio

Irrigation water is being scheduled on the basis of ratio of certain depth of irrigation water to the cumulative evaporation from open pan popularly known as IW/CPE ratio. Irrigation depth is fixed at a value anywhere between 5 to 12 cm depending on soil type, rooting depth and easiness to measure. The ratio is fixed between 0.5 to 2.5 for different crops. Smaller the ratio means irrigation interval is longer and larger interval means irrigation is shorter intervals. For example the wheat crop is irrigated at IW/CPE ratio of 0.5, 0.75, 1.0 and 1.50 with IW = 6.0 cm. this means 6 cm of water is irrigated when open pan evaporation accumulate to values of 12 cm, 8 cm, 6 cm and 4 cm respectively. Depending on evaporation rate the interval will vary.

Determine how much to Irrigate

Enough irrigation water should be applied to replace the depleted plant available water within the root zone and to allow for irrigation inefficiencies. Root depth and root distribution are important because they determine the depth of the soil reservoir from which the plant can extract available water. About 70 per cent of the root mass is found in the upper half of the maximum root depth. Under adequate moisture conditions, water uptake by the crop is about the same as its root distribution. Thus, about 70 per cent of the water used by a crop is obtained from the upper half of the root zone. This zone is referred to as the *effective root depth*. This depth should be used to compute the volume of PAW. Irrigation amounts should be computed to replace only the depleted PAW within the effective root zone.

Scheduling of Irrigation in Paddy

Rice is the largest consumer of water about 2,500 litres are required to produce 1 kg of rice. Water requirement of rice varies from 1190 to 2650 mm depending upon soil, climate, variety and duration. The total water requirement includes land preparation, rising nursery, transplanting and harvest. Most of the water applied to rice field is lost by percolation and surface runoff. In most of the rice growing areas of tropics is 4-5 mm/day during *Kharif* and 10-12 mm/day during summer months. Critical stages for rice are maximum tillering, tiller initiation, primordial initiation, grain filling and flowering are most vital. Lack of adequate moisture at these stages reduced the yield even up to the extent of 50 per cent. For efficient and effective management of irrigation water by adopting improved methods like critical stage approach, intermittent submergence, alternate wetting and drying (AWD), micro irrigation, direct seeded rice, aerobic system of cultivation and saturation maintenance *etc*. Under intermittent submergence during the critical stage of initial tillering and flowering and maintenance of saturation or field capacity during the rest of the crop growth gave yields comparable to those obtained under continuous shallow submergence.

Micro Irrigation in Rice

For good yields of rice it is not necessary to maintain full submergence especially in the *Kharif* season. Use of sprinkler and drip irrigation has been successfully attempted in rice in Tamil Nadu and Andhra Pradesh and it brought out a saving of about 33 per cent in irrigation water. The detailed discussion on micro irrigation in rice has been covered in other chapters of this book chapter.

Table 18.2: Water Requirement of Rice Crop at different Growth Stages

	Avg. Water Requirement (mm)	Per cent of Total Water Requirement
Nursery	50-60	5
Main field preparation	200-250	20
Planting to Panicle initiation	400-550	40
Panicle initiation to flowering	400-450	30
Flowering maturity	100-150	5
Total	1200-1460	**100**

Table 18.3: Depth of Water to be Maintained during different Crop Growth Stages of Transplanted Rice

Stage of Crop	Depth of Water (cm)
At transplanting	2-3
After transplanting (5 to 20 days)	4-5
During tillering (22 to 42 days)	2-3
Reproductive stage,Panicle emergence, Booting, Heading and Flowering	4-5
Ripening stage (21 days after full flowering) Milk stage, Dough stage and Maturity	Drain the field gradually to Saturation Withdraw water 12 days before Harvesting

Scheduling of Irrigation in Maize

About 80 per cent of maize is cultivated during the monsoon season, mostly under rainfed condition, while crop during winter and summer season almost fully irrigated. The overall irrigated maize is only 25 per cent compared to 91 per cent under wheat and 59 per cent under rice. Maize crop is sensitive to both moisture stress and excessive moisture; hence regulate irrigation according to the requirement. In areas where assured irrigation facilities are available water must be supplied at critical growth stages, young seedling, knee high, silking and tasselling, and grain filling stages are the most sensitive stages for moisture stress. Ensure optimum moisture availability during the most critical phase (45 to 65 days after sowing); otherwise yield will be reduced by a considerable extent. Generally 3-5 irrigations may be required in monsoon depending on rainfall distribution. If 5 irrigations are available provide at 6 leaf, late knee high, tasselling, 50 per cent silking and dough stages, if 4 irrigations are available provide at 6 leaf, late knee high, 50 per cent silking and dough stages and if 3 irrigations at available provide at early knee high, tasselling and 50 per cent silking (Rajendra Prasad, 2012). Under fully irrigated condition first irrigation should be applied at 3-4 weeks after germination and subsequent irrigations at 4-5 week intervals upto mid-March and thereafter at 1-2 weeks depending on temperature and rainfall. Maize crop requires about 460-600 mm of water during its life cycle. Water shortage during tasselling and silking even for 2 days can reduce grain yield by about 20 per cent. The shortage for 6-8 days can pull down the yield by 50 per cent (Chhidda Singh *et al.,* 2012).

Table 18.4: Fertigation Schedule and Growth Stages of Maize

Crop stages	Quantity (per cent)			Irrigation Frequency (Days)	Kc Values
	N	P	K		
Vegetative stage (6 – 30 days)	25	25	25	5	0.4
Reproductive stage (30 – 60 days)	50	50	50	5	0.8
Maturity stage (60 – 75 days)	25	25	25	3	1.15
Total	100	100	100		0.7

Table 18.5: Effect of Drip Fertigation on Grain Yield of Maize

Treatments	Grain Yield (kg/ha)
Drip fertigation- 75 per cent RDF	5885
Drip fertigation 100 per cent RDF	6321
Drip fertigation 75 per cent RDF (50 per cent P and K - WSF)	6578
Drip fertigation 100 per cent RDF (50 per cent P and K- WSF)	7309
Drip irrigation 100 per cent RDF	5386
Surface irrigation 100 per cent RDF	4720
CD (p=0.05)	235

Source: Anitta Fanish and Muthukrishnan (2013).

Table 18.6: Number of Irrigations based on the Availability of Water

No. of Irrigation available	Critical Stages
1	CRI
2	CRI+ LJ
3	CRI+ B+ M
4	CRI+ LT+ F+ M
5	CRI+ LT+ LJ+ F+ M
6	CRI+ LT+LJ+ F+ M+ D

CRI- Crown root initiation (21 DAS), LT- Late tillering (42 DAS), LJ- Late Jointing (60 DAS), F- Flowering (80DAS), M- Milk (95 DAS) D- Dough (115 DAS).

Source: Pal *et al.* (1996).

Scheduling of Irrigation in Wheat

Wheat requires about 300-400 mm of irrigation weather depending upon climatic factors. About 90 per cent of area under wheat in India is irrigated although the amount of irrigation water available and applied differs. Climatic parameters especially pan evaporation data can also be used for scheduling irrigation in wheat in the areas where water resources are in abundance. In recent years a ratio of IW/CPE used for scheduling irrigation in wheat ranges between 0.7 to 0.9. Critical stages for soil moisture is presented in Table 21.5.

Scheduling of Irrigation in Sorghum

Eventhough sorghum is considered as drought tolerant crop it responds positively to irrigation and well suited to irrigated condition. Sorghum required 450- 600 mm water for getting higher yield. Roots can extract moisture from 2m of soil depth. *Kharif* season crop grown in rainfed condition, it was very oftenly affected by water stress. However, post rainy season sorghum is grown on residual soil moisture with limited irrigation facilities. The critical stages of sorghum in relation to water requirement are:

1. Initiation of grand growth stage   20-25 DAS
2. Flag leaf stage or boot stage   50-55 DAS
3. Flowering stage   70-75 DAS
4. Grain filling stage   90-100 DAS

Scheduling of Irrigation in Pearlmillet

The crop is grown primarily in rainfed areas. Water requirement of this crop is much lower than maize, sorghum and fingermillet. Eventhough it is rainfed crop irrigation at anthesis or flowering stage is beneficial. It has been estimated that crop requires on an average 140-150 mm of water per metric tonne of grain. A rainfall of 500 mm received during the crop season would be able to produce the grain yield upto 5 t/ha, however it depends on rainfall distribution.

Scheduling of Irrigation in Oilseed Crops

In India the major commercially exploited oilseed crops are groundnut, sunflower, soybean, mustard, safflower, castor, linseed. In general, flowering and pod development stages are sensitive stages for moisture stress.

Groundnut is largely grown in *Kharif* season but in recent years considerable emphasis has been given to rabi and summer irrigated groundnut. Providing irrigation at 75 per cent available soil moisture depletion (ASWD) approach, IW/CPE at 0.4 to 0.6 in *Kharif* season, whereas frequent irrigation in Vertisols has deleterious effect on pod formation and maturity. In *rabi* groundnut improvement in pod yield was observed up to 25 per cent ASWD and IW/CPE ratio of 1.2 in various regions of India. For Rapeseed and mustard providing one or two irrigation depending upon evaporative demand and residual soil moisture gave significantly higher seed yield. At Navsari mustard crop responded positively to 40 per cent ASWD. Sunflower is an exhaustive water requiring grown in *Kharif, rabi* and summer seasons. Scheduling irrigation based on IW/CPE at 0.6 to 1.0 was significantly influence on water seed yield. It was found superior over critical stage approach.

Scheduling of Irrigation for Pulses

Pulses are generally grown under rainfed conditions. Under moisture stress condition pulses respond well to limited irrigation. Some of the pulses like frenchbean, mungbean, urdbean *etc.* are cultivated under irrigated condition. Various approaches such as critical stage approach during crop growth stage, IW/CPE ratio, cumulative pan evaporation have been used in scheduling irrigation. Most of the pulses are sensitive to irrigation at flowering and pod development stages. Providing one irrigation at early pod filling stage of lentil was found most effective. Frequent irrigation to chickpea has no beneficial effect except during vegetative stage. Sprinkler irrigation for greengram increased yield by 39.7 per cent over surface irrigation and resulted in water saving of 49.8 per cent (Velayutham and Chandrasekaran, 2002). Drip fertigation in pigeonpea has profound impact on grain yield and water saving. Studies conducted in *Vertisols* of Raichur, Karnataka revealed that drip irrigation at 5 days interval and nutrient supply through water soluble fertilizers to pigeonpea improved grain yield, water saving and nutrient use efficiency. Another study at IIPR, Kanpur also confirmed drip fertigation at branching and pod development as most efficient in yield improvement as well as higher nutrient use efficiency (Praharaj and Kumar, 2011).

In chickpea irrigation at flowering or at pod formation gave higher yield than rainfed. By using meteorological observations, IW/CPE ratio of 0.6 to 0.8 which required 3-4 irrigations was found to be optimum for higher yield.

Table 18.7: Water Requirement of different Pulse Crops

Crop	Water Requirement (mm)	Critical Stages
Pigeonpea	400-500	Flowering and pod development
Chickpea	150-200	Flowering and pod development
Greengram	250-300	Flowering and pod development
Blackgram	200-250	Flowering and pod development
Fieldbean	250-300	Flowering and pod development
Horsegram	150-200	Flowering and pod development
Minor pulses	200-250	Flowering and pod development

Summary

The primary objective of the irrigation management was to rely on achieving higher water productivity by precise water use. Scheduling of irrigation will give information on when to irrigate and how much irrigate by appropriate method. Removal of excess water from cultivated field is as important as application of irrigation water. Providing irrigation water as and when required by plant grown in soil-plant-atmosphere continuum. There are three factors soil, plant and atmosphere in this continuum govern scheduling of irrigation. The quantity of irrigation water supply has to fulfil evaporative demand and crop water requirement. The major soil indicators are visual observance, soil moisture estimation, depletion of available water *etc*. Among plant related indicators visual symptoms, critical stages, canopy temperature, relative water content, leaf water potential, and mini plot techniques *etc*. While atmosphere parameters include evaporation estimation by USWB evaporimeter, IW/CPE ratio, water balance method are efficient methods to take decision on scheduling of irrigation.

References

Chhidda Singh, 2012. Modern techniques of raising field crops. Eds. Chhidda Singh, Prem Singh and Rajbir Singh. Pub: Oxford and IBH Publishing Company Pvt. Ltd., New Delhi, pp. 1-582.

Reddy, S.R., 2007. Irrigation agronomy. Pub: Kalyani Publishers, New Delhi, pp. 463.

Sankara Reddi, G.H., Yellamanda Reddy T., 2011. Efficient use of irrigation water. Pub: Kalyani Publishers, New Delhi, pp. 1-489.

Lenka, D., 2013. Irrigation and drainage. Pub: Kalyani Publishers, New Delhi, pp. 1-315.

19

Precision Irrigation Water Management in Field Crops

U.K. Shanwad and M.R. Umesh

Traditionally, producers treat the entire field as if it were a homogeneous unit, even though there were variations in soil types, soil fertility, and yield potentials. They apply average rates of inputs over the entire field. As a result, some areas were under applied while others were over applied, resulting in lower profits and chemical and nutrient losses to surface and ground water. Application of information technology in agriculture, precision farming is a feasible approach for sustainable agriculture. Precision farming makes use of remote sensing to macro-control of GPS to locate precisely ground position and of GIS to store ground information. It precisely establishes various operations, such as the best tillage, application of fertilizer, sowing, irrigation, spurting *etc.*, and turns traditional extensive production to intensive production according to space variable data. Precision farming not only utilizes fully the resources, reduce investment, decrease pollution of the of the environment and get the most of social and economic efficiency, but also makes farm products, the same as industry, become controllable, and be produced in standards and batches. However, precision farming has been confined to developed countries.

Land tenure system, smaller farm size (<1ha) level, This will be a stupendous task and a threatening challenge to space and agricultural scientists alike who are currently remotely placed from the ground truth of Indian farming. Currently 65 per cent of agriculture in India is rain dependent (Singh and Kumar, 2009 and Singh *et al.*, 2003). There are extreme variations in rainfall, the western most part getting less than 100 mm annually and the eastern most part receiving 100 times more. Floods and droughts can strike the country simultaneously at different places. For better use of water in agriculture in water-limited environments, efforts are needed from different research disciplines: agronomists, plant breeders, plant physiologists, plant biotechnologists, water engineers and others, to develop new approaches in water conservation. The paper mainly highlights the precision water management, present status in India, obstacle and, strategies for implementation to harvest the fruitful results.

Population growth is expected to increase, and the world population is projected to reach 10 billion by 2050, which decreases the per capita arable land and water. More intensive agricultural production will have to meet the increasing food demands for this increasing population, especially because of an increasing demand for land area to be used for biofuel. These increases in intensive production agriculture will have to be accomplished amid the expected environmental changes attributed to Global Warming. During the next four decades, soil and water conservation scientists will encounter some of their greatest challenges to maintain sustainability of agricultural systems stressed by increasing food and biofuels demands and Global Warming. We propose that Precision Conservation will be needed to support parallel increases in soil and water conservation practices that will contribute to sustainability of these very intensively-managed systems while contributing to a parallel increase in conservation of natural areas.

The original definition of Precision Conservation is technologically based, requiring the integration of a set of spatial technologies such as global positioning systems (GPS), remote sensing (RS), and geographic information systems (GIS) and the ability to analyze spatial relationships within and among mapped data according to three broad categories: surface modelling, spatial data mining, and map analysis. In this paper, we are refining the definition as follows: Precision Conservation is technologically based, requiring the integration of one or more spatial technologies such as GPS, RS, and GIS and the ability to analyze spatial relationships within and among mapped data according to three broad categories: surface modelling, spatial data mining, and map analysis. We propose that Precision Conservation will be a key science that will contribute to the sustainability of intensive agricultural systems by helping us to analyze spatial and temporal relationships for a better understanding of agricultural and natural systems. These technologies will help us to connect the flows across the landscape, better enabling us to evaluate how we can implement the best viable management and conservation practices across intensive agricultural systems and natural areas to improve soil and water conservation.

Present Status

Precision farming aims to manage production inputs over many small management zones rather than on large zones. It is difficult to manage inputs at extremely fine scales, especially in the case of the plantations (sugarcane, tea, coffee *etc.*) irrigation system. However, in real sense we expect site-specific irrigation approach to potentially improve the overall water management in comparison to irrigated farms of hundreds of acres. A critical element of the irrigation scheduling and management is the accurate estimation of irrigation supplies and its proper allocation for the irrigation of structures based on the actual planted areas. All irrigation scheduling procedures consist of monitoring indicators that determine the need for irrigation. The final decision depends on the irrigation criterion, strategy and goal. Irrigation scheduling is the decision of when and how much water to apply to a field. The amount of water applied is determined by using a criterion to determine irrigation need and a strategy to prescribe how much water to apply in any situation.

The right amount of daily irrigation supply and monitoring at the right time within the discrete irrigation unit is essential to improve the irrigation water management of a scheme (Rowshon and Amin, 2010). Many computerized tools have been used for scheduling irrigation deliveries and improving the irrigation project management. One such tool is a Geographical Information System (GIS). Its use in irrigation management with their large volumes of spatially and temporally distributed data is most beneficial. The GIS capability to integrate spatial data from different sources, with diverse formats, structures, projections or resolution levels, constitute the main characteristics of these systems, thus providing needed aid for those models that incorporate information in which spatial data has a relevant role (Adnan *et al.*, 2010). This explained about the capability of GIS for decision-making. The possibility of GIS for easily creating and changing scenarios allows the consideration of multiple alternatives of irrigation scheduling, including the adoption of crop specific irrigation management options. Scenarios may include different irrigation scheduling options inside the same project area applied to selected fields, crops, or sub-areas corresponding to irrigation sectors. This allows tailoring irrigation management according to identified specific requirements (Fortes *et al.*, 2005).

The irrigation scheduling alternatives are evaluated from the relative yield loss produced when crop evapotranspiration is below its potential level. Examples of those successful applications are presented by (Zairi *et al.*, 2003) for surface irrigation in the European region. Irrigation scheduling is the farmers decision process relative to "when" to irrigate and "how much" water to apply at each irrigation event. It requires knowledge of crop water requirements and yield responses to water, the constraints specific to the irrigation method and respective on farm delivery systems, the limitations of the water supply system relative to the delivery schedules applied, and the financial and economic implications of the irrigation practice (Pereira *et al.*, 2003). Irrigation scheduling models are particularly useful to support individual farmers and irrigation advisory services. Zairi *et al.* (2003) reported about an irrigation model embedded within the GIS.

The data were correlated to digital data sets on soils, agro-climate, land use, and irrigation practice to produce tabular and mapping outputs of irrigation need (depth) and demand (volume) at national, regional, and catchment levels. The GIS approach allows areas of peak demand to be delineated and quantified by sub-basins. The GIS-based modelling approach is also currently being used to administer irrigation needs for irrigated crops. Map and tabular output from the GIS model can provide licensing staff with the information necessary to establish reasonable abstraction amounts to compare against requested volumes on both existing and new license applications for spray irrigation. The main value of models results from their capabilities to simulate alternative irrigation schedules relative to different levels of allowed crop water stress and to various constraints in water availability. The main limitation of simulation models is that some model computations are performed at the crop field scale for specific soil, crop, and climate conditions, which characterize that crop field and the respective cropping and irrigation practices.

When the computation procedure is applied at the region scale it becomes heavy and slows due to the need to consider a large number of combinations of field and crop characteristics to be aggregated at sector or project scales. Proper irrigation scheduling can reduce irrigation demand and increase productivity. A large number of tools are available to support field irrigation scheduling, from in-field and remote sensors to simulation models. Irrigation scheduling models are particularly useful to support individual farmers and irrigation advisory services (Rowshon *et al.*, 2003). Ortega *et al.* (2005) outlined applications of GIS-based modelling. Irrigation water requirements are determined by mapping the spatial distribution of water requirements based on soil and crop distributions. GIS-based modelling approaches are used to establish irrigation scheduling based on water-balance modelling. GIS databases for irrigation include coverage for crops, irrigation methods, and soils. These data are coupled with agro-climatic data to provide information on growing-season and water-use requirements.

As previously stated that the economic benefits of having such a system are: reducing monitoring costs, improving the speed of decision making by supporting the decision-makers with real time information, ability to access by everyone and everywhere over the Internet, reducing time and minimizing effort to reach data, high speed, security and high rate of error handling with new Internet technologies, having centralized database that provides a single source of common information which provides standardization and faster retrieval and selective modification of information, and finally, the ability to produce reports based on user specified parameters (Montgomery and Schuch, 1993).

Case Studies in India and Abroad

Irrigation System in Yaqui Valley in Mexico

The Yaqui Valley, an intensively managed wheat based agricultural region, is located in Sonora State, Mexico, between the Sierra Madre Mountains and the Gulf of California. The Valley consists of approximately 2,25,000 ha of irrigated agricultural fields. Referred to as the birthplace of the Green Revolution for wheat, it is one of the country's most productive breadbaskets. Using a combination of irrigation, high fertilizer rates and modern cultivars, Valley farmers produce some of the highest wheat yields in the world. The farmers applied an average of 5-6 irrigations during the crop cycle. The majority of farmers receive water from canals, which are cleaned an average of twice during the growing cycle, either manually (29 per cent), mechanically (22 per cent), chemically (12 per cent), or using a combination of methods. Only 2 per cent of the farmers used pump irrigation. The most common method of irrigation was through beds and corrugations (79 per cent), followed by melgas (18 per cent). Farmers who planted on beds in clay soils had a lower average number of irrigations. The normal rainfall here is lower than 400 mm annually. However, the Farmers Cooperative Irrigation System taking care that it operates 498 bore wells automatically through GIS program and also monitors how much water of bore well should be added to the canal water at each point (as underground water pH is towards acidic).

Precision Agriculture Farmers Association in Bangalore

Precision Agriculture Farmers Association in Bangalore are working towards precise tuning of agriculture inputs to the horticulture crops like papaya, banana through sensors including irrigation water management.

Direct Seeded Rice (DSR) in TBP command in Raichur District

Direct seeded rice continues to remain in focus due to falling water storage in the reservoirs and unscheduled release and closer of canal supplies. This is largely due to vagaries and insufficient monsoonal climate and rains received. Early wetting rains are very essential for the success of DSR for timely planting and introduction of second crops in DSR fallows.

Surface and Sub-surface Drip Irrigation in TBP Command in Raichur District

Water resources are going to be limited in future and developing technology that uses less water to produce higher output have to be devised and developed. Conventionally rice is raised under flooded condition. A newer technology which requires less water and could be grown successfully under limited water resources are need of the hour. During the year we laid 4 demonstrations in farmers field which involved surface and subsurface irrigation. Earlier we have clearly demonstrated how the adoption of DSR could save irrigation water (38-40 per cent) and attracted the large number of farmers. To improve water productivity in DSR, demonstration on micro irrigation initiated in the irrigation command.

Obstacles

There are many obstacles to adopt precision farming in developing countries in general and India in particular. Some are common to those in other regions but the ones specific to Indian conditions are as follows.

   a. Culture and perceptions of the users

   b. Small farm size

   c. Lack of success stories

   d. Heterogeneity of cropping systems and market imperfections

   e. Land ownership, infrastructure and institutional constraints

   f. Lack of local technical expertise

   g. Knowledge and technical gaps

   h. Data availability, quality and costs

nsors Placed at 1 Feet Depth
ge in Haveri District.

Despite the many obstacles listed earlier, business opportunities for precision farming technologies including GIS, GPS, RS and yield monitor systems are immense in many developing countries. The scope for funding new hardware, software and consulting industries related to precision agriculture is gradually widening. In Japan, the market in the next 5 years is estimated at about US $ 100 billion for GIS, and about US $ 50 billion for GPS and RS (Srinivasan, 2001). Punjab and Haryana states in India, where farm mechanization is more common than in others, may be the first to adopt precision farming on a large scale. Recently, the governments of certain Asian countries initiated special efforts to promote precision farming. In Japan, the Ministry of Agriculture has allocated special funds for research on remote sensing applications of precision farming. A quasi-governmental institute "Bio-oriented Technology Research Advancement Institute (BRAIN)" is also funding research on precision farming. In Malaysia, the Malaysian Agricultural Research and Development Institute (MARDI) is promoting research on precision farming of upland rice. In other countries, the private sector, which holds or leases a large acreage, is likely to adopt precision farming sooner than the small holders. Precision farming is useful in many situations in developing countries but only a few priority topics will be discussed here in detail. Rice, wheat, sugar beet, onion, potato and cotton among the field crops and apple, grape, tea, coffee and oil palm among horticultural crops are perhaps the most relevant. Some have a very high

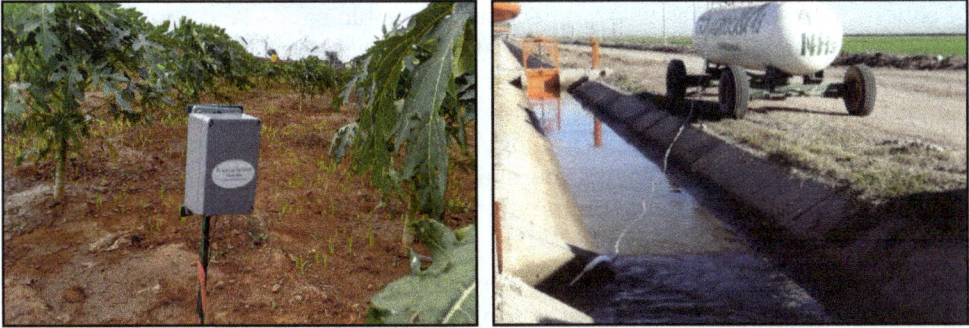

**Nodes installed across farm having Moisture and Team sensors.
Zigbee module used to send data wirelessly to the main unit.
Mechanical enclosure is water and dust proof (IP 65 compliant)**

**Irrigation System in Yaqui Valley in Mexico Anhydrous Ammonia (80 per cent N)
Mixing with Irrigation Water in Mexico.**

value per acre, making excellent cases for site-specific management. For all these crops, yield mapping is the first step to determine the precise locations of the highest and lowest yield areas of the field.

Researchers at Kyoto University recently developed a two-row rice harvester for determining yields on a micro plot basis (Iida *et al.*, 1998). Precision farming can bring several benefits to the sugar beet industry in Hokkaido, where the marketing system was changed in 1986 from a quantity (fresh weight) to quality (sugar yield) basis. Heavy N fertilization, a common practice here, results in excessive N levels and a decreased sugar concentration. It is now possible to estimate sugar and amide

N concentration in leaves using reflectance in visible bands, and root yield using reflectance in visible and infrared bands (Okano *et al.*, 1997). Incorporation of such data into a GIS alongwith precise positioning of non-uniform areas using GPS can be used to vary fertilizer dose within a field, thereby improving productivity. Although weighing conveyor technology has been known for some time, effective yield measurement still remains the main problem in crops such as sugar beet, onion and potato. Further, the preparation of product quality maps for these crops is as important as yield maps.

In India, a few researchers in the private sector initiated studies on precision agriculture in high value crops like cotton, coffee and tea. In cotton, remote sensing coupled with GIS can assist in improved precision of insect pest management and harvesting. In Sri Lanka, researchers at the Tea Research Institute are examining precision management of soil organic carbon. In so far as dairy farming in Asia is concerned, precision farming techniques can help in improving efficiency of methods, timing and rate of application of animal wastes leading to high application efficiency and low environmental pollution. While considering soil and climatic conditions. For instance, factors determining the risk of NO_3 leaching, release of N_2O through denitrification and contamination of surface and ground water by runoff can be mapped and analysed. Likewise, poorly managed areas in grass lands can be identified and the optimum period for cutting on a plot basis determined. Nutrient stress management is another area where precision farming can help Indian farmers. Most cultivated soils in India are acidic and spatial variation in pH is high. Detecting nutrient stresses using remote sensing and combining data in a GIS can help in site-specific applications of fertilizers and soil amendments such as lime, manure, compost, gypsum and sulphur.

This in turn would increase fertilizer use efficiency and reduce nutrient losses. In semi arid and arid tropics, precision technologies can help growers in scheduling irrigation more profitably by varying the timing, amounts and placement of water. For example, drip irrigation, coupled with information from remotely sensed stress conditions (*e.g.*, canopy temperature) can increase the effective use of applied water from 60 to 95 per cent there by, reducing runoff from 23 to 1 per cent and deep percolation from 18 to 4 per cent. Pests and diseases cause huge losses to Indian crops. If remote sensing can help in detecting small problem areas caused by pathogens, timing of applications of fungicides can be optimized. Recent studies in Japan show that pre-visual crop stress or incipient crop damage can be detected using radio-controlled aircraft and near-infrared narrow-band sensors. Likewise, airborne video data and GIS have been shown to effectively detect and map black fly infestations in citrus orchards, making it possible to achieve precision in pest control. Perennial weeds, which are usually position-specific (Wilson and Scott, 1982) and grow in concentrated areas are also a major problem in developing countries. Remote sensing combined with GIS and GPS can help in site-specific weed management. Although cost-benefit analysis has not been done yet, the possible use of precision technologies in managing the environmental side effects of farming and reducing pollution is appealing.

Strategies

Precision farming is still only a concept in many developing countries and strategic support from the public and private sectors is essential to promote its rapid adoption. Successful adoption, however, comprises at least three phases including exploration, analysis and execution. Data on crop yield, soil variables, weather and other characteristics are collected and mapped in the exploratory stage, which is important for increasing the awareness among farmers of long term benefits. The approaches to data collection and mapping must, therefore, reflect local needs and resources. In the analysis stage, factors limiting the potential yield in various areas within a field and their interrelationships are examined using GIS-based statistical modelling. Sadler *et al.* (1998) showed that quantitatively important yield variation may occur over distances as short as 10 m, however, only some factors such as soil structure, water status, pH, nutrient levels, weeds, pests and diseases can be controlled but not the others (soil texture, weather, topography).

After determining the significance of each source of variability to profitability of a particular crop and relative importance of each controllable factor, management actions can be prioritized. It must be remembered that in some low yielding areas, the reason for poor yields may be the lack of sufficient soil nutrients in the first place. In such cases, application beyond just replenishment is necessary. Lastly, execution phase includes variable application of inputs or cultural operations. However, it is not always necessary and/or possible to use variable rate applicators. Efforts must, therefore, initially focus on limiting indiscriminate use of inputs in conventional methods. Once the economic and environmental benefits are known widely, variable rate technology would be rapidly implemented at least in high value crops. To spur adoption of precision farming methods in developing countries, pilot demonstration projects must be conducted at various growers' locations by involving farmers in all stages of the project. The pilot projects must attempt to answer the grower's needs and emphasize the operational implementation of technology and complete analysis of the costs and savings involved. Documentation of pilot projects would help in examining the operational weaknesses and identification of remedial measures.

The projects can be used to train innovative farmers and early adopters, expose the neighbouring non-participating farmers to the new technologies, and show the usefulness of the technology for short and long-term management. The role of agricultural input suppliers, extension advisors and consultants in the spread of these technologies is vital. For instance, public agencies should consider supplying free data such as remotely sensed imagery to the universities and research institutes involved in precision farming research. Also, professional societies of agronomy, agricultural informatics, and engineering must provide training guidance in the use of technologies. The involvement of inter/disciplinary teams is essential in this. Small farm size will not be a major constraint, if the technologies are available through consulting, custom and rental services.

The role of agricultural cooperatives is important in dissemination of precision farming technologies to small farmers. If precision farming is considered a series of discrete services: map generation, targeted scouting, it is possible to fit these services within the structure of a progressive agricultural cooperative in each developing country. Changes in agricultural policies are also necessary to promote the adoption of precision farming. There are basically two policy approaches: regulatory policies and market based policies. The former refer to environmental regulations on the use of farm inputs and later refer to taxes and financial incentives aimed at encouraging growers to efficiently use farm inputs. In most developing countries the lack of penalties for pollutant generation has partly contributed to an excessive use of inputs. Subsidies on inputs and outputs and mechanisms that prevent the price system from rationing limited resources are also common. The latter include state-guaranteed crop prices, tariffs, import quotas, export subsidies. Inputs such as water and fossil fuels are usually sold at prices that are well below the real resource cost of their use, which consists not only production costs but also includes scarcity value and costs of pollution. In such cases, the formulation of policies that reflect the real scarcity value of natural resources and penalize pollution and policies such as green payments for farmers adopting techniques that would lower environmental costs can promote the adoption of precision farming technologies (Branden *et al.*, 1994). In most developing countries, the pollution effects of agriculture have been largely ignored so far because of inability to effectively monitor such effects. The advent of precision farming, and the computerization of input and output flows, will now enable such monitoring. Higher taxes on pollution farms are often recommended, but there is strong opposition to the implementation of the polluter-pays-principle concept in most countries including India. At the same time, some consumers in India would like to see a drastic reduction in the use of pesticides and fertilizers, and are willing to pay as much as 4 to 6 times the normal price for produce such as organic vegetables, soybean and wheat. When the price elasticity of input use is low and the input costs are only a small part of the total production expenditure, as in the case of fertilizers and pesticides. Very high taxes are required to reduce their use adequately.

Given the unfeasibility of such high taxes, a hybrid policy may be implemented for controlling pollution. A tax-free quota of N can be combined with taxes on additional N use. At the research level, many issues remain to be resolved. Although some progress has been made at Space Application Center, Ahmadabad, yield monitors for small farm conditions are yet to be developed. The development of standards for the hardware and software (image transfer formats and GPS transfer formats, map projection formats) is another issue. Crop models and decision support systems must be improved by considering local resources. Data for calibration of models must be made available to increase their accuracy and/or predictability. The ability to finance a creative information venture in agriculture will affect the speed of diffusion of precision farming technologies. Commercial banks, as well as other sources of funding, have to be educated regarding the potential of precision farming. In many developing countries, it may be worthwhile to develop programmes of subsidized credit to enable R and D activities on precision farming.

Conclusion

It's true that the necessity of achieving sustainable management of available water resources for irrigation supplies will determine the development of up-to-date and competitive agriculture. Precision farming in many developing countries including India is in its infancy but there are numerous opportunities for adoption. Although it is recognized that agriculture is a major polluter of the environment in many developing countries, farmers will not adopt precision farming unless it brings in more or at least similar profit as compared to traditional practice. It must be remembered that not all elements of precision farming are relevant for each and every farm. For instance, introduction of variable rate applicators is not always necessary or the most appropriate level of spatial management in Indian farms. Likewise, not all farms are suitable to implement precision farming. Some growers are likely to adopt it partially, adopting certain elements but not others. Precision farming cannot be convincing if only environmental benefits are emphasized. On the other hand, its adoption would be improved if it can be shown to reduce the risk. The adoption of precision farming also depends on product reliability, the support provided by manufacturers and the ability to show the benefits. Effective coordination among the public and private sectors and growers is, therefore, essential for implementing new strategies to achieve fruitful success.

References

Adnan, M., Singleton, A. D., Longley, P. A., 2010.Developing Efficient Web-Based GIS Applications.*CASA Working Papers Series*, Paper 153.

Fortes, P. S., Platonov, A. E., Pereira, L. S., 2005. GISAREG: A GIS-Based Irrigation Scheduling Simulation Model to Support Improved Water Use. *Agricultural Water Management*, 77, 159-179. http://dx.doi.org/10.1016/j.agwat.2004.09.042

Montgomery, G. E., Schuch, H. C., 1993.GIS Data Conversion.*GIS Data Conversion Handbook*, 27-45. http://dx.doi.org/10.1002/9780470173244.ch2

Ortega, J. F., De Juan, J. A., Tarjuelo, J. M., 2005.Improving Water Management: The Irrigation Advisory Service of Castilla-La Mancha (Spain).*Agricultural Water Management*, 77, 37-58. http://dx.doi.org/10.1016/j.agwat.2004.09.028

Ozdilek, O., Seker, D.Z., 2004. A Web-Based Application for Real-Time GIS.http://www.isprs.org/proceedings/XXXV/congress/yf/papers/934.pdf

Pereira, L. S., Teodoro, P. R., Rodrigues, P. N., Teixeira, J. L., 2003. Irrigation Scheduling Simulation: The Model ISAREG. In: Rossi, G., Cancelliere, A., Pereira, L.S., Oweis, T., Shatanawi, M., Zairi, A., Eds., *Tools for Drought Mitigation in Mediterranean Regions*, Kluwer, Dordrecht, 161-180. http://dx.doi.org/10.1007/978-94-010-0129-8_10

Rowshon, M. K., Amin, M. S. M., 2010.GIS-Based Irrigation Water Management for Precision Farming of Rice.*Int. J. Agric. Bio. Engg.*, 3, 27.

Rowshon, M. K., Kwok, C. Y., Lee, T. S., 2003. GIS-Based Scheduling and Monitoring of Irrigation Delivery for Rice Irrigation System: Part I. Scheduling. *Agril Water Mang.*, 62, 105-116.

Singh, S., Bhushan, L., Ladha, J.K., Gupta, R. K., Naresh R.K., Singh. P.P., 2003, Weed Management in Zero-Till Direct Seeded Rice: Some Promising Developments, *Rice-Wheat Information Sheet*. 47(12):7-8.

Zairi, A., E l Amami, H., Slatni, A., Pereira, L. S., Rodrigues, P. N., Machado, T., 2003. Coping with Drought: Deficit Irrigation Strategies for Cereals and Field Horticultural Crops in Central Tunisia. In: *Tools for Drought Mitigationin Mediterranean Regions*, Springer, Netherlands, 181-201. http://dx.doi. org/10.1007/978-94-010-0129-8_11

20

Dry Direct Seeded Rice: Efficient Water Use in Irrigation Commands

P.H. Kuchanur, G.W. VanLoon, Yogeshkumar Singh,
R.H. Rajkumar, S.R. Anand and S.G. Patil*

Rice is one of the most important food crops in the world, and is staple food for more than half of the global population. Under transplanting, even though a good crop can be grown by creating a hard pan below the plough zone of the soil by puddling to reduce the permeability of water. This process, however, leads to high losses of water through percolation and surface evaporation. Around the world, including India, water resources, both surface and underground, are shrinking and water has become a limiting factor in rice production. The looming water crisis and the water intensive nature of rice cultivation alongwith escalating costs of cultivation and energy requirements are stimulating a search for alternative methods so as to make rice cultivation more profitable and to conserve both natural and financial resources.

Rice Ecosystems in India

Rice is grown within different eco-systems on a variety of soils under varying climatic and hydrological conditions. Within India, the different ecosystems include the following:

Irrigated Rice

Rice is grown under irrigated conditions in the states of Punjab, Haryana, Uttar Pradesh, Jammu and Kashmir, Andhra Pradesh, Telangana, Tamil Nadu, Sikkim, Karnataka, Himachal Pradesh and Gujarat. In these states 50-90 per cent of rice area is under irrigation.

Rain-fed Rice

Rain-fed rice is further classified as upland or lowland. Upland rice is cultivated in Assam, Bihar, Eastern MP., Orissa, Eastern U.P., West Bengal and North Eastern Hill region. For rain-fed upland rice there is no standing water after a few hours of cessation of rain. Low land rice is cultivated mostly in Assam, West Bengal, Bihar, Orissa, Eastern M.P., and Eastern U.P. Low land rice is further classified into shallow water (<50cm), semi-deep water (50-100cm) and deep water (>100 cm) depending on the standing depth of water in the field. Coastal areas are frequently subjected to salinity problems and these areas are situated in West Bengal, Orissa, Andhra Pradesh, Tamil Nadu and Kerala. Cold or hill areas are located in Jammu and Kashmir, Uttarakhand and North Eastern hill states. The major problems found in these areas are cold injuries, blast, drought and the very short duration of the cropping season (Rice- A status paper. 2014, Directorate of Rice Development, GOI, Patna, Bihar, 1-16).

Irrigated Rice Cultivation in India

Under irrigation, rice can be established by four principal methods: Dry direct seeding (Dry-DSR), wet direct seeding (wet-DSR), water seeding and transplanting. These methods differ from each other either in the land preparation (tillage) or crop establishment methods or in both. Dry-DSR, wet-DSR and water-seeding are the practices, wherein seeds are sown directly in the main field instead of planting in a nursery and then transplanting rice seedlings, are commonly referred to as direct seeding methods. Direct seeding is actually the oldest method of rice establishment in India. Prior to the 1950s,direct seeding was most common, but was gradually replaced by puddled transplanting (Rao *et al.*, 2009). Dry direct seeded rice is the technology which is most efficient in terms of water, labor and energy and has eco-friendly characteristics that make it an alternative to transplanted rice that is especially appropriate in the 21st century.

Direct Dry Seeded Rice: The Need of the Hour

The majority of current rice cultivation systems consume two to three times more fresh water than for other cereals. Rice consumes about 50 per cent of the total irrigation water used in Asia and accounts for about 34–43 per cent of the world's irrigation water and 24–30 per cent of the total withdrawals of freshwater for all purposes around the world. A grim water scenario for agriculture together with the highly inefficient rice production technologies currently adopted by a majority of farmers globally warrants the exploration of alternative rice production methods, which inherently require less water and are more efficient in water use. Current research points to the virtues of DSR in responding to the water crisis. There, on an average, 25-30 per cent of water is saved in wet and dry direct seedling methods. Furthermore, direct seeding drastically reduces the adverse effects of puddling on soil physical properties and on the succeeding non-rice crop.

Both Dry- and Wet-DSR are more water efficient and they have other advantages over transplanted rice (CT-TPR). For one thing, CT-TPR cultivation is highly labour intensive. Both land preparation (puddling) and transplanting for crop establishment in CT-TPR require a large amount of labour. Besides the cost, rapid economic growth in Asia has increased the demand for labour in non-agricultural sectors, resulting in reduced labour availability for agriculture. Direct seeding can cut the amount of labour time required during the cropping period by approximately half.

Indian farmers are ready to take up this challenge; among many, there is increasing awareness and interest towards conservation agriculture technologies (Kumar and Ladha 2011).

A survey carried out during 2012-13 involving 30 farmers who practiced both dry direct seeding and transplanted rice revealed that there was considerable saving in many categories when using the dry seeding practice. Farmers saved seeds (43.2 per cent), fuel (37.5 per cent), tillage cost (37.4 per cent), fertilizer cost (30.0 per cent) and others without affecting the yield levels as compared to transplanting (Table 20.1). Further, they realized greater net returns as they incurred reduced expenditures for dry seeding practice compared to transplanting method.

Table 20.1: A Comparison between Direct Dry Seeded Rice (DSR) and PTR Practices (as per the survey of 30 farmers in TBP and UKP – 2012, variety: BPT 5204)

Particulars of Operations	DSR	TPR	Saving/Benefit in DSR	
			Quantity	Per cent
Seed rate (kg ha^{-1})	34.5	60.7	26.2	43.2
Fuel (Lha^{-1})	32.7	52.3	19.6	37.5
Tillage cost (R. ha^{-1})	5445	7600	2155	27.4
Fertilizer(NPKkg ha^{-1})	213:137:62	311:195:85	98:58:23	32:30:27
Fertilizer cost(Rs ha^{-1})	14010	19908	5898	30.0
Grain yield (t ha^{-1})	7.21	6.59	0.62 Incr.	9.4 Incr.
Expenditure (R. ha^{-1})	46465	66500	20035	30.0
Gross returns(Rs ha^{-1})	12597	114709		
Net returns (Rs ha^{-1})	79132	48209	30923	
B:C ratio	2.73	1.74		
Water consumption (m^{3}ha^{-1})	8130	12500		
Water consumption (mm)	813	1250		
Virtual water value (VWV) (Lkg^{-1})	1130	2160		

Being able to forgo the nursery stage eliminates the energy consumption and financial expense associated with heavy fuel use in some aspects of land preparation, especially puddling and planking. Add to this the removal of a transplanting step, and labour costs are also significantly less.

Clearly, reduced water requirement is a big plus connected with DSR. And this advantage could be multiplied many fold for the country as a whole where currently about 80 per cent of water consumption is directed to agriculture. Each year there are increasing demands for industrial, commercial and domestic uses as well as in other parts of the agricultural sector. The virtual water value (VWV) is measure of the water required to produce a kilogram of grain, and direct seeding can bring these values nearer to some of the dryland cereals, while maintaining productivity that is comparable to TPR (Table 22.1).

Dry direct seeded rice also offers flexibility in sowing time as compared to transplanted rice in tail end commands of the Tungabhadra (TBP) and upper Krishna Project (UKP), where canal water becomes available later. Especially in the TBP, farmers have to complete transplanting operations on a war footing (within 15 days); otherwise they need to wait another 15 days for the next supply of water needed to transplant the crop. Hence, using DSR the farmers are able to make use of early monsoon rains for sowing and this brings its own benefits such as even germination and uniform establishment of crop.

Requirements for Successful Dry Direct Seeded Rice in Irrigated Areas

Laser Land Levelling

Precise land levelling, achieved by laser land leveller, is a pre requisite for growing a successful crop of DSR. In comparison to traditional land levelling, laser levelling facilitates a healthy crop stand, enabling the farmer to apply uniform irrigation and leading to improved weed control and nutrients use efficiency. Other benefits include savings in irrigation water and higher crop productivity (Jat *et al.*, 2009). It facilitates seed planters to place seeds at even spacing and depth and the uniform distribution of water results in a uniform crop stand (Kumar and Ladha, 2011). Undulation in traditionally level fields often leads to poor establishment of DSR due to uneven depth of seedling and uneven water distribution. Hence, precision in land levelling is first step for better DSR crop.

Soil Type

Direct seeding (DSR) can be practiced in almost all type of soils suitable for rice. Local experience has indicated that black cotton soils and red soils give good results but medium textured soils are more suited to DSR. For conventional till DSR, the field should be pulverized to maintain good soil moisture and to maximize soil-to-seed contact but for ZT DSR, existing weeds should be killed using herbicides such as paraquat (0.5 kg ai/ha or phyphosate (1.0 kg ai/ha) (Kumar and Ladha, 2011; Gopal *et al.*, 2010).

Planting Date

To optimize the use of monsoon rain, the optimum time for sowing DSR is about 10- 15 days prior to onset of the monsoon. In the tail end of the Tunga Bhadra command areas in Karnataka, farmers practice dry sowing as well as sowing (followed by seed covering) after the receipt of soaking rains.Further, in the case of a delayed monsoon, farmers have used early duration varieties for sowing upto the first week of September and they were able to harvestgood yields. A good rice crop can be established by dry sowing followed by light irrigation during both *kharif* and*rabi*/summer seasons.

Cultivar Selection and Crop Establishment

Among the existing varieties and hybrids which were bred for puddled rice, some hybrids and basmati varieties have been found suitable for DSR. An extensive survey of farmers who practiced dry direct seeded rice as well as puddled rice indicated that the varieties released for transplanting also performed well under DSR (Table 20.2).

Table 20.2: Performance of different Varieties under Dry Seeding and Transplanting

	Yield (t ha⁻¹)		Per cent Increase over PTR
	DSR	TPR	
BPT 5204 (30)	7.21	6.59	9.4
Kaverisona (11)	6.87	-	-
MTU 1010 (5)	6.60	-	-
JK 3333 (2)	9.00	-	
JGL 11470 (2)	7.97	7.50	6.3
Gangavatisona (1)	7.13	7.13	0
Sri Ram Gold (2)	6.75	6.56	2.9
IR 64 (5)	6.11	-	-
Mean	7.2	6.9	4.6

Table 20.3. Rice varieties suited for DSR at different regions

Sl.No.	Regions	Genotypes suitable for DSR
1	Bihar (India)	Satyam, Rajendra, Mahsuri-I, NDR-359, Prabhat, Birsa dhan-101, Birsa dhan-104
2	Eastern Uttar Pradesh (India)	Aditya, NDR-359, Sarjoo-52, Mahsoori, Swarna, Moti, Pusa-44, KRH-2
3	Haryana, Punjab, Western U.P. (India)	Pusa-1121, Pusa Sugandh-5, PRH-10, Pusa Basmati-1, Pant Dhan-12, Sharbati, PHB-71, Kanchan, Kalinga-3, HeeraPathra, Sneha, Sahbhogi, Birsa dhan-101, 104, 105, 201 and 202, Saket-4, VLK dhan, Kranti, Satya
4	Tarai of Uttarakhand	Nidhi, Narendra-359, Sarvati, PR-113, Sarjoo-52
5	Cambodia	Koshihikari, W42 (tuong 2008)
6	Nepal	SonaMasuli, Hardinath, Radha-4, Radha-11, Chaite 2
7	Thailand	IR 57514- PMI-5- B-1-2, IR 20
8	Japan	RS-15, RS -20

Potential Constraints/Risks for Direct Seeding

The successful DSR crop establishment under rainfed conditions depends on evenly distributed timely rain. In northern India, proliferation of weedy or red rice alongwith new and non paddy weed flora like *Leptochloa, Eragrostis, Dactyloctenium* are increasing the dependence on herbicides. The other adverse effects include reduced availability of soil nutrients such as N, Fe, and Zn especially in Dry-DSR, increased incidence of new soil-borne pests and diseases such as nematodes, enhanced nitrous oxide emissions from soil and relatively more soil carbon loss due to frequent wetting and drying. Even though management with DSR is faced with the aforementioned risks, nevertheless the overall benefits outweigh the risks or constraints associated with it (Kamboj *et al.*, 2012).

Future Research Needs

Direct seeded rice is slowly making inroads into the rice ecosystem of the country and is slowly gaining acceptance among farmers. However, the large scale implementation of the technology will be possible only after several research gaps are addressed by appropriate research programmes. Much research and many adoptive evaluations carried out during the past decade have provided management options, including improved drills to precisely place seed and fertilizer and integrated approaches for managing weeds. However, additional research is needed in weed management, including (1) monitoring shifts in weed flora, (2) developing management strategies for emerging problems of weedy rice, (3) identifying new herbicides/tank mixtures with wide-spectrum weed control ability, (4) identifying vulnerabilities in weed life cycles through analysis of weed population dynamics under reduced till/ZT conditions, and (5) developing integrated strategies to minimize/avoid/delay in the development of herbicide resistance in weed populations. Although refinements in agronomy and management will continue to be important, targeting varietal improvements in rice under DSR is likely to be crucial for improving the potential of direct seeding (Kumar and Ladha 2011).

References

Gopal, R., Jat, R. K., Malik, R. K., Kumar, V., Alam, M. M., Jat, M. L., Mazid, M. A., Saharwat, Y. S., McDonald, A., Gupta, R. 2010, Direct dry seeded rice production technology and weed management in rice based systems. *Technical Bulletin. International Maize and Wheat Improvement Centre,* New Delhi, India. pp28.

Kamboj, B, R., Kumar, A., Bishnoi, D. K., Singla, K., Kumar, V., Jat, M. L., Chaudhary, N., Jat, H. S., Gosain, D. K., Khippal, A., Garg, R., Lathwal, O.P., Goyal, S. P., Goyal, N.K., Yadav, A., Malik, D.D., Mishra, A., Bhatia, R. 2012, Direct seeded rice technology in western Indo-Gangetic plains of India. *CSISA experiences. CSISA, IRRI and CIMMYT.* pp16.

Kumar, V., Ladha, J. K., 2011, Direct seeded rice: recent developments and future research needs. *Advances in Agron.,* 111: 299-391.

21

Herbigation in Field Crops

G.S. Yadahalli, B.M. Chittpaur and
Vidyavathi G. Yadahalli

The shrinking water resources and competition from other sectors, the share of water allocated to irrigation is likely to decrease by 10 to 15 per cent in the next two decades. One of the ways of alleviating water scarcity is by enhancing its use efficiency. Among different methods of irrigation, micro irrigation (drip) method results in maximum water and input use efficiency. Unnecessary applications of water and fertilizer can also allow weeds to flourish in modern agriculture. While irrigation systems are usually designed and managed with a crop of interest in mind; the impact of irrigation on weed growth is an important component of any modern production system.

Weed causing annually 45 per cent of yield loss to the Indian Agriculture (Das., 2008). Weed infestation and competition is more severe in all the field crops. Among the agronomic requirements to improve the yield levels of different crops timely weed control plays an important role. Traditional method of weed management practices are widely adopted for control of weeds in all the crops. These practices are tedious, time consuming, labour intensive, costly and not possible to practice over an extensive area. Further, due to labour scarcity and high labour wages as a result of rapid industrialization and urbanization, traditional weed management practices are being impracticable.

Chemical weed management with appropriate herbicides and their time of application provides ample opportunity for efficient weed management in all crops. The herbicides have proven effective in controlling weeds. Because, the crop and weed seedlings are in similar growth stages (De Datta and Liagas, 1984). Herbicides play a greater role in effective weed management in field crops than other physical/mechanical methods. But, intensive herbicide use can cause environmental contamination and the development of herbicide resistance in weeds (Zhao *et al.*, 2006). Hence, early weed management is essential to avoid yield loss. In this context, application of weedicides alongwith the sprinkler or trickle or drip irrigation to enhance the resource use efficiency need to be considered.

Herbigation is "the process of injecting an approved chemical (herbicide) into irrigation water (drip) and applying through the irrigation system to crop, weed or field". Herbigation is the improved method of application of herbicides through irrigation water and may be superior over conventional spraying mainly by reducing the herbicide loss through run-off and leaching. Reducing the cost and residual effect in soil as well as crop produce. It may also improve the herbicidal activity compared to conventional method of spraying. Mulching and repeated inter cultivation are the other possibilities to minimize evaporation loss and weed menace under aerated soil conditions. Herbigation may increase crop production efficiency by decreasing costs for application equipment, labor, and fuel (Ogg 1986), and by reducing wheel traffic through the field compared to ground application. Runoff is less in fields without wheel tracks because soil in wheel tracks is compacted, which decreases water infiltration (Chesters *et al.*, 1989).

Herbigation in Field Crops

1. Weed Growth and Weed Control Efficiency

Application of EPTC herbicides via overhead micro-sprinkler irrigation systems in potato after 40 days reduced weed population and dry weight of grasses (95.01 per cent and 45.36 per cent, respectively) and broad leaved weeds (87.3 per cent and 41.67 per cent, respectively) compared to conventional method of weed control (Saito and Santos, 1980). Charlotte *et al.* (2000) reported higher tuber yield of potato in complete weed free condition (43,300 kg ha^{-1}) compared to herbigated plot (42,100 kg ha^{-1}) and no herbicide application treatment (12,900 kg ha^{-1}).

Gruzdev *et al.* (1990) reported the effectiveness of weed control in maize and after harvest of soybean crops when herbicides were applied together with irrigation water in 200 m^3 water ha^{-1} compared to traditional boom sprayer method in 300 litres spray. The results indicated that a wider range of weeds was destroyed in application with irrigation water. Kohansal *et al.* (2010) noticed that reduced herbicides rates of metribuzin (5 litre ha^{-1}) and attrazine (300 g ha^{-1}) besides, herbigation for weed control in corn (*Zea mays*) reduced weed population (90.3 per cent) and weed dry weight (85.8 per cent) as compared unweeded check.

Fourie (1992) noticed that herbigation with Oxadiazon @1.4 kg ha^{-1} and simazine @ 3 kg ha^{-1} recorded lower weed population compared to control both after 30 and 60 days of application. Sujith (1997) noticed that application of alachlor @ 2 kg a.i ha^{-1} through irrigation water recorded lower weed population (monocots, dicots and sedges) and weed control efficiency (80.9 per cent) in groundnut compared to alachlor at same rate as soil spray.

Anon. (2001) reported that lower weed dry matter (8.5 g m^{-2}) in cotton was reported when pendimethalin applied as pre-emergence spray at 1 kg a.i. ha^{-1} followed by one hand weeding and metolachlor application at 1 kg a.i ha^{-1} (30 DAS) followed by one hand weeding (60 DAS) through irrigation water compared to applying the same herbicides in conventional spraying (18.5 g m^{-2}). Similarly, Velayatham *et al.* (2001) reported that pre-emergent herbicide metolachlor (1 kg a.i. ha^{-1}) applied as a herbigation registered lower dry matter production of cotton (33.65 kg ha^{-1}) as compared to unweeded check (8.2 kg ha^{-1}).

Nalayini *et al.* (2004) reported that the weed control efficiency was higher in case of weed control through herbigation (88.4 per cent) compared to conventional spraying of herbicides (86.5 per cent) in cotton. Koumanova *et al.* (2009) reported that micro-sprinkling can be successfully used for the application of soil herbicides with the irrigation water which leads to increases weed control efficiency, economical and does not have a negative effect on the crops and the environment.

3. On Growth and Yield of Crops

Velayatham *et al.* (2001) reported that pre-emergent herbicide metolachlor (1 kg a.i. ha^{-1}) applied as herbigation registered higher grain yield of cowpea (311 kg ha^{-1}) and soybean (443.87 kg ha^{-1}) compared to unweeded check in cotton based cropping system. Nalayini *et al.* (2004) reported that the yield of the cotton in the herbigation (3998 kg ha^{-1}) was higher over conventional spray (3498 kg ha^{-1}).

Abbasi *et al.* (2008) reported that application of trifluralin (1920 g a.i. ha^{-1}) through drip irrigation followed by bentazon (960 g a.i. ha^{-1}) in soybean recorded higher grain yield (2143 kg ha^{-1}), biomass (9410 kg ha^{-1}) and harvest index (23 per cent) as compared to weedy check (1379 kg ha^{-1}, 5905 kg ha^{-1} and 24 per cent, respectively). The highest grain yield (10.6 t ha^{-1}) of corn in the treatment applying of Edadicane (2.kg a.i ha^{-1}) herbigation at second irrigation compared conventional irrigated (7.5 t ha^{-1}) corn (Kesthkar *et al.,* 2010).

4. Soil Micro-organisms

The microbial population was higher in herbigated weed control treatment (105.67 cfu × 104 g^{-1} of dry soil) compared to the conventional sprayed treatment (105.67 cfu × 104 g^{-1} of dry soil) at 90 days after sowing cotton, However, the initial set back in microbial population was recovered over a period of time as the toxicity level decreased with time (Nalayini *et al.,* 2004). Rankova *et al.* (2009) noticed that soil respiration (microbial activity) increased in case of micro sprinkling compared to other methods of herbicide application mainly due to splitting of herbicide with irrigation water acted as a source of nitrogen and carbon for microbes because of higher surface area available. As time advances the microbial activity increases irrespective of the method of application as the herbicidal activity reduced in the soil.

5. Residual Effect

The germination of bio assay crop of greengram raised immediately after the harvest of cotton was not affected by any of the herbicides tested or method of application. Hence, herbigation is found safe to be used in cotton based cropping system (Nalayini *et al.,* 2004). Abdel-Aziz (2006) reported no residues of Butralin herbicide in the soil after 15 days from application under different herbigation as well as conventional spraying treatments.

Limitations of Herbigation

☆ Uniform chemical application depends on uniform water distribution.

☆ High initial cost.

☆ Most herbicide compounds are not approved for application with irrigation water

☆ Lack of research information in respect to rate of application, amount and frequency of application

☆ Operator must be skillful (know calibration procedure).

☆ Economically more viable in orchard or high value crops

Conclusion

Irrigation management is essential to developing a holistic system for weed management in crops. As water resources become costlier, drip irrigation technologies will become more widely utilized by growers worldwide. Although drip irrigation may be adopted due to water savings, the impact of drip irrigation on weed control is noteworthy. Herbigation is the improved method of application of herbicides through irrigation water and superior over conventional spraying mainly by reducing the herbicide loss through run-off and leaching. It also reduces the cost and residual effect in soil as well as crop produce. The ability to reduce soil wetting will allow for improved weed control over surface irrigation systems.

References

Abbasi, R., Alizedeh, H.M., Khanghah, H. Z., Jahroni, K., 2008, Integration of mechanical methods and herbicides in controlling weeds of soybean. *15th Australian Weed Conf.* **5**: 25-63.

Abdel-Aziz, A. A., 2006, The use of chemigation technique for weed control and minimizing environmental pollution with herbicides in newly cultivated lands. *Misr J. Eng.*, **23**(2): 571-592.

Anonymous, 2001, *Annual Report,* Central Institute of Cotton Research, Nagpur, p.38.

Charlotte, V. E., Bradley, A. K., Mary, J. G., 2000, Evaluating an automated irrigation control system for site-specific herbigation. *Weed Tech.*,**14**(1): 182-187.

Chesters, G., Simsiman, G. V., Levy, J., Alhajjar, B. J., Fathulla, R. N., Harking, J.M., 1989. Environmental fate of alachlor and metolachlor. *Rev. Environ. Contam. Toxicol.* 110:1–74.

Das, T. K., 2008, Weed evolution, concept, definition, nomenclature and distribution, In *Weed Sci. Basic Appl.*, Jain Brothers, New Delhi. Pp.1.

De Datta, S. K., Liagas, M. A., 1984, Weed problems and weed control in upland rice in tropical Asia; in an overview of upland rice research. *Proceeding of 1982 upland rice workshop*, IRRI, Los Banos, Philippines. pp: 321-341.

Kesthkar, G., Alizodeh, H., Abbasi, F. Zorch, A., 2010, The effectiveness of applying eradicane through irrigation on grain yield and yield attributes of corn. *Proc. of 3rd Iranian Weed Sci. Cong.*, **2**:55-59.

Kohansal, A., Majab, M., Kuhnavarda, F., Hasseini, M, 2010, Evaluating the efficiency of metribuzin and attrazin for controlling corn (*Zea mays*) weeds on the herbigation application. *Proc. of 3rd Iranian Weed Sci. Cong.*, **2**:17-18.

Nalayini, P., Shankaranarayanan, K., Velmourougane, K., 2004, Herbigation in cotton (*Gossypium spp.*): Effects on weed control, soil microflora and succeeding greengram (*Vigna radiata*). *Indian J. Agric. Sci.*, **83** (11):1144-1148.

Ogg, A. G., JR. 1986. Applying herbicides in irrigation water—a review. *Crop Prot.* 5:43–65.

Rankova, Z., Koumanov, K. S., Kolev, K., Shilev, S., 2009, Herbigation in a cherry orchard- translocation and persistence of Pendimethalin in the soil. *Acta. Hort.*, **825**: 305-312.

Saito, S. Y., Santos, H. N. G., 1980, Herbigation-a new method of applying herbicides. *Resumos Congrosso.* **2**:111-113.

Sujith, C. M., 1997, Effect of irrigation scheduled and methods of herbicide application on growth and yield of groundnut. *M. Sc. (Agri.) Thesis,* Uni. Agri. Sci., Bengaluru.

Velayatham, A., Mohammad, A., Veerabhadran, V., Sanbavalli, S., 2001, Impact of herbicides and their application techniques on yield and residues in cotton based intercropping systems. *Acta Agron. Hungarica,* **49**(3):283-292.

Zhao, D. L., Atlin, G. N., Bastiaans, L.and Spiertz, J. H. J., 2006, Comparing rice germ plasm groups for growth, grain yield and weed suppressive ability under aerobic soil conditions. *Weed Res.*, 46: 444-452.

22

Concepts and Design of Drip Irrigation in Arable Crops

N. Ananda and M.R. Umesh

India has the second largest net irrigated area in the world, after China. The irrigation efficiency under canal irrigation is not more than 40 per cent and for ground water schemes, it is 69 per cent. The net irrigated area in the country is 53.5 Mha, which is about 38 per cent of the total 'sown area. Although considerable area has been brought under irrigation since independence; there is not much scope for its expansion in the future. Irrigation water for agriculture finds competition from domestic use, industrial and hydroelectric projects. At present, the efficiency of the irrigation systems adopted is less than 30 per cent. As such as 50 per cent of the water release at the project head is lost in transmission of the canal outlet. Additional loss occurs in water courses which is directly proportional to their length and duration of water flow. Considerable scope exists for enhancing the water use efficiency to bring additional area under irrigation. Scientific management of irrigation water is necessary to improve crop productivity and alleviate irrigation related problems such as shortage of irrigation water, water logging, salinity *etc.*

The average grain yield in the country however is about 2 t ha^{-1} in irrigated and 1.5 t ha^{-1} in rainfed areas. With appropriate management practices, achieving a target of more than 450 Mt by the year 2050 AD is a distinct possibility. Even all the water resources have been tapped for irrigation; almost 50 per cent area will still remain rainfed. But, whether it is irrigated or rainfed agriculture water holds the key for enhancing and sustaining agricultural production. Since, sustainability and enhanced productivity are the need of the hour, the focus has to shift from crops to cropping systems that are more input use efficient going with resource conservation technologies. Out of the 250 cropping systems in India, 30 are the most common ones and out of them, several are well fitted under drip and sprinkler irrigation system.

There is an immense scope for conservation, distribution and on farm utilization of water and attaining higher water use efficiency through micro irrigation system; yield can be maximized significantly with a limited amount of water. Modern irrigation techniques like sprinkler and drip should be promoted where water is scarce and the topographic and soil condition do not permit conventional methods of irrigation.

Irrigation in India has been practiced since Maurya's time who contributed the most in building ancient irrigation system in India. Irrigation through drip is a newly introduced system in the country and little work has been done on application and evaluation of drip for cotton and cotton-based cropping system in the country. Although there are various ways of irrigating crops, drip and sprinkler irrigation systems considered as the best in bringing about water and fertilizer use efficiency alongwith improved crop productivity. In this system, water is directly delivered to the root zone of the individual plants by network of tubing. The tubing can be moved around different locations, topography and slopes as per plan and convenience to deliver water at desired pressure through emitters/micro tubes to the plants.

History of Micro-Irrigation

Earlier attempts were made by the researchers in Germany during 1860 by simply pumping the irrigation water into the clay pipes through underground drainage system. The first work on MIS (Micro- irrigation systems) was initiated at Colorado in 1913 and it was concluded that drip system was too expensive. Later on an important breakthrough was made in Germany in 1920 when perforated pipes were used for irrigating the crops. However, in 1930, the peach growers in Australia, pumped water through 5 cm GI pipes laid along the tree rows with water emitting points made on the pipe as small triangular holes. In early 1940, Symcha Blass observed that a tree near a water leaking point exhibited vigorous growth as compared to other trees in the area. This led to the concept of MI (Micro-irrigation) where water is applied in very small amounts as drop by drop. Later on, a remarkable breakthrough was made in the material science, when poly ethylene, a crack resistant and cheaper alternative was accidentally produced in a British laboratory. Later LDPE (Low density poly ethylene) gave place to HDPE (High density poly ethtylene) and in 1977, LLDPE ((Low lenior density poly ethlene) was introduced. Thus, micro-irrigation systems really got off the ground with the developments in plastic industry. Later on the orifice emitters were developed to improve the consistency of "holes drilled into the pipes" and gradually sophisticated water emission small diameter plastic tubes and microtubes were developed. Turbulent flow emitters were also developed which are being used at present.

Why Micro-Irrigation?

  ☆ High water use efficiency (30-60 per cent)
  ☆ High quality and higher yields (10-60 per cent)

☆ Minimized fertilizer loss and soil erosion

☆ Can be laid out in undulating fields

☆ Moisture within the root zone can be maintained

☆ Efficient weed control (30-90 per cent)

☆ Water soluble fertilizers can be applied through MI

☆ Low labour cost

☆ Day and night irrigation - Low interference with cultivation

According to a recent estimate, thirty four countries in the world will be facing water scarcity by 2025 AD indicating that per capita availability of fresh water supplies will be less than 100 m^3 person^{-1} year^{-1}. A country with renewable water availability on an annual per capita basis exceeding about 1700 m^3 will suffer only occasional or local water problems. Below this threshold, countries begin to experience periodic or regular water stress. India (1400 m^3) and China (1700 m^3) will 'come first into this category in the year 2025 AD, while USA will have more than 7000 m^3 person^{-1} year^{-1} and will not face any scarcity. Rising demand for urban and industrial water supplies in the world poses a serious threat to irrigated agriculture. The allocation of water for agriculture will come down to 50 per cent from the present level of 70 per cent. However, to achieve required food and fibre production with increasing population, India has to enhance the current irrigation potential of 91 M.ha to 160 M ha. However, to fulfil the additional requirement of the irrigation with improved technologies for water harvesting, excess runoff collection, storage and recycling for precision water application by economizing the available amount of irrigation water needs to be adopted.

The major problem associated with decreasing amount of fresh water for irrigation is conveyance losses, reducing the net utilization of irrigation water to 46 per cent only. The net utilization of irrigation water in drip system is 90 per cent and through sprinkler system, it is 82 per cent. In view of the same, micro-irrigation is having paramount importance with brighter future prospects.

Why Modern Irrigation Technologies are needed?

☆ The productivity of irrigated land is low compared to its potential

☆ The productivity per unit water is very low

☆ Water available for irrigation is becoming scarce

☆ Cost for generating water source is ever increasing

☆ The predominance of soils with low water retention capacities and very low hydraulic conductivities make the Arid and Semi-arid regions an ideal case for light and frequent irrigations through micro-irrigation

☆ Micro-irrigation will increase the irrigation cover using the existing available water

☆ Micro-irrigation with fertigation will enhance production per unit input in these nutrients poor, shallow and sloppy soils. Micro-irrigation is a co-ordinated and controlled water management system where water is made to flow under pressure through a net work of pipes of varying diameters, the main-line, the sub-main lines and the lateral lines with appropriately placed emitters along the length of the latter through which water is discharged to the root zone.

Need for Micro-Irrigation

To achieve required food production with increasing population, India has to enhance the current irrigation potential of 91 M.ha to 160 M.ha. But, the total water resources estimated are 230 M.ha.m will have to cater the need to the non-agricultural uses also. The country is likely to be water stressed in the coming years. Therefore hand in hand with technologies for water harvesting and storage, technologies for precision water application methods need to be adopted.

Advantages of Micro-Irrigation System

☆ Saving of ample irrigation water

☆ Low water application rate

☆ Uniformity of water application around the plant

☆ Precision placement of water

☆ Efficient fertilizer and chemical application

☆ Better control of root zone environment

☆ Significant yield enhancement

☆ Improved quality of the farm produce

☆ Improved disease control

☆ Discourages weed growth

☆ Saving of power due to lesser use

☆ Reduces labour cost

☆ Being light in weight, the system can be shifted without any problem

☆ It can be moved on undulating topography

☆ It can be put to use during night also

Water for Agriculture

Agriculture is the biggest user of water. In Africa and Asia, 85-90 per cent of water used is for agriculture. Rising demand from urban and industries water supplies in the world pose a serious problem or threat to irrigated agriculture. Allocation of water to agriculture comes down from 69 per cent to 50 per cent.(Table 22.1). The water used by the different agricultural produce is presented in Table 22.2.

Table 22.1: Water Withdrawal from different Sectors

Withdrawal	World (per cent)	India (per cent)
Agriculture	69	88
Industry	23	7
Domestic	8	5
TOTAL	100	100

Table 22.2: Water Use by Agricultural Produces

Crops	Water Use
Rice	3500 lit/kg
Groundnut	1300 lit/kg
Banana	100 lit/kg fruit
Silk	80000 lit/kg of silk

Water Demand in India

☆ 60 million ha area under irrigation

☆ The current irrigated area is 40 per cent of total cultivated area

☆ Present water availability-1086 billion cubic metres/year

The Total anticipated water demand in India is presented in Table 22.3.

Table 22.3: Total Anticipated Water Demand in India (Billion cubic metres)

Year	Normal Requirement	With Improved Management
2010	813	710
2025	1093	843
2050	1447	1180

The fresh water demand in India and per capita availability of water is presented in Tables 22.4 and 22.5.

Table 22.4: Freshwater Demand in India

Purpose	Water Use in the Year					
	2000		2025		2050	
	BCM	Per cent of Total	BCM	Per cent of Total	BCM	Per cent of Total
Domestic	44	6.6	77	6.2	93	5.6
Irrigation	520	78.5	909	72.8	1072	64.7
Energy	27	4.1	70	5.6	212	12.8
Industries	30	4.5	120	9.6	199	12.0
Others	41	6.2	72	5.8	80	4.8
Total	662	100.0	1248	100.0	1656	100.0

Assessment of water availability and requirement – Ministry of water resources, Govt. of India

Table 22.5: Per Capita Water Availability (m³)

Countries	Per Capita Water Availability (m³)
Japan	65000
USA	62000
Russia	17500
World	7420
Asia	3240
India	2025

Water Challenges

☆ Degradation of existing water supplies

☆ Degradation of irrigated crop land

☆ Groundwater depletion

☆ Increasing pollution/declining water quality

☆ Poor cost recovery

☆ Trans boundary water disputes

☆ Increasing costs of new water

☆ Strategies

Supply Management

New structures – system improvement

Demand Management

☆ Crop diversification

☆ Improving pump efficiency

☆ Watershed management

☆ Artificial recharge - abandoned wells/ponds

☆ Efficient water management practices

What is Micro-irrigation?

Micro irrigation is the system that provides precise quantity of water in and around root zone of plant with the help of irrigation pipe net work and emitters.

Micro irrigation is the frequent application of small quantities of water on, above or below the soil surface, by surface drip, subsurface drip, micro sprayers or micro sprinklers.

The highest area under drip irrigation is in Maharashtra and sprinkler irrigation is in Rajasthan (Table 22.6).

Table 22.6: Indian Scenario: Area under Micro-irrigation

State	Drip (ha)	Sprinkler (ha)	Total(ha)
Rajasthan	10025	554708	564733
Haryana	4258	503877	508135
Maharashtra	341848	153507	495355
Andhra Pradesh	155441	124510	279951
Karnataka	114433	157028	271461
West Bengal	110	150020	150130
Gujarat	53707	96374	150081
Tamil Nadu	116665	26332	142996
Madhya Pradesh	6483	100000	106483
Orissa	2036	20220	22256
Punjab	5101	10000	15101
Uttar Pradesh	4609	10000	14609
Kerala	10562	1548	12110
Sikkim	80	10030	10110
Chattishgarh	1979	3765	5744
Himachal Pradesh	116	581	696
Grand Total	**829067**	**1927009**	**2756076**

Table 22.7: The Comparative Advantages of Drip over the Flood Irrigation

Variable	Drip Method	Flood Method
Water saving	High 40-100 per cent	Less due to evaporation
Irrigation efficiency	80-90 per cent	30-50 per cent
Input cost: Fertilizer, Pesticides and Tilling	less in labor	Comparatively higher
Weed problem	Almost nil	High
Suitable water	Even saline water Can be used	Only normal water can be used
Diseases and pest Problems	Relatively less	High
Water logging	Nil	About 8.5 mh under water logging
Water control	high and easy	Less
Evaporation and Transportation	Very low	High seepages leackage *etc.*
Efficiency of Fertilizer use	Very high and regulated supply	Heavy losses due to leaching
Increase in yield	20-100 per cent	Less compared to drip

Surface Drip Irrigation

Drip or trickle irrigation is one of the latest methods of irrigation. It is suitable in water scarcity and salt affected soils. In drip irrigation method, water is applied frequently and at low volume so that it approaches the consumptive use of the plants and minimizes losses such as deep percolation, runoff and soil water evaporation. The system applies water slowly to keep the soil moisture within the desired range for plant growth.

Components

A drip irrigation system consists of a main line, submains, laterals and emitters. The mainline delivers water to the submains and the submains into the laterals. The emitters which are attached to the laterals distribute water for irrigation. In mains, submains and laterals are usually made of black PVC (poly vinyl chloride) tubings. The emitters are also made of PVC material. PVC material is preferred for drip system as it can withstand saline irrigation water and is not affected by chemical fertilizers.

The ancillary components include a valve, pressure regulator, filters, pressure gauge, fertilizer application components *etc.*

There are three main connections in a drip irrigation system. They are:

1. Main - sub main connection and sub main manifold
2. Sub main - lateral connection
3. Lateral end arrangements

The main sub main connection is usually a 'Tee" either thread or slip. A sub-main is used to deliver water into laterals and also used as a controller so that the field can be irrigated separately under a desired water pressure at any selected time. All the specific items used for control are installed in the sub main which is called a sub main manifold.

Pump

The pressure necessary to force water through the components of the system including the fertilizer tank, filter unit, mainline, lateral and the nozzle is obtained by a pump. Centrifugal pump operated by engines or electric motors are commonly used. The laterals may be designed to operate under pressures as low as 0.15 to 0.2 kg/cm^2 and as large as 1 to 1.75 kg/cm^2. The water coming out of the emitters is almost at atmospheric pressure.

Fertilizer Tank

A fertilizer tank is provided at the head of the drip irrigation systems for applying fertilizers in solution directly to the field alongwith irrigation water.

Filter

It is an essential part of drip irrigation system. It prevents the blockage of holes and passages of drip nozzles. The filter system consist of valves and a pressure gauge for regulation and control. A two stage filter unit is usually provided. It consists of one coarse filter and a fine filter.

Emitters

Drip nozzles commonly called drippers or emitters are provided at regular intervals on the laterals. They allow the water to emit at very low rates usually in trickles. There are 3 general types of trickles. The amount of water dripping out of each nozzle in a unit time will depend mainly upon the pressure at the nozzle, size of the opening and frictional resistance due the length and size of the water passage in the drip nozzle (emitter). The discharge rate of emitters usually ranges from 2 to 10 litres per hour.

Microtubes are frequently used in a drip lateral. They are used mainly in the following ways (1) as emitters (2) as connectors, (3) as pressure regulators.

Advantages

1. Water saving - losses due to deep percolation, surface runoff and transmission are avoided. Evaporation losses occurring in sprinkler irrigation do not occur in drip irrigation.
2. Uniform water distribution
3. No land leveling required
4. No soil erosion
5. Better weed control
6. Nutrient preservation
7. Use of unsuitable supply. Highly saline soil or saline or brackish water source, normally considered unsuitable for agriculture can be utilised.

Disadvantages

☆ High initial cost

☆ Drippers are susceptible to blockage

☆ Trees grown may develop shallow confined root zones limiting the area of extraction of soil nutrients resulting in poor anchorage of roots. Toppling of trees under strong wind.

Adaptability

☆ It is adaptable to all vegetables and most crops and climatic variations.

☆ It is adaptable to highly permeable soils like sandy soils, sandy loam and less permeable soils like clay, clay loam, silky clay or shallow soils underlain with less permeable strata.

☆ However, it is most suited to coarse sandy soils under the adverse condition of high salinity of soil and water, high temperatures and low relative humidity.

Sub Surface Drip Irrigation

This system with laterals buried below soil surface is gaining importance as problems with clogging have been reduced. Advantage of this system includes freedom from necessity of anchorage of tubing at the beginning and removing at the end of growing season, little interference with cultural operations and longer economic life.

Subsurface drip irrigation is defined by American Society of Agricultural Engineers as "Application irrigation water below the soil surface through emitters, with discharge rates generally in the same range as drip irrigation (Camp, 1998).The high labour requirement in spreading and collecting laterals every season and the deterioration of exposed drip lines limits the further expansion of drip irrigation. Subsurface location of trucklers may solve these problems. Furthermore, it would position the supply of nutrients in the centre of root system, where the water content is relatively high. Nutrients introduced via subsurface trickles can move in a spherical volume around the emitter, while transport in surface application is bound with in a hemisphere below the point source.

Sprinkler Irrigation System

The sprinkler or overhead irrigation system consists of conveying water to the field by aluminium or polyvinyl chloride (PVC) pipes and allowed to sprinkle over the field under pressure through a system of nozzles. This system is designed to distribute the required depth of water uniformly which is not possible in surface irrigation. Here water is applied at a rate less than the infiltration rate of the soil hence the runoff from irrigation is avoided.

A sprinkler system usually consists of the following parts:

1. A pumping unit
2. Tubing – mains, sub-mains and laterals (pipelines)
3. Couplers
4. Sprinklers
5. Other accessories such as filter, valves, bends, plugs and risers

Pumping Unit

A high speed centrifugal or turbine pump can be installed for pumping water under pressure.

Tubings or Pipelines

Pipelines are generally of two types. Main, sub-main and lateral. Main pipelines carry water from the pump to many parts of the field. In some cases, sub main lines are provided to take water from the mains to laterals. The lateral pipelines carry the water from the main or sub main pipe to the sprinklers. The pipelines may be either permanent, semi-permanent or portable.

Couplers

Provides connection between two tubings and between tubings and fittings.

Sprinklers

Sprinklers may rotate or remain fixed. The rotating sprinklers can be adapted for a wide range of application rates and spacings. They are effective with pressure of about 10 to 70 m head at the sprinkler. Pressures ranging from 16-40 m head are considered the most practical for most farms.

Fixed head sprinklers are commonly used to irrigate small lawns and gardens.

Other Accessories/Fittings

1. Water meters - It is used to measure the volume of water delivered.
2. Flange, coupling and nipple - For proper connection to the pump and suction delivery.
3. Pressure gauge - It is necessary to know whether the sprinkler is working with the desired pressure in order to deliver the water uniformly.
4. Bends, tees, reducers, elbows, hydrants, butterfly valves, end plugs and risers
5. Fertilizer applicators. These are available in various sizes. They inject fertilizers in liquid form to the sprinkler system at a desired rate.
6. Filters: This is needed when water is obtained from streams, ponds, canals or other surface supplies to filter the debris like sand, weed seeds, leaves, sticks, moss and other trash that may otherwise plug the sprinklers. It is not required when water is pumped from wells.

Types of Sprinkler System

Based on arrangement for spraying irrigation water, sprinkler systems are classified into.

1. Rotating head (or) revolving sprinkler system
2. Perforated pipe system

Rotating Head (or) Revolving Sprinklers

This can again be divided into 3 categories.

1. Conventional system/small rotary sprinklers
2. Boom type and self propelled sprinkler system
3. Mobile raingun/large rotary sprinklers

Perforated Pipe System

There are three types of spraying systems

1. Stationary
2. Oscillating
3. Rotating

Based on the portability, sprinkler systems are classified in to the following types.

1. **Portable system:** It has portable mainlines and laterals and a portable pumping unit

2. **Semi portable system:** A semi portable system is similar to a fully portable system except that the location of the water source and pumping plant are fixed.

3. **Semi permanent system:** A semi permanent system has portable lateral lines, permanent main lines and sub-mains and a stationery water source and pumping plant. The mainlines and sub mains are usually buried, with risers for nozzles located at suitable intervals.

4. **Solid set system:** A solid set system has enough laterals to eliminate their movement. The laterals are placed in the field early in the crop season and remain for the season.

5. **Permanent system:** It consists of permanently laid mains, sub mains and laterals and a stationary water source and pumping plant. Mains, sub mains and laterals are usually buried below plough depth. Sprinklers are permanently located on each riser.

Advantages

1. Economical use of water - with the available water 2-3 times more area can be irrigated.

2. Lower water loss – water loss is estimated as 18 per cent as against 54 per cent in surface irrigation with lined channels and 71 per cent in a system in unlining.

3. Effective water management. Water application is controlled - Over or under irrigation is avoided.

4. Saving in land - It helps to conserve water up to 70 per cent and can irrigate 2-3 times the area compared to surface irrigation.

5. Saving in fertilizers - even distribution and avoids wastage.

6. Land levelling not necessary.

7. Soil is conserved.

8. Frost control - protect crops against frost and high temperature.

9. Free aeration of root zone.

10. Drainage problems eliminated.

Disadvantages

1. High initial cost

2. Efficiency is seriously affected by windy weather

3. Higher evaporation losses in spraying water

4. A stable water supply is needed for the most economical use of the equipment

5. Higher power requirement

6. Excessive use of saline water on foliage is inadvisable

Suitability

1. It is suitable to regions of water scarcity

2. For close spaced crops in well irrigated areas and also in tank and canal commissioned area this method is suitable.

3 Sprinkler irrigation is suitable for all types of soils, more particularly coarse, sandy and gravelly soils and for almost all crops like wheat, sorghum, cotton, potato, tobacco, groundnut, vegetables, ragi *etc.*

4. It is particularly suitable for production of high yielding crops or for having continuous and quick growth of valuable crops

5. It is not recommended for rice and jute

Opportunities in Micro-Irrigation and Fertigation

☆ Waste-land and fallow lands making up a total of 26.3 m ha

☆ Undulating terrains and hilly slopes can be irrigated

☆ Waste water reuse and saline water use for cultivation

☆ Opportunity to research about chemigation techniques

Constraints in Micro Irrigation

☆ High initial cost

☆ Maintenance requirements (emitter clogging, *etc.*)

☆ Lack of training to extension officials and farmers

☆ Difficulties in getting subsidies

☆ Poor quality drip components and pipes

☆ After care by irrigation firms

☆ Lack of confidence on the drip technology with farmers

Reason for Slow Adoption of Micro-irrigation

☆ Inadequate knowledge with farmers as well as dealing officers on cost benefit of MI.

☆ High initial investment.

☆ Water and energy usage in the farming sector not priced properly.

☆ Inadequate and timely availability of assistance.

☆ Rigid credit framework of financial institutions.

☆ Poor after sales service and quality assurance by suppliers.

☆ Cumbersome procedures for assistance disbursal

☆ Inadequate availability of crop and area specific package of practices.

Case Studies

Drip irrigation is used widely in fruit and vegetable crops and their production and productivity has increased significantly with increased fertilizer and water use efficiency. The work done at different places in the country on different crops is highly encouraging with the main objective to economise the irrigation water with higher yield and economic return per unit of land and water. Crop wise salient features are as given below:

Drip Irrigation

Sugarcane

Drip system resulted in cane yield of 171.4 t ha^{-1} as compared to conventional surface irrigation (86.9 t ha^{-1}). Application of 80 per cent recommended dose of fertilizer by drip gave the highest cane yield (182.84 t ha^{-1}) and superiority was recorded in increasing WUE over conventional method (Shelke *et al.*, 2002).

Cotton

Drip irrigation improved seed sprouting by 66 per cent within 6 days, while only 46 per cent germinated under flood irrigation during the same period resulting in 74.9 per cent germination in the conventional system as compared to 93.5 per cent in drip system (Nalayini and Shanmugham, 2002). Water use efficiency ranges from 16.3 to 35 kg ha^{-1} for drip as against 4.9-8.3 kg ha^{-1} for flood irrigation system. Drip irrigation favoured the growth of summer cotton at Coimbatore and resulted in saving of 50 per cent water besides increasing seed cotton yield by 34.5 per cent as compared to conventional (flood irrigation) method.

Summer Groundnut

Application of 20 kg N + 40 kg P$_2$O$_5$ ha^{-1} as drip fertigation significantly improved the nutrient availability in the soil and uptake by crop and resulted an increase in pod yield by 20.7 per cent as compared with fertilizer applied in soil under surface irrigation (Devidayal and Malviya, 2002). The treatment drip fertigation of 100 per cent RDF as WSF recorded maximum plant height (42.0 cm), number of branches (8), leaf area (1411.1 cm^{-2} plant^{-1}), leaf area index (4.70) and dry matter production (29.48 g plant^{-1}) at maturity. This was mainly due to continuous constant availability of nutrients from WSF which resulted in better translocation of photosynthates and more carbohydrate synthesis and better yield. Higher nitrogen availability that in turn increased leaf area, chlorophyll content and photosynthetic activity for longer period (Sanju, 2013).

Maize

Drip irrigation helps to save the water up to 32 to 43 per cent compared to surface irrigation in maize. Irrigation given through drip alongwith a fertigation of

100 per cent RDF with 50 per cent P and K as WSF (Water soluble Fertilizer) had a higher WUE of 23.7 kg ha^{-1} mm^{-1}. Drip fertigation with 100 per cent RDF in which 50 per cent P and K as WSF increased the grain yield to the tune of 15.5 per cent as compared to drip fertigation of 100 per cent RDF with normal fertilizer.

Chickpea

Deolankar and Berad (1999) reported that N fertigation through drip irrigation improved fertilizer use efficiency through saving of 25 per cent of fertilizer with 11.07 per cent increase in chickpea yield over surface irrigation with soil application of fertilizers at Rahuri in Maharashtra.

Subsurface Drip Irrigation

Tomato

Marketable yield was 22 per cent greater for plants grown with subsurface drip irrigation than furrow irrigation (Bogle *et al.*, 1989).

Sprinkler Irrigation

Sprinkler irrigation maintained superiority in case of groundnut and chilli and increased yield significantly over irrigation in furrow, border or check. Irrigation through drip in different vegetable crops such as tomato, bhendi, radish brinjal, *Ipomea batata* and sugarcane crops maintain its superiority over surface irrigation. Similarly, drip recorded higher water use efficiency in all these crops over surface irrigation.

Application of Fertilizers

Assume 20 kg N is to be fertigated through urea

↓

Amount of urea required = 20/0.46 = 43.5 kg

↓

Solubility of urea (summer) is 1,100 g/litre

↓

Amount of water required for dissolution = 43.5/1.1 = 40 litres

↓

Take 50 litres capacity and filled with 40 litres of water and then dissolve the 43.5 kg urea by mixing it slowly

↓

Siphon the clear solution for injecting through the fertigation applicator

Worked Example

Design a Drip Irrigation System for the following Data (By Muhammad Ashraf, ICARDA, 2012)

☆ Area 30 acre= 400 m x 300 m

☆ Tophography – flat

☆ Crop: Citrus

☆ Spacing 6.1 m x 6.1 m (3225 plants)

☆ Water source: Tube well at the center of the field

☆ Suction Lift : 3m

☆ Delivery lift: 3 m

☆ Tube well discharge 15 lps

☆ Emitter discharge- 4 lph

☆ Total emitters (4 per plant): 12900

☆ Total flow rate: 51599 lph= 14.3 lps

☆ Divided the area into 4 blocks- (7.5 acre x 4)

☆ lateral Length- 75 m

☆ Laterla inside diameter" 16mm

☆ No. of emitter/lateral:49

☆ Discharge of emitters: 197 lph= 0.05 lps

☆ Head loss in lateral (0.91 m/100m): 0.68m

☆ Sub main-1

☆ Length – 200 m

☆ Diameter- 62.5 mm

☆ No. of latrals on the sub main: 66

☆ Total discharge of the sub main: 12984 lph= 3.6 lps

☆ Head loss in sub main (2.78 m/100m): 5.56 m

☆ No. of sub mains: 4

☆ Total discharge of main line (4 sub mains): 51934 lph: 14.4 lps

☆ Diameter of main line: 100mm

☆ Length of main line: 150m

☆ Head loss in main line (2.67 m/100m): 4m

☆ Operating pressure: 10m

Irrigation Scheduling

☆ Crop: Citrus

☆ Area: 30 acres

☆ Root zone depth: 80 cm

☆ Maximum allowable deficit (MAD): 40 per cent

☆ ET0 = 8mm/day

☆ Kc: 0.8

☆ Soil Texture: Loamy clay

☆ Bulk Desnity: 1.4 g/ml

☆ Field capacity: 32 per cent

☆ Wilting point : 15 per cent

☆ Available mositure: 17 per cent

☆ Daily peak season water demand: 7.2 mm/day

☆ Gross daily demand (mm/day), assuming 90 per cent efficiency: 8 mm/day

☆ Available mositure by volume : 0.24 cm^3 of water/cme of soil

☆ Total available mositure: 19.04 cm

☆ Water content at 40 per cent MAD: 7.62 cm

☆ No. of days after irrigation is due: 10 days

Irrigation Scheduling

Canopy Diameter (m)	Gross Crop Water Requirement (litres/day)	Time of Irrgation with 4 Emitters (hrs/day)
2	25	1.3
3	57	2.8
4	100	5.0

Fertigation in Tomato

☆ Concentration of NPK fertilizers: 180-50-250 kg/ha

☆ Type of fertilizers available: Ammonium nitrate (33.5-0-0) NH_4NO_3; Diammonium phosphate DAP (16-48- 0); $(NH_4)_2HPO_4$; Potassium chloride (0-0-60) K_2O

☆ System flow: 23,000/hrs

☆ Irrigation dosage: 18000 litres

☆ Duration of application: 1.5 hours.

Phosphate and potassium are given in oxides, therefore they are converted into P and K elements by multiplying by 0.4364 and 0.8302 respectively. Calculation of the amounts of fertilizers needed in grams per 1000 litres of water

K = 250 x 100 ÷ (60 x 0.8302) = 0.502 kg K_2O

P = 50 x 100 ÷ (48 x 0.4364) = 0.239 kg $(NH_4)_2HPO_4$

This amount also provides 38 g of N.

N = (180-38) x 100 ÷ 33.5 = 0.424 kg NH_4NO_3

Thus, for 18 m^3 of water, which is the irrigation dosage, the exact quantities are:

0.502 kg x 18 = 9.036 kg K_2O

0.239 kg x 18 = 4.30 kg $(NH_4)_2HPO_4$

0.424 kg x 18 = 7.63 kg NH_4NO_3

The amount of water needed for the dilution of the above quantity of fertilizers is estimated by taking into account the solubility of the fertilizers:

9.036 kg K_2O x 3 litres 27.0 litres

4.30 kg Ca (H_2PO_4) x 2.5 litres 10.75 litres

7.63 kg NH_4NO_3x 1 litre 7.63 litres

Minimum amount of water needed 45 litres

If the fertilizers are diluted in 60 litres of water and the duration of the irrigation is 1.5 h (1 h 30 min), then the injection rate should be about 40–45 l/h in order to complete the fertigation in approximately 1 h 25 min.

References

Bogle, C.R., Hartz, T.K., Nunez, C., 1989. Comparison of subsurface trickle and furrow irrigation on plastic mulched and bare soil for tomato production. *J Am. Soc. Hort. Sci.*, 114(1):40-43

Camp, C.R., 1998. Subsurface drip irrigation a review.*Trans Am Soc Agric.Engr.*41 (5)1353-1367.

Devidayal, Malviya, D.O., 2002. Effect of drip fertigation on nutrient availability, nutrient uptake and yield of summer groundnut (*Arachis hypogaea*) in medium black calcarious soils. Extended summaries, Vol. 2. 2nd International Agronomy Congress, Nov. 26-30, New Delhi,India. pp. 1346-1347.

Deolankar, K. P., Berad, S. M., 1999. Effect of fertigation on growth, yield and water use efficiency of chickpea (*Cicer arietinum*). *Indian J. Agron.*, 44(3): 581-583.

Sanju, H. R., 2013. Effect of precision water and nutrient management with different sources and levels of fertilizer on yield of groundnut. *M.Sc. (Agri.) Thesis, Univ. Agric. Sci.*, Bangalore (India).

Shelke, D.K., Digrase, L.N., Sondge, V.D., 2002. Optimization of irrigation water and fertilizers for seasonal sugarcane (*Saccharum officinarum*) through drip irrigation system.Extended summaries, Vol. 2. 2nd International Agronomy Congress, Nov. 26-30, New Delhi, India. pp. 1340-1341.

Nalayini, P., Shanmugham, K., 2002. Efficacy of drip irrigation for summer cotton (*Gossypium* species). Extended summaries, Vol. 2. 2nd International Agronomy Congress, Nov. 26-30, New Delhi, India. pp. 1343-1344.

23

Irrigation Water on Incidence and Management of Plant Diseases

K.R. Shreenivasa and D. Rekha

The presence of a pathogen against a particular plant will generally not cause serious disease unless the environmental conditions are favorable. This includes the aerial environment and the soil (*edaphic*) environment. Human attempts at controlling disease usually involve manipulating the environment in some way. For example, breeding wheat cultivars to tolerate dry conditions allows Australian farmers to plant the crop in areas that are not favorable for pathogens such as powdery mildew and leaf rust. Properties of the aerial environment that influence disease development include moisture levels, temperature and pollution.

Moisture is particularly important to pathogenic bacteria and fungi. Rain splash plays an important role in the dispersal of some fungi and nearly all bacteria, and a period of leaf wetness is necessary for the germination of most airborne spores. By using water for dispersal, propagules are dispersed at a time when they are likely to be able to germinate as well. Because the process of germination and infection takes time, the duration of leaf wetness also influences the success of the infection. The duration necessary for infection varies with temperature. Usually, a longer period of leaf wetness is needed to establish an infection in cooler temperatures, as germination and infection are generally accelerated in warmer conditions.

Temperature also affects the incubation, or latent, period (the time between infection and the appearance of disease symptoms), the generation time (the time between infection and sporulation), and the infectious period (the time during which the pathogen keeps producing propagules). The disease cycle speeds up at higher temperatures, resulting in faster development of epidemics. The period of leaf wetness, combined with temperature information can be used to predict outbreaks of some diseases (infection periods) and be used to time preventative treatments, such as spraying. A recently recognized aspect of the aerial environment that can influence disease in plants is air pollution. A high concentration of pollutants can affect disease development and, in extreme cases, damage the plants directly by causing acid rain.

The edaphic (soil) environment affects soil-borne diseases, largely by determining the amount of moisture available to pathogens for germination, survival and motility. Germination and infection success also rely on the temperature of the soil. The fertility and organic matter content of the soil can affect the development of disease. Plant defences are weakened by nutrient deficiency, although some pathogens, such as rusts and powdery mildews, thrive on well-nourished plants. Other diseases thrive in soils that are specifically low in organic matter.

The pathogen, the host and the environment interact, usually in ways that are difficult to quantify and predict. Control measures can include sowing of a crop species early, to avoid exposing seedlings to a disease during the time of year that provides the best environmental conditions for the pathogen.

Effect of Moisture

Moisture, like temperature, influences the initiation and development of infectious plant diseases in many interrelated ways. It may exist as rain or irrigation water on the plant surface or around the roots, as relative humidity in the air, and as dew. Moisture is indispensable for the germination of fungal spores and penetration of the host by the germ tube. It is also indispensable for the activation of bacterial, fungal, and nematode pathogens before they can infect the plant. Moisture, in such forms as splashing rain and running water, also plays an important role in the distribution and spread of many of these pathogens on the same plant and on their spread from one plant to another. Finally, moisture increases the succulence of host plants and thus their susceptibility to certain pathogens, which affects the extent and severity of disease.

The occurrence of many diseases in a particular region is closely correlated with the amount and distribution of rainfall within the year. Thus, late blight of potato, apple scab, downy mildew of grapes, and fire blight are found or are severe only in areas with high rainfall or high relative humidity during the growing season. Indeed, in all of these and other diseases, the rainfall determines not only the severity of the disease, but also whether the disease will even occur in a given season. In fungal diseases, moisture affects fungal spore formation, longevity, and particularly the germination of spores, which requires a film of water covering the tissues. In many fungi, moisture also affects the liberation of spores from the sporophores, which, as in apple scab, can occur only in the presence of moisture. The number of infection cycles per season of many fungal diseases is closely correlated with the number of rainfalls per season, particularly of rainfalls that are of sufficient duration to allow establishment of new infections. Thus in apple scab, for example, continuous wetting of the leaves, fruit, and so on for at least 9 hours is required for any infection to take place even at the optimum range (18 to 23°C) of temperature for the pathogen. At lower or higher temperatures the minimum wetting period required is higher, *e.g.,* 14 hours at 10°C and 28 hours at 6°C. Similar conditions are required for the initiation and development of infections in many other diseases. If the length of the wetting period is less than the minimum required for the particular temperature, the pathogen fails to establish itself in the host and fails to produce disease.

Most fungal pathogens require free moisture on the host or high relative humidity in the atmosphere for spore release or for germination of their spores. Most pathogens become independent of outside moisture once they can obtain nutrients and water from the host. Some pathogens, however, such as those causing late blight of potato and the downy mildews must have high relative humidity or free moisture in the environment throughout their development. In these diseases, although spores may be released following a short leaf-wetness period, the growth and sporulation of the pathogen, and the production of symptoms, come to a halt as soon as dry, hot weather sets in. All these activities resume only when it rains again or after the return of humid weather.

Although most fungal and bacterial pathogens of aboveground parts of plants require a film of water to infect hosts successfully, spores of the powdery mildew fungi can germinate, penetrate, and cause infection even when there is only high relative humidity in the atmosphere surrounding the plant. In powdery mildews, spore germination and infection are actually lower in the presence of free moisture on the plant surface than they are in its absence. In some of them, the most severe infections take place when the relative humidity is rather low (50 to 70 per cent). In these diseases, the amount of disease is limited rather than increased by wet weather, as indicated by the fact that powdery mildews are more common and more severe in the drier areas of the world. The relative importance of powdery mildews decreases as rainfall increases. In high rainfall areas and periods, other diseases become more prevalent.

In many diseases affecting underground parts of plants, such as roots, tubers, and young seedlings, *e.g.,* in the *Pythium* damping off of seedlings and seed decays, the severity of the disease is proportional to the amount of soil moisture and is greatest near the saturation point. The increased moisture seems to affect primarily the pathogen, which multiplies and moves (zoospores in the case of *Pythium*) best in wet soils. Increased moisture may also decrease the ability of the host to defend itself through a reduced availability of oxygen in water-logged soil and by lowering the temperature of such soils. Many other soil pathogens [*e.g., Phytophthora, Rhizoctonia, Sclerotinia,* and *Sclerotium*], some bacteria (*e.g., Erwinia* and *Pseudomonas*), and most nematodes usually cause their most severe symptoms on plants when the soil is wet but not flooded. Several other fungi, *e.g., Fusarium solani,* which is the cause of dry root rot of beans, *Fusarium roseum,* the cause of seedling blights,and *Macrophomina phaseoli,* the cause of charcoal rot of sorghum and of root rot of cotton, grow fairly wellin rather dry environments. Apparently that characteristic enables them to cause more severe diseases in driersoils on plants that are stressed by insufficient water. Vascular wilts caused by the fungus *Verticillium* and canker diseases of forest trees and seedlings caused by fungi are significantly more severe when the plants suffer from water stress. Similarly, *Streptomyces scabies*, which causes the common scab of potatoes, becomes most severe in soils drying out after wetting.

Most bacterial diseases, and also many fungal diseases of young tender tissues, are particularly favoured by high moisture or high relative humidity. Bacterial pathogens and fungal spores are usually disseminated in water drops splashed by rain, in rainwater moving from the surfaces of infected tissues to those of healthy ones, or in free water in the soil. Bacteria penetrate plants through wounds or natural openings and cause severe disease when present in large numbers. Once inside the plant tissues, the bacteria multiply faster and are more active during wet weather, probably because the plants, through increased water absorption and resulting succulence, can provide the high concentrations of water that favour bacteria. The increased bacterial activity in wet weather produces greater damage to tissues. This damage, in turn, helps release greater numbers of bacteria onto the plant surface, where they are available to start more infections if the wet weather continues

One of the most important factors influencing attack by fungal pathogens may be the availability of water. Moisture can influence the distribution and spread of many pathogens, as well as affecting the development, longevity, germination, and infectiveness of fungal spores. While some fungal diseases, such as dry rot of beans (caused by *Fusarium solani*), prefer drier environments, the occurrence of many diseases is favoured by heavy rainfall, dew, or extended periods of high relative humidity. For example, such soil-borne "damping off" pathogens as *Pythium* and *Phytophthora* have long been known to be favoured by high humidity and wet, poorly drained soils.

Variation among habitats in soil moisture may result in spatial variation in losses to seed pathogens. In an field study conducted it was found that fungal mortality of seeds was significantly higher in wetland sites than in upland sites. Studies by also suggest that moist, shady microenvironments can increase losses of germinating seeds. Patterns such as these may have profound importance for the distribution of plant species, but few other studies of natural systems have considered interactions between spatial variation in physical conditions and losses of seeds to fungi. Together, these results indicate that seeds of at least some species are at a greater risk of loss to fungal pathogens in wet soils than in dry soils. This applies both to species of dry habitats (*Danthonia*) and to plants of wetter sites (*Glyceria Glyceria*). At the same time, desiccation is one of the primary causes of mortality of germinating seeds Consequently, for seeds of many plants, the habitats that are safest in terms of physical environment may be the riskiest in terms of fungal attack.

Erratic rainfalls and rise in temperature have become more frequent under the changing scenario of climate particularly in semi-arid tropics. As a consequence of it, a drastic shift of chickpea diseases have been recorded throughout the major chickpea growing regions in India and elsewhere. Dry root rot (DRR) caused by *Rhizoctonia bataticola* (Taub.) Butler [Pycnidial stage: *Macrophomina phaseolina* (Tassi) Goid] was found as a potentially emerging constraint to chickpea production than wilt (*Fusarium oxysporum* f. sp. *ciceris*). Increasing incidence of DRR indicate strong influence of climate change variables such as temperature and moisture on the development of disease.

The DRR incidence was significantly affected by high temperature and soil moisture deficit. Out of five temperature regimes (15°C, 20°C, 25°C, 30°C and 35°C) and four moisture levels (40 per cent, 60 per cent, 80 per cent and 100 per cent), a combination of high temperature (35°C) and soil moisture content (60 per cent) predisposes chickpea to DRR. The study clearly demonstrates that high temperature coupled with soil moisture deficit is the climate change variables predisposing chickpea to *R. bataticola* infection, colonization and development.

Codinaea fertilis is a root-rot pathogen of white clover plants which can reduce root and shoot production of this economically important pasture forage legume. This fungus is potentially a contributing factor to poor persistence of clover in North Island pastures. Soil moisture is an important environmental factor which can influence the pathogenicity of the fungus. In dry soils where plants are already stressed, this pathogen could cause increased root damage leading to low pasture productivity and poor persistence.

A study on the effect of soil moisture showed that maximum disease occurred at low soil moisture level (40 per cent) and disease incidence decreased as the moisture level increased (100 per cent). In other words dry soils are more favourable for the disease. The past work carried out by several workers on different crops also support our observation. The susceptibility of sorghum to charcoal rot in Karnataka revealed that most of the activated varieties and hybrid were equally susceptible to the disease when moisture stress condition coupled with soil temperature prevailed during grain filling period.

Charcoal rot, caused by, is one of the most important diseases of sorghum in India. The organism is also known to cause root and stem diseases of many economic crops, notably charcoal rot of soybean, corn *etc.* As soon as the moisture level increased from 40 to 100 per cent average disease rating decrease. At different temperature level's 25, 30, 35 and 40,maximum per cent infection at 40°C (21.3 per cent) while minimum at 25°C (13.8 per cent). Similarly maximum numbers of internodes are crossed at 40°C where disease rating is 1.57 and minimum at 25°C maximum length is also affected at 40°C.

In the southeast, watermelon is typically grown on bare soil and watered by overhead irrigation or on raised beds covered with plastic film mulch and watered with drip irrigation. Due to the occurrence of droughts over the last few years, there is high interest among vegetable growers to increase irrigation water use efficiency. The use of plastic film mulch and drip irrigation has the potential to optimize irrigation water use efficiency. However, over irrigation in many farms is frequent due to inadequate irrigation scheduling. Over irrigation results in water waste and nutrient leaching, and may be conducive to increased incidences of plant diseases such as Phytophthora rot. Phytophthora rot (caused by *Phytophthora capsici* Leonian) is an important disease of several vegetable crops including squash, watermelon, and pepper. This disease can cause complete destruction of an individual crop. The pathogen produces sporangia on roots, stems and fruit of infected plants. These sporangia release zoospores that are dispersed by the rain and irrigation water causing the spread of the disease. Since water is important in the dispersal of the pathogen, the disease is associated with abundant rainfall and high soil moisture condition

References

Mamta Sharma, Suresh Pande, 2013. Unravelling Effects of Temperature and Soil Moisture Stress Response on Development of Dry Root Rot [*Rhizoctonia bataticola* (Taub.)] Butler in Chickpea, *American Journal of Plant Sciences* 4: 584-589

Michelle Schafer, Peter M. Kotanen, 2003. The influence of soil moisture on losses of buried seeds to fungi, Acta Oecologica 24 : 255–263.

Manjeet arora, Savita Pareek, 2013. Effect of soil moisture and temperature on the severity of *macrophomina* charcoal rot of sorghum, *Indian J.Sci.Res.* 4(1): 155-158

Waipara, N.W., DI Menna, M.E., Cole, A.L.J., Skipp, R.A., 1996. Soil moisture effects on root rot of white clover caused By *Codinaea Fertilis Proc. 49th N.Z. Plant Protection Conf. 1996: 216-219.*

24

Use of Remote Sensing and GIS in Irrigation Water Management

N.L. Rajesh and B.M. Chittapur

Population growth and recurring droughts have increased pressure on the available water resources (M.H. Ali, 2010). Land is most important natural resource covering 30 per cent of the earth's surface and not all parts of land are productive and habitable. The U.N. Food and Agriculture Organization (FAO) estimates that more than 850 million people worldwide suffer from hunger and malnutrition today (Bob bell *et. al.* 2005), because of rainfall shortage and poor soil quality. To feed the estimated 9 billion world population by 2050, a comprehensive holistic strategy needs to be adopted to solve the geographical problems of acute rainfall and poor land quality. Geo-spatial technologies such as RS (Remote Sensing) and GIS (Geographical Information System) are the proven robust tools to tackle any Geo-graphic problem, therefore adoption of RS and GIS in estimation of water availability for irrigation and domestic purpose would help plan for judicial usage of water and for sustainable crop plan development to feed the world population.

RS and GIS in Estimation of Water Availability

Today agriculture uses 70 per cent of available water and rest 30 per cent is used for domestic and industrial purposes (FAO). This excess, unscientific and non-judicial use of available water has to be mapped, and assessed to develop plan for reorganizing the usage through implementing best policies for judicial usage of available water for any purpose. The FAO states that the causes for serious public health and environmental problems are due to increased "soil erosion" and "polluted groundwater and surface water" as opined by water experts, which is

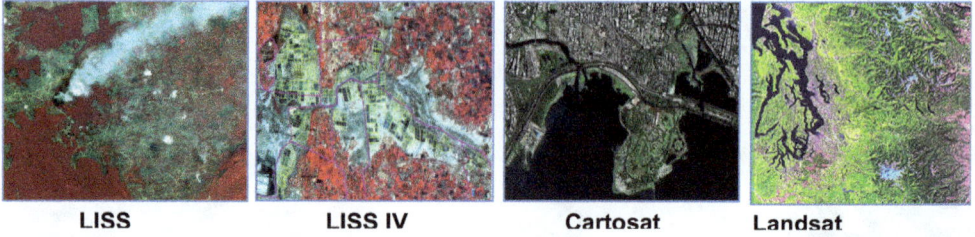

LISS   LISS IV   Cartosat   Landsat

a result of Green Revolution's dependence on technological and chemical inputs. RS data will help interpreting the timeline changes in usage estimation of available water. The advances in remotely sensed data encompass very high spatial, spectral, radiometric and temporal resolutions. The EoS (Earth Observatory Satellites) of IRS images such as LISS III, LISS IV and Cartosat – I, having spatial resolutions of 23.5 m, 5.8 m and 2.5 m respectively and Landsat ETM+ from NASA, USA, having 7 bands with 15 m spatial resolution help identifying significant water signatures of the earth surfaces.

Apart from estimating the surface water, remote sensing technology will also help identifying the water below surface *viz.*, top soil moisture, ground water (through interpreting the associated surface features) and the water depth of

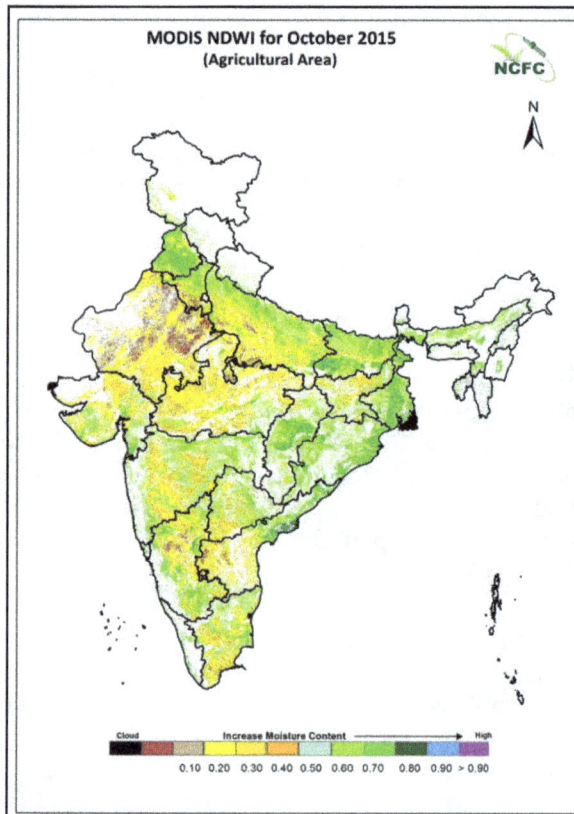

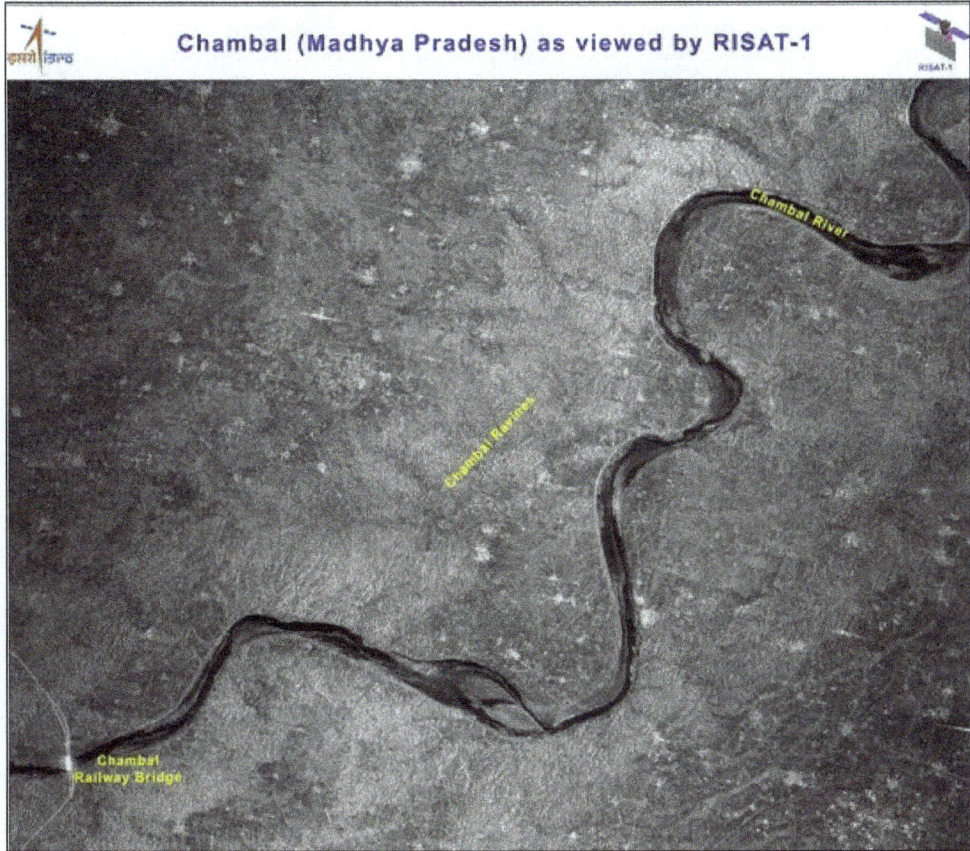

Chambal (Madhya Pradesh) as viewed by RISAT-1

different water bodies. Recent imageries like RISAT which is India's first indigenous RADAR satellite imagery capable of recording radio signal echos of different surface features at different spatial resolutions *viz.*, 50 m (Coarse Resolution mode), 25 m (Medium Resolution mode), 3 m to 12 m (Fine Resolution mode) and Spotlight mode which has 1 m spatial resolution. These RADAR images are active sensors and capable of recoding radio signals even in night and can penetrate the cloud. The RISAT can record the radio signals reflected from depth of few centimetres (approximately up to the depth of 10 cm) of the surface soil. This will help in continuous monitoring of the surface water runoff, soil moisture, and even the floods/inundations. The moisture present in this surface soil is very important for shallow root crops to sustain crop life and yield.

Indices

The pixel level classification of these EoS imageries would identify even the smallest water-bodies on the earth's surface. Alongwith the satellite image interpretation and image classification techniques, the ground truth will improve the accuracy of output of water resources estimation. The NDWI is a remote sensing based indicator sensitive to the change in the water content of leaves (Gao, 1996). NDWI (Normalized Differential Water Index) offers a means to view water bodies based on the given ratio, a range of satellite images right from multispectral to hyper-spectral images can be classified to understand the water availability. According to PRODUCT FACT SHEET: NDWI – EUROPE, (Version 1, 2011), by providing near-real time information on the plant water stress to the stakeholders, water and agricultural management can be much improved, notably by irrigating specifically areas where plant water needs are not fulfilled anymore. During drought event, vegetation canopy can be affected by water stress. This can have major impact on the plant development in general and can cause crop failure or lower crop production in agricultural areas. Early recognition of plant water stress can be critical to prevent such consequences. The Normalized Difference Water Index (NDWI) is known to be strongly related to the plant water content. It is therefore a very good proxy for plant water stress.

NDWI = (Green – NIR)/(Green + NIR)

where,

NIR = Near Infrared

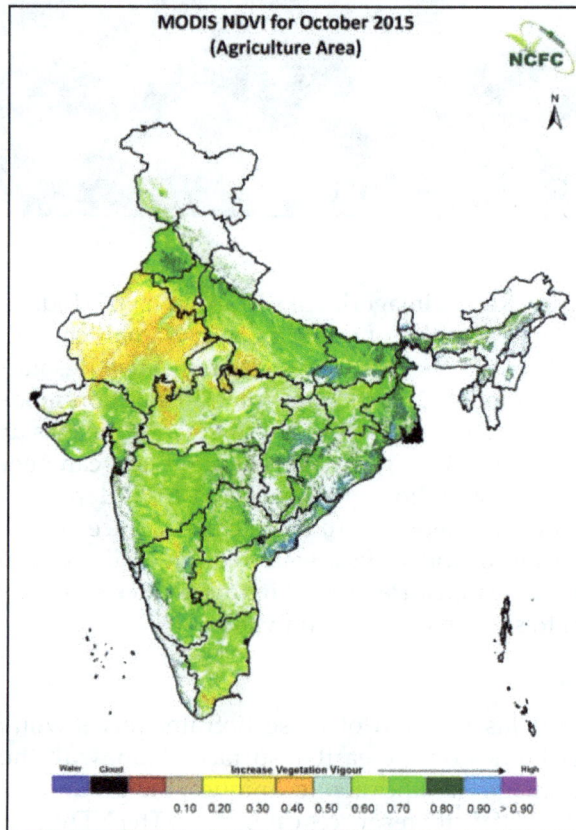

MODIS NDVI for October 2015 (Agriculture Area)

The NDWI products can be used in conjunction with NDVI (Normalized Differential Vegetative Index) change products to assess context of apparent change areas.

NDVI = (NIR − Red)/(NIR + Red)

The range of these indices varies from -1 to +1, if the value is towards negative it indicates stress and if it is towards positive it indicates healthy. According to Gao (1996). NDWI is a good indicator for vegetation liquid water content and is less sensitive to atmospheric scattering effects than NDVI.

RS and GIS in Crop Plan Development

The land resources are under severe strain at present due to various forms of degradation and competing demands of the various land uses. This vast stretch of land is affected by severe soil loss due to erosion, low and uncertain productivity, low rainwater use efficiency, rapid depletion of ground water, low level of technological penetration, acute fodder shortage and poor livestock productivity and poor marketing opportunities. Therefore it is necessary to map, understand and learn to make farm level crop suitability for the sustainability of the farmers and farming system. Remote Sensing and GIS will largely help to generate cadastral level data

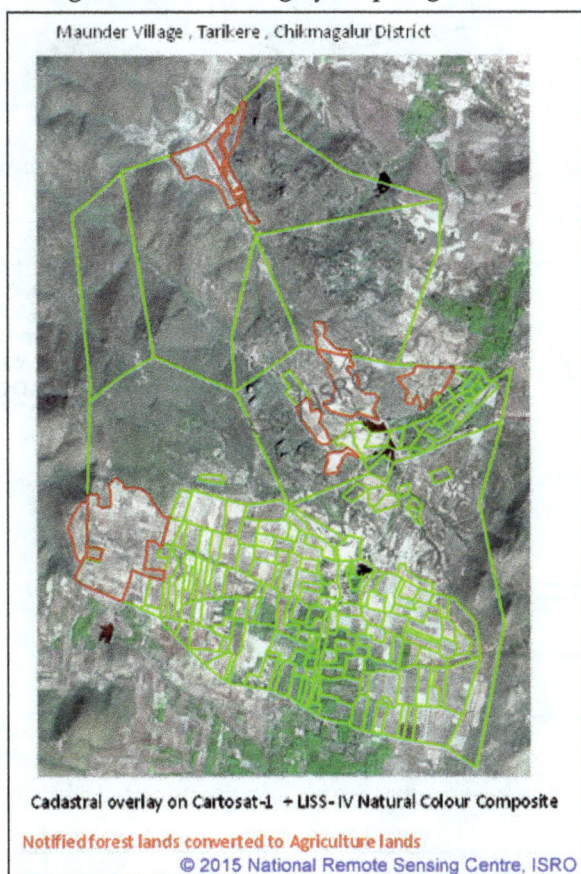

Maunder Village , Tarikere , Chikmagalur District

Cadastral overlay on Cartosat-1 + LISS-IV Natural Colour Composite

Notified forest lands converted to Agriculture lands

© 2015 National Remote Sensing Centre, ISRO

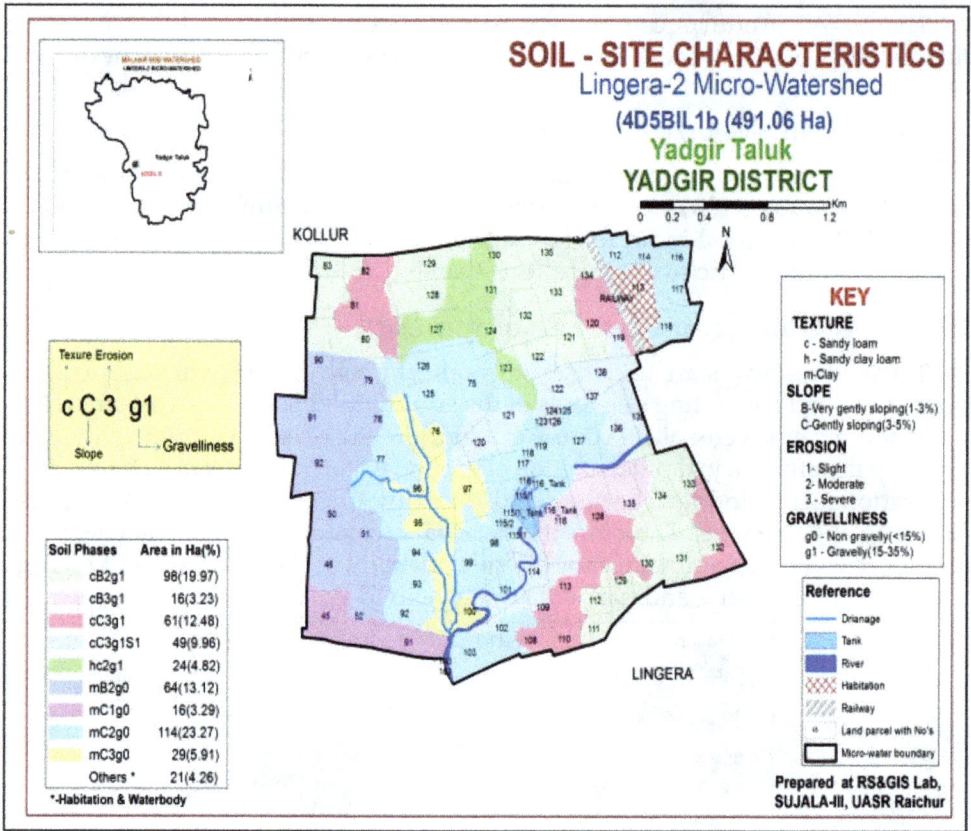

SOIL - SITE CHARACTERISTICS
Lingera-2 Micro-Watershed
(4D5BIL1b (491.06 Ha)
Yadgir Taluk
YADGIR DISTRICT

KEY

TEXTURE
c - Sandy loam
h - Sandy clay loam
m-Clay

SLOPE
B-Very gently sloping(1-3%)
C-Gently sloping(3-5%)

EROSION
1- Slight
2- Moderate
3 - Severe

GRAVELLINESS
g0 - Non gravelly(<15%)
g1 - Gravelly(15-35%)

c C 3 g1

Texure Erosion
Slope — Gravelliness

Soil Phases	Area in Ha(%)
cB2g1	98(19.97)
cB3g1	16(3.23)
cC3g1	61(12.48)
cC3g1S1	49(9.96)
hc2g1	24(4.82)
mB2g0	64(13.12)
mC1g0	16(3.29)
mC2g0	114(23.27)
mC3g0	29(5.91)
Others *	21(4.26)

*-Habitation & Waterbody

Reference
Drianage
Tank
River
Habitation
Railway
Land parcel with No's
Micro-water boundary

Prepared at RS&GIS Lab, SUJALA-III, UASR Raichur

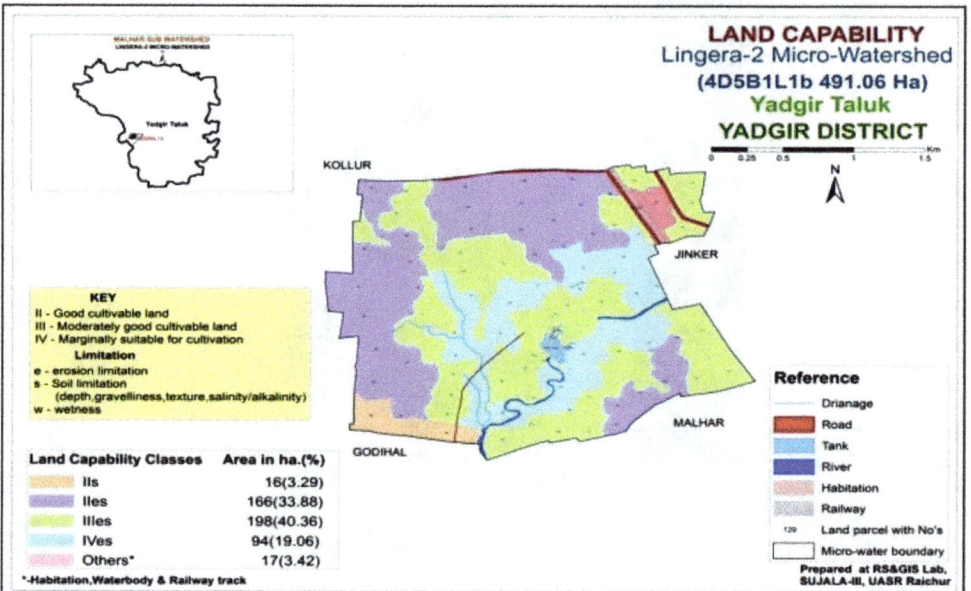

LAND CAPABILITY
Lingera-2 Micro-Watershed
(4D5B1L1b 491.06 Ha)
Yadgir Taluk
YADGIR DISTRICT

KEY
II - Good cultivable land
III - Moderately good cultivable land
IV - Marginally suitable for cultivation
Limitation
e - erosion limitation
s - Soil limitation
(depth,gravelliness,texture,salinity/alkalinity)
w - wetness

Land Capability Classes	Area in ha.(%)
IIs	16(3.29)
IIes	166(33.88)
IIIes	198(40.36)
IVes	94(19.06)
Others*	17(3.42)

*-Habitation,Waterbody & Railway track

Reference
Drianage
Road
Tank
River
Habitation
Railway
Land parcel with No's
Micro-water boundary

Prepared at RS&GIS Lab, SUJALA-III, UASR Raichur

Variable Rate of Application of Fertilizer for in Plot No. 125 (Paddy) Jangamara Kalgudi N

of soil, land use/land cover, hydrology, weather parameters *etc.* which will help for site specific agriculture management. The high resolution Indian remote sensing data like Cartosat – I (2.5 m) merged with LISS – IV is good enough to generate the land resources data at cadastral scale (1:8000).

Detailed database pertaining to the nature of the land resources, their constraints, inherent potentials and suitability for various land based rural enterprises, crops and other uses is a prerequisite for preparing location-specific action plans, which are in tune with the inherent capability of the land resources. Land Resource Inventory provides the required information for farm level planning. For site-specific needs and for developmental works, we need to develop detailed farm level database at 1: 8, 000 scale. The cadastral level resource map should be generated by studying all the site characteristics like slope, erosion, drainage, salinity, rock fragments *etc.* and soil characteristics like depth, texture, colour, structure, consistency, gravels, porosity, soil reaction *etc.* followed by grouping of similar areas based on soil-site characteristics into homogenous (management) units and showing their extent and distribution on the cadastral map.

This can be accomplished effectively by using digital cadastral base in conjunction with remote sensing data products like Cartosat – I imagery. From the database generated for any area, the required thematic outputs can be generated through the use GIS and further any spatial analysis like overlay, buffer, and network analysis can be done to get comprehensive output for management.

The point data of soil samples at parcel level can be interpolated thorough advanced *"Kriging"* techniques to understand the spatial distribution of soil fertility for precise agricultural input management. From the data collected at farm level, viable, sustainable land use options suitable for each and every land holding can be identified with the defined criteria for various horticultural and agricultural crops.

References

Bob Bell, David Kauck, Marianne Leach, Priya Sampath, 2005. Hunger: Facing the Facts, eJOURNAL USA, IIP/PUBJ, U.S. Department of State, Volume 12, Number 9, Page 19

Gao, B. C., 1996. NDWI - A normalized difference water index for remote sensing of vegetation liquid water from space. Remote Sensing of Environment 58: 257-266.

M.H. Ali, 2010. GIS in Irrigation and Water Management, Practices of Irrigation and On-farm Water Management: Volume 2, Affiliated with Agricultural Engineering Division, Bangladesh Institute of Nuclear Agriculture (BINA), pp 423-432.

Online Reference

http://deltas.usgs.gov/fm/data/data_ndwi.aspx

25

Improved Water Management Practices in Horticultural Crops

B.R. Premalatha and B.N. Maruthi Prasad

Water is the vital input of horticultural crop production. Adequate, timely and assured supply of water is essential to maximise the crop yield both in terms of quantity and quality. Though water is the renewable resource, it is most limiting factor in agriculture, horticulture and plantation crop production. Thus, the water management assumes greater importance to obtain higher water use efficiency and to reduce or minimise the wastage of water.

The sources of water that are potentially useful are known as water resources. Over 70 per cent of our Earth's surface is covered by water though it looks abundant, the real issue is the amount of availability of fresh water. Of the total water resources of the World 97.5 per cent of the water on the Earth is salt water and only 2.5 per cent is fresh water. Nearly 70 per cent of this fresh water is frozen in the icecaps of Antarctica and Greenland (Anon., 2008), most of the remainder is present as soil moisture, or lies in deep underground aquifers as groundwater not accessible to human use. Only nearly 1 per cent of the world's fresh water is accessible for direct human usage. This is the water found in lakes, rivers, reservoirs and those underground sources that are shallow enough to be tapped at an affordable cost. Only this available water can be regularly renewed by rain and snowfall, and hence available on a sustainable basis.

Globally, the agricultural sector is by far the biggest user of freshwater with irrigation accounting for 70 per cent of global water withdrawals. The industrial and domestic sectors account for the remaining 20 per cent and 10 per cent, respectively, although the figures vary considerably across the countries. In most of the world's least developed countries, agriculture accounts for more than 90 per cent of water withdrawals (Anon., 2014).

By 2025, agriculture is expected to increase its water requirements by 1.3 times, industry by 1.5 times, and domestic consumption by 1.8 times (Anon., 2008). The demand for water in industrial and domestic sectors is likely to increase as a result greater portion of the fresh water will be diverted towards these sectors and thus smaller quantity of water would be diverted for agriculture sector. At present, hardly 38 per cent of cultivable area is irrigated. Looking into the current level of irrigation, even after exploitation of all the available resources we can cover only 50 per cent of cultivated area and more than 50 per cent area may remain rainfed. Without improved efficiencies, agricultural water consumption is expected to increase globally by about 20 per cent by 2050. Thus, more attention needs be diverted to increase the productivity and water use efficiency.

As per the land use statistics 2011-12, the total geographical area of the country is 328.7 million hectares, which constitutes 2.4 per cent of the global geographical area and only 4 per cent of water resource currently supports about 17.1 per cent of the human and 11 per cent livestock population of the world. The net sown area is 140.8 million hectares and 195.2 million hectares is the gross cropped area with a cropping intensity of 138.7 per cent. The net irrigated area is 65.3 million hectares (Anon., 2014). Water is the key issue to sustain required growth of agriculture in future to attain higher crop intensity. Availability of fresh water for agriculture is expected to decline from current level of 80 per cent to about 70 per cent by 2050 AD due to ever increasing demand of water for industrialization and urbanization (Table 25.1).

Table 25.1: Water Availability Facts at a Glance

Area of the country as per cent of World Area	2.4 per cent
Population as per cent of World Population	17.1 per cent
Livestock as per cent of World livestock Population	11.0 per cent
Water as per cent of World Water	4 per cent
Rank in per capita availability	132
Rank in water quality	122
Average annual rainfall	1160 mm (world average 1110 mm)
Range of distribution	150-11690 mm
Range Rainy days	5-150 days, Mostly during 15 days in 100 hrs
Range PET	1500-3500 mm
Per capita water availability (2010)	1588 m^3

Source: Water Resources at a Glance 2011 Report, CWC, New Delhi. http://www.cwc.nic.in.

Irrigation constitutes the main use of water and is thus focal issue in water resources development. As of now, irrigation use is 84 per cent of total water use. This is much higher than the world's average, which is about 65 per cent. In advanced nations, the figure is much lower. For example, the irrigation use of water in USA is around 33 per cent. In India, the remaining 16 per cent of the total water use accounts for rural domestic and livestock, municipal domestic and public, thermal-electric power plants and other industrial uses.

In India, the surface water potential is about 180 million ha-m and the ground water resource is about 44 million ha-m. With annual precipitation of about 400 million ha-m, the average annual natural flow is about 188 million ha-m. The annual requirement of fresh water is estimated at 105 million ha-m by the year 2025 AD which is nearly equal to the ultimate water resource's level of the country. Out of this, 77 million ha-m has been considered for irrigation purpose. In terms of area, the ultimate irrigation potential of the country has been assessed at 155 million ha (58 million hectare from major/medium projects, 17 million hectares from surface water minor irrigation projects and 80 mha from ground water projects). India has acquired an irrigation potential of about 84.9 mha against the ultimate irrigation potential. Although, India has the largest irrigation system in the world, its water use efficiency has not been more than 40 per cent because of huge conveyance and distribution losses (Rosegrant, 1997; INCID, 1994).

Though India has the largest irrigated area in the world, the coverage of irrigation is only about 40 per cent of the gross cropped area as of today. One of the main reasons for the low coverage of irrigation is the predominant use of flood (conventional) method of irrigation, where water use efficiency is very low due to various reasons (Rosegrant, 1997; INCID, 1994).

Thus, with current level of efficiency, even after exploitation of all the available resources, more than 50 per cent area may still remain rainfed. If it is continued like this, the water crisis would result in reduced production and productivity, which would affect our food and nutritional security. This calls for more productive use of water and more crop yield per drop of water. This scenario will demand increasing water use efficiency and water productivity in agriculture, both under irrigated and rainfed systems.

Vagaries of monsoon and declining water table due to over exploitation have resulted in shortage of fresh water supplies for agricultural use, which calls for an efficient use of this resource. Strategies for efficient management of water for agricultural use involves conservation of water, integrated water use, recycling of water optimal allocation of water depending on the growth stages of crop and enhancing water use efficiency by crops.

India is the second largest producer of fruits and vegetables next to China in the World.

Horticultural crops play a unique role in India's economy by improving the income of the rural people. Cultivation of these crops is labour intensive and as such they generate lot of employment opportunities for the rural population. Fruits and vegetables are for nutritional security as they are rich source of vitamins, minerals, proteins, carbohydrates *etc.* which are essential in human nutrition and address the competing problems of malnutrition. Hence, these are referred to as protective foods. Cultivation of horticultural crops plays a vital role in the prosperity of a nation and is directly linked with the health and happiness of the people.

It is estimated that all the horticulture crops put together cover 23.69 million hectares area with an annual production of 268.9 million tonnes during 2012-13 (Anon. 2014). Horticulture accounts for 30 per cent of India's agricultural GDP from 8.5 per cent of the cropped area.

In India, only 10 per cent of water is used for irrigating horticultural crops and thus most of the water is diverted to agriculture for growing the staple food crops. Therefore the improvement of water use efficiency of horticultural crops has potential of reducing water requirement in terms of providing unit energy, unit protein and other nutrients as well as economic returns. As majority of the horticultural crops are perennial in nature, they invariably have deep and extensive root system, capable of extracting water from deeper layers and large canopy to harvest optimum natural resources. Hence, they have better productivity than field crops.

Water-use Efficiency/Water Productivity

Water Use Efficiency (WUE) or water productivity has emerged from the ideas of drought resistance and drought tolerance (Passioura, 2006). At the beginning of the sixties of the last Century, water use efficiency has been generally defined in agronomy (Viets, 1962) as:

$$WUE = \frac{\text{Crop yield (usually the economic yield) (kg ha}^{-1}\text{cm}^{-1})}{\text{Water used to produce the yield}}$$

The term Water Use Efficiency can be used at wide range of scales; for example, it can be used at the farm, the field, the plant, or down to plant parts level, such as the leaf (Morison *et al.*, 2008).

Water Management in Horticulture Crops

Horticultural crops demand and require significant amounts of water due to their perishable nature. In case of perennial crops stress not only affects the current season's crop, but future crop production as well. Vegetables are also quite perishable and require sufficient amount of water. As these crops are highly productive and income generating crops they have to be raised under protective irrigation conditions.

Water management refers to efficient use of water for best possible crop production. It includes the optimization of water use and maximization of crop yields. The water losses either through deep percolation beyond the root zone and run off is to be minimized or avoided to the maximum extent. Many parameters like type of soil, crop growth stage and its sensitivity to water stress, climatic conditions and water availability in the soil determine when to irrigate or the so-called irrigation frequency. However, this frequency depends upon the irrigation method and therefore, both irrigation scheduling and the irrigation method are inter-related.

The management practices to be followed for efficient utilization of the available water are:

1. Adoption of Soil and Moisture Conservation Techniques

Contour cultivation, contour trip cropping, mixed Cropping, tillage, mulching, zero tillage, are some of the agronomical measures for the in-situ soil moisture conservation. Mechanical measures like contour bunding, graded bunding, bench terracing, vertical mulching *etc.* also need to be followed for effective soil and moisture conservation in dry lands. Another technology for efficient utilization of run-off is water harvesting and recycling. Rainwater harvesting includes collecting runoff water into dug out ponds, farm ponds or tanks in small depressions, gullies and into storage dams of earth or masonry structures. Rain water harvesting is possible in areas having rainfall as little as 500 to 800 mm. Depending on the rainfall and soil characteristics, 10-50 per cent of the runoff can be collected in farm pond. Surface run off thus collected in a farm pond can be used to provide protective irrigation or life saving irrigation at critical stages of crop during the periods of prolonged dry spell.

2. Enhancing Soil Organic Matter Content

If you take care of our soil, the soil will take care of our plants. Continuous efforts are required to improve the soil organic carbon. Incorporation of crop residues, vermicompost, green manures and farm yard manure to soil improves the organic matter status, improves soil structure and soil moisture storage capacity and soil microflora. Organic matter content of the soil can also be improved by following multiple cropping systems like alley cropping, crop rotation and agro forestry. Vegetable being short duration crop and having faster growth phases, the available organic matter needs to be properly composted. Vermi composting can be followed for quicker usage of available organic matter in the soil and improving the soil moisture holding capacity. In case of perennial crops, for proper growth and development the maintenance of soil health is very essential and hence the application of organic matter is most essential.

3. Mulching Practices

The technique of covering the soil with natural crop residues or plastic films for soil and water conservation is called mulching. Mulching can be practiced in fruits, plantation, flowers and vegetable crops using crop residues and other organic material available on the farm. Recently plastic mulches have come into use due to the inherent advantages of efficient moisture conservation, weed suppression and maintenance of soil structure. Presently different colour plastic films are used as mulches such as black, red, yellow and dual colored like silver-black, white-black *etc.* Wide variety of vegetables can be successfully grown using mulches. In addition to improved yield and quality, soil and water conservation, suppression of weed growth, mulches can improve the use efficiency of applied fertilizer nutrients and also use of reflective mulches are known to minimize the incidence of pest and virus diseases. For vegetable production generally polyethylene mulch film of 30µ thick and 1 to 1.2 m width are used. Generally raised bed with drip irrigation system is followed while laying the mulch film. Whereas for fruit crops and perennial plants generally 100µ, medium duration crops like papaya, pine apple -50µ, monocot weeds 100µ and dicot weeds 30-50µ thickness mulches are recommended.

4. Wind Breaks, Hedges, Green Manure Crops and Intercropping

To overcome the adverse effect of high temperature and dry winds, tall growing trees need to be planted all along the boundary of the farm. In recently established orchards inter cropping of vegetable crops, green manure crops or short duration pulses of the area can be practiced during *kharif* months. Maize/sorghum can be grown all along the border of the vegetable plot to mitigate the effect of desiccating winds. In most of plantation crops in areas where soil erosion is a problem green manure crops like cowhage, thornless mimosa, stylosanthus are grown and incorporated into the soil before flowering when enough moisture is available.

5. Knowledge about the Soil Type, Effective Root Zone Depth, Water Requirement and Moisture Sensitive Periods of Crops

The amount of water used by a particular crop depends on a number of factors, including variety, crop growth stage and environmental conditions (temperature, wind, relative humidity). The speed at which soil moisture is depleted depends on crop use and the soil type (sand, clay, *etc.*). Applying adequate amounts of moisture requires a basic understanding of soils and the general water use of the crop. Moisture stress/excess can influence crop yield and survivability (over-wintering) (Tables 25.2 and 25.3).

Table 25.2: Effective Rooting Depth of Selected Vegetables

Shallow (6-12")	Moderate (18-24")	Deep (>36")
Beet	Cabbage, Brussels Sprouts	Asparagus
Broccoli	Cucumber	Lima Bean
Carrot	Egg plant	Pumpkin
Cauliflower	Muskmelon	Sweet Potato
Celery	Pea	Water melon
Greens and Herbs	Potato	Squash winter
Onion	Snap Bean	
Pepper	Squash Summer	
Radish	Sweet Corn	
Spinach	Tomato	

Most of the active root system for water uptake may be in the top 6"-12".

Table 25.3: Critical Soil Moisture Periods of Crops

Crop	Critical Soil Moisture Stage of Crops
Chillies	Tenth leaf to Flowering and fruit development and after periodical harvests
Potato	Stolon formation, Tuberization and tuber enlargement
Onion	Bulb formation and bulb enlargement
Tomato	Flowering and fruit development and each harvest
Peas	Flowering and pod development

Crop	Critical Soil Moisture Stage of Crops
Cabbage	Head formation and enlargement
Cauliflower	Curd formation and enlargement
Brinjal	Flowering and fruit development and after each harvest
Cucumber	Flowering and fruit development
Bhendi	Flowering and pod development
Leaf vegetables	Entire crop duration
Fruit crops	
Citrus	Flowering, fruit setting, fruit growth
Banana	Early vegetative period flowering and fruit formation
Mango	Start of fruiting to maurity
Pine apple	Vegetative growth
Grape	Vegetative growth. Frequent irrigation during vegetative stage may cause rotting of friuts
Guava	Period of fruit growth
Ber	A drought resistant plant ; irrigation is required during fruit growth

Table 25.4: Water Requirements of Horticultural Crops

Sl.No.	Crop	Water Requirement	
		Initial Stage	Later Stage
1.	Tomato	8 lt/m² (first month)	12 lt/m² (subsequent months)
2.	Capsicum	8 lt/m² (first month)	12 lt/m² (subsequent months)
3.	Cucumber	8 lt/m² (first month)	12 lt/m² (subsequent months)
4.	Brinjal	8 lt/m² (first month)	12 lt/m² (subsequent months)
5.	Cabbage	8 lt/m² (first month)	12 lt/m² (subsequent months)
6.	Cauliflower	8 lt/m² (first month)	12 lt/m² (subsequent months)
7.	Knol-khol	8 lt/m² (first month)	12 lt/m² (subsequent months)
8.	Chilli	8 lt/m² (first month)	12 lt/m² (subsequent months)
9.	Bhendi	8 lt/m² (first month)	12 lt/m² (subsequent months)
10.	Water melon	8 lt/m² (first month)	12 lt/m² (subsequent months)
11.	Musk melon	8 lt/m² (first month)	12 lt/m² (subsequent months)
12.	Bottle gourd	8 lt/m² (first month)	12 lt/m² (subsequent months)
13.	Bitter gourd	8 lt/m² (first month)	12 lt/m² (subsequent months)
14.	Ridge gourd	8 lt/m² (first month)	12 lt/m² (subsequent months)
15.	Pumpkin	8 lt/m² (first month)	12 lt/m² (subsequent months)
		Through drip irrigation for 30 minutes during initial stages	Through drip irrigation for 45 minutes at subsequent stages

Sl.No.	Crop	Water Requirement	
		Initial Stage	Later Stage
		Flower crops	
16.	Jasmine	3-4 lt/plant/day (6 months)	6-8 lt/plant/day
		Through drip irrigation for 30 minutes	Through drip irrigation for 45 minutes
17.	Marigold	8 lt/m² (first month)	12 lt/m² (subsequent months)
		(30 minutes)	(45 minutes)
18.	China Aster	8 lt/m² (first month)	12 lt/m² (subsequent months)
		(30 minutes)	(45 minutes)
19.	Gladiolus	8 lt/m² (first month)	12 lt/m² (subsequent months)
		(30 minutes)	(45 minutes)
20.	Chrysanthemum	8 lt/m² (first month)	12 lt/m² (subsequent months)
		(30 minutes)	(45 minutes)
		FRUITS	
	Banana	1200-2200	
	Citrus	900-1200	
	Grapes	500-1200	
	Pineapple	700-1000	
	Coconut	80-100 (lit/plant/day)	
	Mango	30-40 (lit/plant/day)	
	Guava	22-30 (lit/plant/day)	
	Banana	8-12 (lit/plant/day)	

Published by: Precision Farming Development Centre (PFDC), Division of Horticulture, UAS, GKVK, Bangalore

Irrigations Methods

Irrigation methods are classified under surface, subsurface and pressurized irrigation.

- ☆ Surface irrigation can be used for all types of crops.

- ☆ Sprinkler and drip irrigation, because of their high capital investment per hectare, are mostly used for high value crops but the quantity and quality of the produce help them to recover the spent amount within a short span of time and moreover the government is providing subsidy.

- ☆ Drip irrigation is suited to irrigating individual plants. It is not advisable for closely spaced crops.

Surface Irrigation Methods

Border, furrow and/or basin are the most common irrigation methods practiced to till date. It refers a scientific and systematic approach to spread the water over the surface crop area as uniform as possible to satisfy the demand of replenishable soil moisture and bring it to the field capacity maintaining the gradient, design discharge and duration or opportunity time to infiltrate the water. They are excellent systems but often result in inefficient use of water due to overwatering beginning and ending segments and under watering the middle segments. Depending on the system used and design, these systems have a 55-75 per cent efficiency rating; can be better if properly designed. Many times inefficiency is made worse due to the failure of water reaching the tail end. Length of runs should be dictated by soil types and not field size. Among these methods, under limited water situations the switch over to furrow system of irrigation results in 30-60 per cent saving of water and significant increase in productivity, especially in case of widely spaced crops.

Studies conducted on methods of irrigation have indicated that from improved planting methods (sunken and raised bed, raised bed furrow irrigated system (FIRB), broad bed furrows (BBF) and by adopting alternate-furrow irrigation and widely-spaced furrow irrigation in capsicum, tomato, okra and cauliflower saved 35 to 40 per cent of irrigation water without adversely affecting yield. In addition to low water use efficiency, these irrigation methods often require a much higher labour input - for construction, operation and maintenance - than sprinkler or drip irrigation, accurate land levelling, regular maintenance and a high level of farmers' organization to operate the system.

Micro-irrigation Methods

Drip and sprinkler methods of irrigation are treated as Micro Irrigation methods. Both the methods differ in terms of flow rate, pressure requirement, wetted area and mobility (Kulkarni, 2005). Drip method is also known as Trickle irrigation where it supplies water directly to the root zone of the crop through a network of pipes with the help of emitters. Sprinkler irrigation is a method of applying irrigation water which is similar to rainfall. Water is distributed through a system of pipes usually by pumping under high pressure of 5 kg/cm². It is then sprayed into the air through nozzles and irrigates entire soil surface which subsequently breaks up into small water drops.

In case of drip irrigation method the water is applied directly to the root zone of the crop, instead of land like in flood irrigation method and thus the water losses occurring through evaporation and distribution are completely absent (INCID, 1994, Narayanamoorthy, 1996; 1997; Dhawan, 2002). The on-farm irrigation efficiency of properly designed and managed drip irrigation system is estimated to be about 90 per cent, while the same is only about 35 to 40 per cent for surface method of irrigation (INCID, 1994). As water is applied over the entire field of the crop in sprinkler irrigation method, water saving is relatively low (up to 70 per cent) as compared to drip irrigation. As per the data provided by the National Committee on Plasticulture Applications in Horticulture, by 2012, it is speculated that around 6 M ha were under micro-irrigation in India. Out of this, 2.5 M ha were under drip irrigation and remaining 3.5 M ha under sprinkler.

The experiments conducted at different locations of the country on different crops have clearly shown that there is large amount of water saving, yield increase and thus the water use efficiency. Thereby it also helps to reduce the cost of cultivation. INCID (1994) report presents the results of various crops carried out at different locations in the country (Table 25.5). It shows that the productivity of different crops is significantly higher under drip irrigation method (DIM) when compared to flood irrigation method (FIM) apart from excellent quality.

Table 25.5: Water Saving and Productivity Gains under Drip Method of Irrigation: India

Crop's Name	Water Consumption (mm/ha)		Yield (tons/ha)		Water Saving over FIM (per cent)	Yield Increase over FIM (per cent)	Water Use Efficiency (yield/ha)/ (mm/ha)	
	FIM	DIM	FIM	DIM			FIM	DIM
Vegetables								
Ash gourd	840	740	10.84	12.03	12	12	0.013	0.016
Bottle gourd	840	740	38.01	55.79	12	47	0.045	0.075
Brinjal	900	420	28.00	32.00	53	14	0.031	0.076
Beet root	857	177	4.57	4.89	79	7	0.005	0.023
Sweet potato	631	252	4.24	5.89	61	40	0.007	0.023
Potato	200	200	23.57	34.42	Nil	46	0.118	0.172
Lady's finger	535	86	10.00	11.31	84	13	0.019	0.132
Onion	602	451	9.30	12.2	25	31	0.015	0.027
Radish	464	108	1.05	1.19	77	13	0.002	0.011
Tomato	498	107	6.18	8.87	79	43	0.012	0.083
Chillies	1097	417	4.23	6.09	62	44	0.004	0.015
Ridge gourd	420	172	17.13	20.00	59	17	0.041	0.116
Cabbage	660	267	19.58	20.00	60	2	0.030	0.075
Cauliflower	389	255	8.33	11.59	34	39	0.021	0.045
Fruit crops								
Papaya	2285	734	13.00	23.00	68	77	0.006	0.031
Banana	1760	970	57.50	87.50	45	52	0.033	0.090
Grapes	532	278	26.40	32.50	48	23	0.050	0.117
Lemon	42	8	1.88	2.52	81	35	0.045	0.315
Watermelon	800	800	29.47	88.23	Nil	179	0.037	0.110
Mosambi*	1660	640	100.00	150.00	61	50	0.060	0.234
Pomegranate*	1440	785	55.00	109.00	45	98	0.038	0.139

Notes: * - yield in 1000 numbers: Sources: INCID (1994) and NCPA (1990).

In case of vegetables, it was noticed that the productivity was over 40 per cent in vegetable crops such as bottle gourd, potato, onion, tomato and chillies, whereas the same is noticed over 70 per cent in many fruit crops. The water saving ranged from 12 per cent to 84 per cent per hectare over the conventional method of irrigation for vegetable crops. Similar trend was noticed in case of fruit crops where it varied from 45 per cent to 81 per cent per hectare (Table 25.5). Thus, it resulted in higher water use efficiency when compared with conventional method of irrigation.

The main reasons for water saving and increased yield in drip irrigation method are

☆ Water is supplied through a network of pipes, the evaporation and distribution losses of water, percolation below root zone are very minimum or completely absent.

☆ Water is supplied depending on growth stage and as per the crop requirement and thus over irrigation is avoided.

☆ Water is supplied only to the root zone of the crop and not to entire field.

Micro Irrigation has following advantages

1. Water savings: 40-60 per cent saving in water apart from improved quality produce.

2. Energy savings: This type of irrigation system requires a smaller power unit and consumes less energy.

3. Weed and disease reduction: This type of irrigation system is helpful in inhibiting growth of weeds as it keeps limited wet areas. Under this condition the incidence of disease is also reduced up to major extent.

4. Can be automated. Fertilizers and chemicals can be applied with water through micro irrigation system. This systems can be automated which reduces labor requirements.

5. Improved production on marginal land. On hilly terrain, micro-irrigation systems can operate with no runoff and without interference from the wind. The fields need not be levelled.

6. In a short period of time required quantity of water can be given to entire large area at a time.

The Micro Irrigation System has following disadvantages

1. Management: Micro-irrigation systems normally have greater maintenance requirements. Soil particles, algae, or mineral precipitates can clog the emission devices. The impurities are

   ☆ Physical impurities: soil particles, algae, organic plant debris and mica.

   ☆ Chemical impurities: carbonates and bicarbonates minerals and fertilizer precipitation

   ☆ Biological impurities: algae, bacteria *etc.*

2. Potential for damage. Animals, rodents and insects may cause damage to some components. The drip and bubbler irrigation systems need additional equipment for frost protection.

3. High initial cost. Micro-irrigation systems are ideal for high value installations such as orchards, vineyards, greenhouses, and nurseries where traditional irrigation methods may not be practical. However, the

investment cost can be high that can be compensated by increased yield and quality produce and by availing subsidy.

Fertigation and Chemigation in Horticultural crops

Fertilizers influence yield as quality of horticultural crops, particularly colour, shape, size, taste, shelf life and processing characteristics. Fertilizers also influence the physiology of plant and thereby determine the composition of fruits and vegetable produce and the resistance of these plants to environmental stress. The conventional method of fertilizer application has low nutrient use efficiency and thus improvements were made in fertilizer application, which is known as **"Fertigation"**.

Fertigation was opted for two reasons:

1. Application of fertilizer in small doses spread across the entire growing season which helps to meet the nutrient requirement of crops, to improve nutrient uptake efficiency, minimize losses, thus to maximize the returns per unit amount of fertilizer used.

2. Minimizes the leaching of nutrients especially nitrate nitrogen which can have negative impact on raising its concentration in the groundwater above the maximum contaminant limit that is recommended for drinking water quality.

Application of fertilizers through irrigation system is referred to an 'fertigation'.

Fertigation through micro-irrigation system provides a technique of application of water and nutrients to an area of the soil where most of the roots are present to coincide with the timing of nutrient requirement of crops. Therefore, fertigation is expected to increase the nutrient uptake efficiency, thereby minimizing leaching losses compared with the conventional method of fertilizer application.

Fertilizer Efficiencies of Various Application Methods

Nutrient	Fertilizer Use Efficiency (per cent)	
	Soil Application	Fertigation
Nitrogen	30-50	95
Phosphorous	20	45
Potassium	50	80

Source : http://agritech.tnau.ac.in/agriculture/agri_nutrientmgt_fertigation.html.

The advantages of localized soil fertigation include: combined application of water, fertilizers and pesticides with high precision and uniformity; improved distribution and control of water and nutrients in the soil and the potential for application of water and nutrients in accordance with the demands of the plant.

Advantages of Fertigation

☆ Simple, saves time, labour and energy

☆ Nutrients can be applied to all points uniformly

☆ Nutrients can be applied as per the crop growth stage and as per the demand of the crop.

☆ Nutrients and water are supplied near the active root zone through fertigation which results in greater absorption by the crops.

☆ As water and fertilizer are supplied evenly to all the crops through fertigation there is possibility for getting 25-50 per cent higher yield.

☆ Fertilizer use efficiency through fertigation ranges between 80-90 per cent, which helps to save a minimum of 25 per cent of nutrients.

☆ By this way, alongwith less amount of water and saving of fertilizer, time, labour and energy use is also reduced substantially.

Promoting Greenhouse and Plasticulture

The green-house technology and use of plasticulture using drip system of irrigation alongwith fertigation is one of the most modern technologies at present to grow high value crops with remarkable saving in water use. However, the design of green house has to be location specific. Due to controlled environmental conditions, the high value crops and nurseries of off-season vegetables, flowers and fruits can be grown throughout the year under protected conditions with water economy of 40-50 per cent.

Crops that can be grown under green houses are:

☆ **Vegetables:** Cherry tomato, Colored capsicum, European cucumber, Pole beans, Musk melon, Water melon, Braccoli, Lettuce, Brussels sprouts

☆ **Flowers:** Rose, Gerbera, Carnation, Orchids, Anthurium, Lilliums, Chrysanthemum

☆ **Fruits:** Grapes, Cherry, Strawberry

Crops under shade net houses:

☆ **Vegetables:** Broccoli, Ridgegourd, Bottlegourd, Cucumber, Chilli, Capsicum, Lettuce, Brussels sprouts

☆ **Flowers:** Gerbera, Anthurium, Rose

Conclusions

Globalization, urbanisation and industrialisation has resulted in reduced per capita availability of water. Water is a national asset. Declining water availability to agriculture has been a matter of discussion in the last two decades and therefore water management has received focus for realizing high water productivity. Out of several available water management strategies, providing irrigation at the most sensitive stages of growth, use of micro-irrigation, well designed surface, sub-surface system of irrigation, mulching, fertigation, plasticulture, rain water harvesting, collection and reuse would help in increasing water productivity.

References

Anonymous 2008. VITAL WATER GRAPHICS, An Overview of the State of the World's Fresh and Marine Waters - 2nd Edition - 2008. http://www.unep.org/dewa/vitalwater/article48.html.

Anonymous 2014. Hand Book on Horticulture Statistics 2014, Government of India. Ministry of Agriculture, Department of Agriculture and Cooperation, New Delhi.

Anonymous, 2014a. World Water Development Report 2014, (http://www.unwater.org/statistics/en).

Anonymous, 2014b. Annual Report 2014-15, Department Of Agriculture and Cooperation Ministry Of Agriculture, Government Of India Krishi Bhavan, New Delhi-110001.

Dhawan, B. D., 2002. Technological Change in Indian Irrigated Agriculture: A Study of Water Saving Methods, Commonwealth Publishers, New Delhi.

INCID, 1994. Drip Irrigation in India, Indian National Committee on Irrigation and Drainage, New Delhi.

Kulkarni, S.A., 2005. "Looking Beyond Eight Sprinklers", Paper presented at the National Conference on Micro-Irrigation, G. B. Pant University of Agriculture and Technology, Patnagar, India, June 3-5, 2005.

Morison, J.I., Baker, N.R., Mullineaux, P.M., Davies, W.J., 2008. Improving water use in crop production. Philo. Trans. R. Soc. London B Biol. Sci., 12: 639-658.

Narayanamoorthy, A., 1996. Evaluation of Drip Irrigation System in Maharashtra, Mimeograph Series No. 42, Agro-Economic Research Centre, Gokhale Institute of Politics and Economics, Pune, Maharashtra.

Narayanamoorthy, A., 1997. "Drip Irrigation: A Viable Option for Future Irrigation Development", Productivity, Vol.38, No.3, October-December, pp. 504-511.

NCPA, 1990. Status, potential and approach for adoption of drip and sprinkler irrigation systems, National Committee on the Use of Plastics in Agriculture, Pune, India.

Passioura, J., 2006. Increasing crop productivity when water is scarce: From breeding to field management. Agric. Water Manage., 80: 176-196.

Rosegrant, W. Mark, 1997. Water Resources in the Twenty-First Century: Challenges and Implications for Action, Food and Agriculture, and the Environment Discussion Paper 20, International Food Policy Research Institute, Washington D.C., U.S.A., March.

Viets, F.G., 1962. Fertilizers and the efficient use of water. Adv. Agron., 14: 233-264.

26

Alternative Use of Agro-based Industries Effluents for Crop Production

S. Bhaskar, M.S. Dinesha and C.A. Srinivasamurthy

In many arid and semi-arid countries water is becoming an increasingly scarce resource and planners are forced to consider any source of water which might be used economically and effectively to promote further development. At the same time, with population expanding at a high rate, the need for increased food production is apparent. The potential of irrigation to raise both agricultural productivity and the living standards of the rural poor has been widely recognized. Whenever good quality water is scarce, water of marginal quality will have to be considered for use in agriculture. Although there is no universal definition of 'marginal quality' water, for all practical purposes it can be defined as water that possesses certain characteristics which have the potential to cause problems when it is used for an intended purpose. For example, brackish water, agro-based industrial effluents are marginal quality water for agricultural use because of their high dissolved salt content and municipal wastewater is a marginal quality water because of the associated health hazards. From the viewpoint of irrigation, use of a 'marginal' quality water requires more complex management practices and more stringent monitoring procedures than when good quality water is used.

Effluent use in agriculture is much more commensurating than many believe. At present, approximately 20 million hectares of arable land worldwide are reported to be irrigated with effluent. The unreported use of wastewater in agriculture can be expected to be significantly higher. It is particularly common in urban and peri-urban areas of the developing world, where insufficient financial resources and institutional capacities constrain the installment and operation of adequate facilities for proper wastewater collection and treatment. Effluent use in agriculture has certain benefits by providing water and nutrients for the cultivation of crops, ensuring food supply to cities and reducing the pressure on available fresh water resources. India is the seventh largest nation in terms of industries. Majority of the industries are agro-based and utilize large volumes of good quality water and other raw materials and generate almost entire quantity of water as effluent and appreciable quantities of solid wastes. Since industrialization and pollution are complementary to each other, measures have to be adopted to render the pollutants less harmful to biosphere. These wastes can be profitably made use of in agriculture after subjecting the wastes to preliminary treatments.

1. Utilization of Coffee Pulp Effluent for Banana (*Musa paradisiaca* L.) Production

It has been estimated that about 75,000 to 80,000 litres of waste water is generated for curing one ton coffee beans and to process 2.23 lakh tons of coffee through wet processing, 8.4 million cubic metres of waste water is generated. The by-products of coffee processing are mainly coffee pulp, pulp effluent, parchment husks and coffee husk.

Properties of Coffee Pulp Effluent Used for Irrigation

The values presented in Table 28.1 indicate that the raw and treated effluents differ in chemical properties and hence it is always advisable to use treated effluent. The pH of raw effluent (3.94) and microbial treated effluent (4.27) was found to be acidic in nature, whereas lime treated effluent (7.16) and microbial and lime treated effluent (7.59) were neutral in reaction. It is clear that lime treatment at the rate of 0.5 per cent is helpful in neutralizing the pH of the effluent. Further, microbial treatment with *Pleuratus* has added advantage in increasing the pH. The electrical conductivity ranged from 1.091 to1.366 dSm^{-1}. Higher amount total solids (suspended and dissolved solids) are found in raw effluent (16.108 g l^{-1}) followed by lime treated effluent (12.338 g l^{-1}), microbial treated effluent (9.466 g l^{-1}) and microbial and lime treated effluent (8.403 g l^{-1}). Again both lime and microbial treatment caused reduction in total solids.

The concentration of BOD and COD varies with the treatment of effluent. Raw effluents always have maximum values (16500 and 27700 mg l^{-1}, respectively) and microbial and lime treated effluents (7800 and 14900 mg l^{-1}, respectively) have minimum values. There is no much difference with respect to the major and micronutrient contents of treated and raw effluent. As the nutrient concentrations are not sufficient for supplementing large quantity particularly the major nutrients, the effluent can be used as irrigation water during pulping season and coffee growing areas during the pulping season do not receive any rainfall and pulping units will be in need of water in large quantity for pulping purpose.

Table 26.1: Chemical Composition of Raw and Treated Coffee Pulp Effluent

Parameters	Raw Effluent	Microbial Treated Effluent	Lime Treated Effluent	Microbial and Lime Treated Effluent
pH	3.94	4.27	7.16	7.59
EC (dSm^{-1})	1.366	1.091	1.343	1.112
Suspended solids (g l^{-1})	7.843	4.512	5.766	3.614
Dissolved solids (g l^{-1})	8.265	4.954	6.572	4.789
Total solids (g l^{-1})	16.108	9.466	12.338	8.403
BOD (mg l^{-1})	16500	10200	13600	7800
COD (mg l^{-1})	27700	20400	24200	14900
Chlorides (meq l^{-1})	5.84	5.21	4.63	5.42
Bicarbonates (meq l^{-1})	6.72	7.04	6.37	6.82
Total nitrogen (per cent)	0.105	0.094	0.099	0.112
Total phosphorus (per cent)	0.0023	0.0028	0.0037	0.0032
Total potassium (per cent)	0.058	0.0583	0.0613	0.0501
Iron (ppm)	24.49	25.02	23.17	23.33
Zinc (ppm)	0.696	0.762	0.667	0.621
Copper (ppm)	1.793	1.833	2.162	1.810
Manganese (ppm)	0.586	0.531	0.494	0.511

Response of Banana (*Musa paradisiaca* L.) and other Crops to Coffee Pulp Effluent Irrigation

The results of field experiments conducted on farmers fields have clearly indicated that alternate irrigation with lime treated coffee pulp effluent and fresh water with microbial culture provided higher bunch yield (75.1 t ha^{-1}) in banana plant crop (Table 26.2). The greater bunch yields in those studies were mainly attributed to improvement in yield parameters in the same treatment and it is interesting to note the yields are on par with RDF with fresh water irrigation suggesting that treated effluent could be used safely for irrigating crop like banana during pulping season in coffee growing areas. Besides banana other short duration crops like baby corn, fodder maize and any perennial fodder grass can be raised and irrigated with treated coffee effluent.

Influence of Coffee Effluent Irrigation on Soil Properties

Low soil pH values are seen with raw coffee pulp effluent irrigation without microbial culture (5.86) when compared to the application treated effluent with lime and microbial culture. Hence it is imperative that only treated effluent can be utilized for irrigation purpose. Application of both raw and lime treated coffee pulp effluent with microbial culture increases electrical conductivity (0.463dSm^{-1}) and organic carbon content (0.83 per cent). In contrast to this, lower soil electrical conductivity and organic carbon content occur in fresh water irrigation (0.186 dSm^{-1} and 0.51 per cent). Addition of organic matter through effluent aided in increasing organic carbon while lime added to neutralize the pH contributes for higher salt content in soils. However, the EC is found to be well within the limits as far as crop growth is concerned. Application of raw or treated coffee pulp effluent with or without microbial culture will have higher amount of soil available nitrogen (258.9 to 303.8 kg/ha) and potassium (336.5 to 402.1 kg ha^{-1}) when compared to fresh water irrigation (216.3 and 302.0 kg ha^{-1}) which recorded lower soil available nitrogen and potassium. With respect to P, again alternate irrigation with lime treated coffee pulp effluent and fresh water with microbial culture contributes for maximum soil available phosphorus (52.8 kg ha^{-1}) when compared to lower soil available phosphorus with fresh water irrigation (38.6 kg ha^{-1}).

Table 26.2: Effect of Coffee Pulp Effluent Irrigation and Microbial Culture on Bunch Yield of Banana Plant Crop Soil Properties after Harvest of Banana Crop

Treatments	Bunch Yield (t ha^{-1})	pH	EC (dS m^{-1})	OC (Per cent)	Avail. N (kg ha^{-1})	Avail. P$_2$O$_5$ (kg ha^{-1})	Avail. K$_2$O (kg ha^{-1})
Fresh water irrigation	70.7	6.25	0.186	0.51	216.3	38.6	302.0
Raw CPE irrigation without microbial culture	38.6	5.86	0.430	0.83	303.8	41.7	402.1
Raw CPE irrigation with microbial culture	51.3	6.04	0.463	0.78	287.2	44.9	381.5
Lime treated CPE irrigation without microbial culture	61.8	6.51	0.429	0.76	285.0	42.7	394.7
Lime treated CPE irrigation with microbial culture	66.6	6.78	0.394	0.74	287.6	48.2	385.0
Alternate irrigation with lime treated CPE and fresh water without microbial culture	71.0	7.12	0.303	0.70	262.3	49.2	343.6
Alternate irrigation with lime treated CPE and fresh water with microbial culture	75.1	7.07	0.305	0.68	258.9	52.8	336.5
1:1 ratio irrigation with lime treated CPE and fresh water without microbial culture	70.1	7.13	0.283	0.66	264.4	48.4	342.1
1:1 ratio irrigation with lime treated CPE and fresh water with microbial culture	70.5	7.19	0.316	0.61	260.4	50.5	335.5
S. Em±	2.44	0.15	0.028	0.04	11.58	2.39	12.73
CD at 5 per cent	7.30	0.44	0.084	0.12	34.72	7.17	**38.17**

Note: Recommended dose of fertilizer and FYM is common for all the treatments.

2. Utilization of Distillery Effluent in Agriculture

Molasses based distilleries are considered to be one of the most polluting agro-based industries due to generation of large amount of foul smelling, brown coloured waste water also called as spent wash with very high BOD and COD levels. However, it is very rich in plant nutrients containing not only major ones but also all the micronutrients in appreciable amounts. These nutrients taken up by sugarcane and in organic form if recycled could save a lot on expenditure of fertilizers. Approximately 40 million m^3 of distillery spentwash are discharged annually from 285 alcohol distilleries in India which can be productively used as a nutrient source for crop production. 1 m^3 of spentwash contains - 1.0 kg N, 0.02 kg P_2O_5 and 10.0 kg K_2O. In addition, it supplies around 2 kg Ca, 1.5 kg Mg and 1.5 kg S. Hence, it been regarded as liquid fertilizer by sugar and alcohol industries. Spent wash is nothing but waste water generated by distilleries during the distillation of fermented molasses to ethyl alcohol using specific strains of yeast. It is a dark brown coloured liquid containing residual nutrients from sugarcane and yeast cells, it does not contain any heavy metals or other toxic residues. Distillery spentwash is categorized into raw spentwash that comes out from the distillation unit which is reddish brown in colour characterized by low pH, high BOD and COD values. The spentwash that leaves bio-methanation plant after the anaerobic digestion is referred to as primary spentwash which is dark brown in colour with neutral to alkaline pH and has relatively lower BOD and COD values than the raw spentwash. The distillery spentwash is also a source of valuable plant nutrients such as N, K, Ca, Mg, S and micronutrients. Because of its non-toxic nature and containing plant nutrients, it can be considered as a liquid fertilizer.

Use of Spent Wash through Drip Fertigation in Banana

Drip fertigation with water soluble fertilizers always provides higher number of hands (10.9 bunch^{-1}) mean bunch weight (31.4 kg) in plant crop over surface irrigation with application of recommended dose of fertilizers to soil (9.4 bunch^{-1} and 27.0 kg, respectively) and surface fertigation with normal fertilizers (9.4 bunch^{-1} and 27.7 kg, respectively). This contributes for greater total bunch yield in drip fertigation with water soluble fertilizers (78.6 t ha^{-1}). The studies show that surface fertigation with water soluble fertilizers (73.7 t ha^{-1}), drip fertigation with distillery spentwash (73.4 t ha^{-1}), 73.0 t ha^{-1} each with surface fertigation with distillery spent wash and drip fertigation with normal fertilizers could also provide on par yields both in plant and in ratoon crop (Table 26.3). From these findings, one can infer that treated and filtered spent wash can be safely used for fertigation through drip system hence it is possible to save exorbitant cost being incurred on fertilizers.

One time land application of distillery effluent is beneficial in dry land where it is advocated to be applied once in 2 or 3 years and such application would bring down the cost on fertilizers without causing any decline in yields. This is really a boon for dry land farmers who are resource poor farmers and not in position to go for balanced application of nutrients. Since distillery spent wash is known to contain all the 16 essential nutrients, the fertility of nutrient hungry soils can be maintained. This would also be handy for distilleries in safe utilization of effluent without causing pollution of water and/or land resources.

Table 26.3: Number of Hands per Bunch, Mean Bunch Weight and Bunch Yield of Banana as Influenced by Distillery Spentwash Fertigation Method at Harvest

Treatments	Main Crop			Ratoon Crop		
	Number of Hands per Bunch	Mean Bunch Weight (kg)	Total Bunch Yield (t ha⁻¹)	Number of Hands per Bunch	Mean Bunch Weight (kg)	Total Bunch Yield (t ha⁻¹)
Surface fertigation with normal fertilizers	9.9	27.7	69.2	9.2	25.8	64.4
Surface fertigation with water soluble fertilizers	10.8	29.5	73.7	10.0	27.4	68.5
Surface fertigation with distillery spentwash	10.6	29.2	73.0	9.9	27.2	68.1
Drip fertigation with normal fertilizers	10.5	29.2	73.0	9.9	27.2	67.9
Drip fertigation with water soluble fertilizers	10.9	31.4	78.6	10.2	29.0	72.6
Drip fertigation with distillery spentwash	10.6	29.3	73.4	10.0	27.3	68.3
Surface irrigation with application of recommended dose of fertilizers to soil	9.4	27.0	67.6	8.7	25.1	62.7
C.D. (P=0.05)	0.81	2.26	5.65	0.66	1.89	**4.71**

Bio-compost : 15 t ha⁻¹ or 6 kg plant⁻¹ was common for all the treatments,

Recommended dose of fertilizer: Main crop: 405:245:507 kg N, P_2O_5 and K_2O ha⁻¹,

Ratoon crop: 250:270:500 kg N, P_2O_5 and K_2O ha⁻¹

Influence of Distillery Spent Wash on Soil Parameters

Soil pH does not differ due to fertigation with distillery spent wash after harvest of plant and ratoon crop of banana (Figure 26.1). Higher EC and OC are noticed with application of drip fertigation with distillery spent wash (T₆: 0.63 dSm⁻¹ and 0.63 per cent) when compared to lower EC and OC in surface irrigation with application of recommended dose of fertilizers to soil (0.54 dSm⁻¹ and 0.57 per cent, respectively) but it was on par with rest of the treatments after the harvest of plant and ratoon crop.

Although application of distillery spent wash is known to increase EC slightly but the raise is well within the critical limits from the point of plant growth. While the improvement in organic carbon content is desirable as it adds to soil organic carbon.

Available Nitrogen, Phosphorus and Potassium in Soil

Higher available nitrogen and P_2O_5 content in soil after harvest of banana has been reported in drip fertigation with distillery spent wash (421.0 and 42.8 kg ha⁻¹)

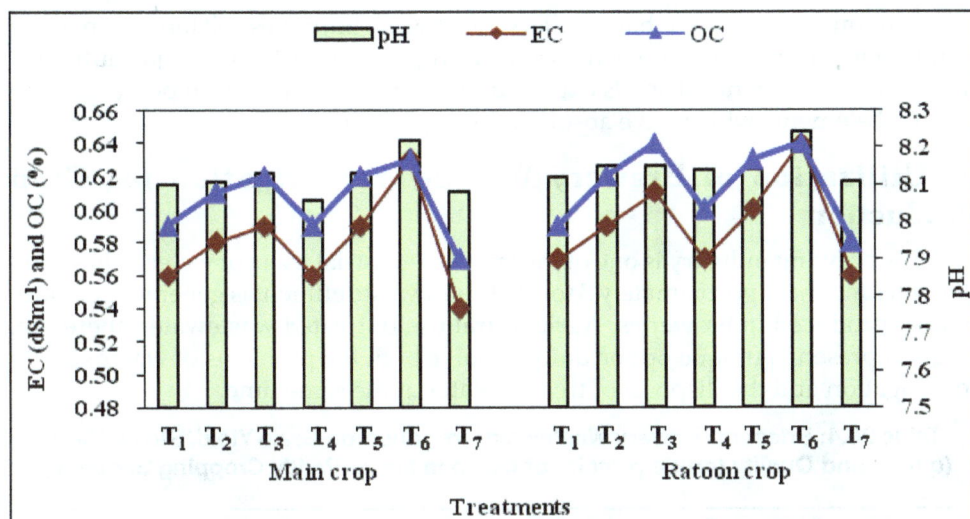

Figure 26.1: Soil pH, EC and OC in Soil after Harvest of Crop as Influenced by Fertigation Methods with Distillery Spent Wash Treatments.

compared to lower available nitrogen in soil with surface irrigation with application of recommended dose of fertilizers to soil (387.7 and 38.6 kg ha^{-1}, respectively) after plant and ratoon crop (Figure 26.2). Higher content of available potassium is seen in surface fertigation with distillery spent wash (649.0 kg ha^{-1}). Application of surface irrigation with application of recommended dose of fertilizers is not beneficial in improving available potassium in soil (280.6 kg ha^{-1}) both in plant and ratoon crops.

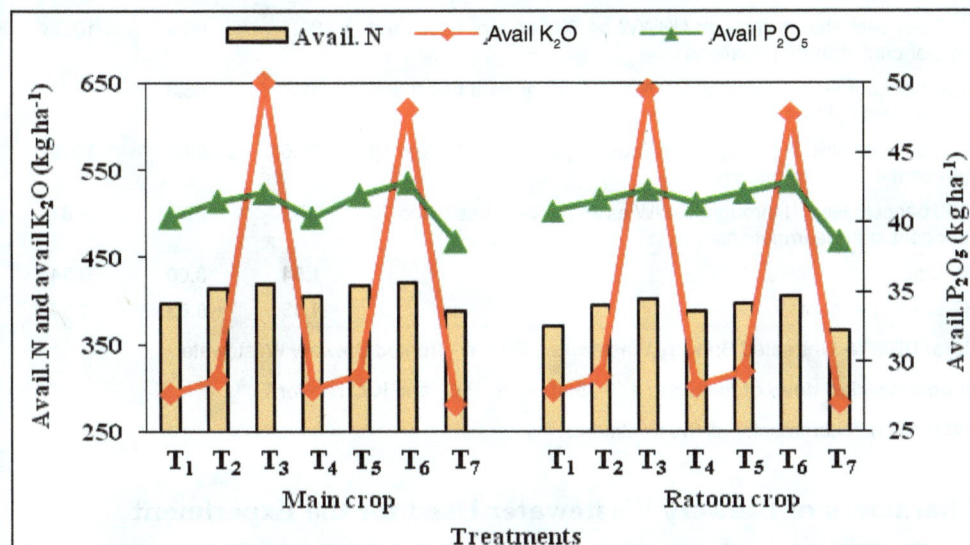

Figure 26.2: Available Nitrogen, Phosphorus and Potassium (kg ha^{-1}) in Soil after Harvest of Crop as Influenced by Fertigation Methods with Distillery Spentwash Treatments.

These findings clearly hint that distillery spent wash improves soil fertility in terms of nitrogen and potassium when used as fertigation. Apart from major nutrients, micronutrient contents of soil is expected to increase as spent wash possess all the essential elements which have absorbed by sugarcane.

3. Utilization of Brewery Wastewater Irrigation for Crop Production

The brewing industry is one of the largest industrial users of water. It has been documented that approximately 3 to 10 litres of waste effluent is generated per litre of beer produced in breweries. Agricultural use of treated wastewater, therefore, might represent a unique opportunity to solve both the problems of water supply for irrigation and the disposal of treated water at the same time.

Table 26.4: Effect of Brewery Wastewater Irrigation on Grain Yield, Stover Yield (q ha⁻¹) and Quality (crude protein) of Maize in Maize-Paddy Cropping Sequence

Treatments	Grain Yield	Stover Yield	CP (per cent)
Fresh water + RDF	30.5	51.0	6.99
Rec. N through UBWW as 50 per cent basal and 50 per cent in three irrigations	32.3	53.5	7.70
Rec. N through UBWW as 25 per cent basal and 75 per cent in three irrigations	30.7	52.1	7.11
Rec. N through TBWW as 50 per cent basal and 50 per cent in three irrigations	32.5	55.2	7.89
Rec. N through TBWW as 25 per cent basal and 75 per cent in three irrigations	31.2	54.6	7.03
150 per cent Rec. N through UBWW as 50 per cent basal and 50 per cent in three irrigations	39.1	69.2	10.38
150 per cent Rec. N through UBWW as 25 per cent basal and 75 per cent in three irrigations	37.8	66.3	9.71
150 per cent Rec. N through TBWW as 50 per cent basal and 50 per cent in three irrigations	39.6	69.9	10.62
150 per cent Rec. N through TBWW as 25 per cent basal and 75 per cent in three irrigations	38.2	67.6	9.81
S. Em±	**1.54**	**3.00**	0.54
C. D. at 5 per cent	**4.43**	**8.63**	1.54

Note: UBWW- Untreated Brewery Wastewater, **TBWW-** Treated Brewery Wastewater.

Recommended dose of fertilizer: 150:75:40 kg N, P_2O_5 and K_2O ha⁻¹ for T_1

FYM: 10 t ha⁻¹ commonfor all the treatments.

Characters of Brewery Wastewater Used for the Experiment

The pH of treated brewery waste water is slightly alkaline (7.34) and untreated is slightly acidic (6.34). Electrical Conductivity of treated wastewater (2.83 dSm⁻¹) and untreated wastewater is relatively high (3.72 dSm⁻¹). Treated wastewater will have higher total nitrogen (0.27 per cent) compared to untreated (0.19 per cent) sample. The P content ranges from 0.002 to 0.004 per cent and K from 0.035 to 0.05 per cent.

The COD of treated and untreated brewery wastewater was 1450 and 2500 (mg l⁻¹), respectively. Chloride contents of treated and untreated brewery wastewater are 52.71 and 61.72 mg l⁻¹, respectively. The micronutrients content of the treated and untreated brewery wastewater was low.

Response of Maize to Brewery Wastewater Irrigation in Cereal - Cereal Cropping Sequence

Application of 150 per cent recommended N through TBWW (50 per cent as basal and 50 per cent in three irrigations) is beneficial in giving higher grain yields of maize (39.6 q ha⁻¹) compared to lower grain yield (30.5 q ha⁻¹) in plots receiving fresh water + RDF (Table 26.4). Results have amply showed that there was no deleterious effect of either treated or untreated effluent on yielding potentials of both grain and stover. Further because of N content effluent, the crude protein content of grain also improves.

Table 26.5: pH, EC (dSm⁻¹), available Nitrogen, Phosphorus and Potassium (kg ha⁻¹) in Soil after Harvest of Maize as Influenced by Brewery Wastewater Irrigation of Maize on Maize-Paddy Cropping Sequence (mean of two years)

Treatments	pH	EC	Nitrogen	Phosphorus	Potassium
Fresh water + RDF	7.38	0.18	168.7	21.8	144.7
Rec. N through UBWW as 50 per cent basal and 50 per cent in three irrigations	6.44	0.17	177.2	27.8	146.7
Rec. N through UBWW as 25 per cent basal and 75 per cent in three irrigations	6.40	0.15	178.6	28.4	147.6
Rec. N through TBWW as 50 per cent basal and 50 per cent in three irrigations	7.41	0.22	177.3	27.0	146.5
Rec. N through TBWW as 25 per cent basal and 75 per cent in three irrigations	7.40	0.22	178.1	28.4	147.4
150 per cent Rec. N through UBWW as 50 per cent basal and 50 per cent in three irrigations	6.38	0.19	184.0	33.3	151.9
150 per cent Rec. N through UBWW as 25 per cent basal and 75 per cent in three irrigations	6.39	0.18	186.3	35.0	152.3
150 per cent Rec. N through TBWW as 50 per cent basal and 50 per cent in three irrigations	7.37	0.24	183.6	33.3	151.1
150 per cent Rec. N through TBWW as 25 per cent basal and 75 per cent in three irrigations	7.38	0.24	185.4	33.9	152.3
S. Em±	0.06	0.03	1.55	1.57	1.08
C. D. at 5 per cent	0.17	0.09	4.47	4.51	**3.11**

Influence on Soil Properties

When untreated effluent is used, the soil reaction and salt content are below the neutral range (Table 26.5) due to its acidic nature. While treated effluent being alkaline in nature can slightly enhance the pH and salt content and the increase in these parameters could be comparable to the plots applied with RDF alongwith fresh water. Plots receiving150 per cent recommended N through UBWW as 25 per cent basal and 75 per cent in three irrigations contributes for higher available nitrogen, P_2O_5 and K_2O (186.3, 35.0 and 152.3 kg ha^{-1}, respectively). By using either treated or untreated effluent, the nutrient status could be increased thus there is saving on cost of fertilizers.

Conclusion

All the effluents used for crop production in general and in banana and maize in particular have showed that there is a great potential to recycle these liquid wastes for production purposes in agriculture. Besides improving the productivity, the effluents are also beneficial in preventing environmental pollution without affecting the soil health. Even the microbial analysis carried out in these studies have clearly shown that the beneficial micro flora was in fact either unaffected or some cases improved their status. Economically the use of effluents would bring down cost of production to be incurred on costly input like fertilizers.

27

Irrigation Scheduling and Techniques in Fruit Crops

N. Jagadeesha and L. Madhu

Banana in India is known as "Instant energy provider", is the second most important fruit crop in India next to mango. Its year round availability, affordability, varietal range, taste, nutritive and medicinal value make it the favorite fruit among all classes of people. Banana requires large quantity of water during its life cycle. Water is becoming increasingly scarce worldwide due to various reasons. The problem of water scarcity is expected to be aggregated further. In spite of having the largest irrigated area in the world, India too has started facing sever water scarcity in different regions. Due to various reasons the demand for water for different purposes has been continuously increasing in India, but the probable water available for future use has been waning at a faster rate (Saleth, 1996). The agricultural sector (irrigation), which currently consumes larger portion of the available water in India, continues to be one of the major water-consuming sectors due to the intensification of agriculture. Banana (*Musa* sp.) requires constant water supply throughout their growth, hence water management is an important technique for efficient utilization of available water for maximum production.

Irrigation Scheduling in Banana

☆ For the first 6 months use drip and basin irrigation.

☆ Put enough water in each basin (25 liters of water every 2-3 days).

☆ The amount of water can be decreased or increased depending on the weather conditions and the soil type.

☆ Irrigation should be given immediately after planting and life irrigation after 3-4 days; subsequently irrigations are to be given once in a week for irrigated plantations of garden lands and once in 10-15 days for wet lands.

☆ The field should be sufficiently irrigated after every manuring.

☆ After 6 months some other method of irrigation like drip irrigation so that the whole ground is watered, can be adopted.

☆ Irrigating for 6 to 8 hours every three to four days would be sufficient depending on weather conditions.

☆ If the use of the hose is to be continued, then watering should no longer be restricted to the basin because the roots would have spread beyond the basin.

☆ Flooding can be used on flat land.

☆ However, flooding on slopes is discouraged because it can cause soil erosion.

☆ Drip irrigation can be advantageous to reduce water loss in conveyance and improving efficiency. Drip systems are especially useful in salt affected soil. Nearly 40 per cent savings in water is noticed when drip systems are used.

Drip Irrigation System

In Banana, micro-irrigation is introduced primarily to save water and increase the water use efficiency in agriculture. However, it also delivers many other economic and social benefits to the society. Drip irrigation refers to frequent application of small quantities of water on or below the soil surface as drops. It embodies the philosophy of irrigating the root zone instead of entire land. Banana being a succulent, evergreen and shallow rooted crop requires large quantity of water of increasing productivity. Water requirement of banana has been worked out to be 1800-2000 mm per annum. In winter, Irrigation is provided at an interval of 7-8 days while in summer it should be given at an interval of 4-5 days. However, during rainy season irrigation is provided if required as excess irrigation will lead to root zone congestion due to removal of air from soil pores, thereby affecting plant establishment and growth. In all, about 70-75 irrigations are provided to the crop. Application of drip irrigation and mulching technology has reported to improve water use efficiency. There is saving of 58 per cent of water and increase yield by 23-32 per cent under drip. Raw bunch gets matured earlier by 30-45 days and yield is increased by 15-30 per cent and 58-60 per cent of water is saved on.

Concerned Problems in Adoption of Drip Irrigation

☆ Irregular and insufficient availability of electricity

☆ Low price of canal water

☆ Fear about drip system clogging

☆ Water availability by drip system is not satisfactory

☆ Services availability of drip system

☆ Unaffordable establishment cost for the drip system

☆ Insufficient awareness of its benefit

☆ Lack of training and extension facilities for farmers.

Suggestions and Recommendation

☆ Training facilities for farmers are essential to increase the adoption of drip irrigation. All registered suppliers under Govt. schemes should provide facilities for training of farmers in operation and maintenance of the system.

☆ It is observed that farmers are very well aware with the benefits of drip irrigation system but they are hesitant to adopt this technology because they are not having proper information. Benefit –cost ratio should be demonstrated to farmers.

☆ It is observed that banana is being irrigated by flooding water although the drip system is installed in the field. The practice of drip irrigation should be encouraged in the banana fields. Since the farmers are getting water for the low cost from the public irrigation system, they are least interested to adopt this technology.

Pomegranate

The pomegranate is native from Iran to the Himalayas in northern India and has been cultivated and naturalized over the Mediterranean region and the Caucasus region of Asia since ancient times. Pomegranate adapts to all kinds of soil and climate; it is tolerant of drought, salt, iron chlorosis and active calcium carbonate. It is widely cultivated throughout Iran, India, the drier parts of south East Asia, Malaya, the East Indies, and dry, hot areas of the United States and Latin America. It typically grows below 1000 m in altitude, is mainly confined to the tropics and subtropics and grows well in arid and semi-arid climates. Favorable growth takes place where winters are cool and summers are hot. It has the ability to withstand frosty conditions, but below $-10°C$ will not survive long. A temperature of $38°C$ and a dry climate during fruit development produces the best quality fruits. Areas with high relative humidity or rain are totally unsuitable for its cultivation, as fruits produced under such conditions tend to taste less sweet and are prone to cracking.

Pomegranate is grown mainly in arid or semiarid regions with limited irrigated water sources; hence water management is an important technique for efficient utilization of available water for maximum production. The water requirement of pomegranate crop depends on age, season, location and management strategies.

A. Water Management Techniques

Covering the soil/ground plant canopy with inorganic or organic mulches during dry months after the rainy season conserves soil moisture and saves irrigation water, creates favorable conditions for plant growth, development and efficient crop production, prevents the direct evaporation of moisture from the soil. Increases soil temperature during winters, reduces soil compaction and fruit production is earlier. Organic mulches in addition increase water and nutrient retention capacity, improve soil oxygen, root growth and also supply nutrients on decomposition.

1. Types of Drip Irrigation System

☆ **Sub – Surface drip irrigation:** The drippers and lateral lines lay below the ground level plant root zone.

☆ **Surface drip irrigation:** The drippers and the lateral are laid on the soil surface. Based on the types of laterals or the emitting devices used, the drip irrigation systems can be classified as:

☆ **On line drip irrigation system:** The drippers or emitters are fixed on the lateral pipes by punching suitable holes on the drip lateral pipes at location specific to the crop being irrigated. Single dripper 15 cm away plant may be used till 6-12 months, 2 drippers till 2-3 years and 4-6 drippers on 2 laterals after 3rd year depending on plant height and spread

☆ **In line drip irrigation system:** The drippers are factory installed within or on the drip lateral at regular intervals. This is suitable for close spaced crop.

2. Water Requirement

The water requirement of Pomegranate in different stages and seasons is give in Tables 27.1–27.3.

3. Precautions for Trouble Free Irrigation

☆ Additional flushing must be conducted at the end of the irrigation season

☆ Flushing the manifold and mainline ends flushing laterals to remove sediments that accumulates at the drip lateral ends

☆ Clean filters at regular intervals based on the water quality and content

☆ Check pressure at each of the system's stations, head, valves, laterals, beginning and end

☆ Checking the laterals flow random drippers

☆ Leaks can occurs unexpectedly as results of insects, animals or farming tools. So, monitor lines for any physical damage regularly.

Easy Way to know Water Requirement of your Crop

Water requirement depends on several factors-climates and whether, soil type organic content, plant age, size, plant stage and load, hence, will vary from orchard to orchard. Therefore, best and easy way to standardize water for your orchard is:

1. Irrigation the crop say for 1 hour.

Table 27.1: Water to be Applied (lit/day) for One Five Year Old Pomegranate Tree during Mirg bhar

Month	MW	1st	2nd	3rd	4th	5th	Months	MW	1st	2nd	3rd	4th	5th
June	23	3	5	6	8	11	2nd week of September to October	39	2	9	16	25	31
	24	3	5	6	8	13		40	2	8	15	24	30
	25	3	5	7	11	16		41	2	8	16	24	30
	26	3	5	8	13	17		42	3	9	16	25	32
	27	3	6	9	15	20		43	3	9	18	27	34
								44	3	9	17	26	33
July to 1st week of September	28	3	6	10	16	22	November to 2nd week of January	45	3	9	17	26	33
	29	2	7	11	18	23		46	3	8	16	25	32
	30	2	7	10	17	23		47	3	8	15	24	30
	31	2	7	12	19	25		48	3	7	15	23	29
	32	2	8	13	21	27		49	3	7	14	23	29
	33	2	8	14	22	29		50	3	6	13	21	27
	34	2	8	14	23	29		51	3	6	13	21	26
	35	2	8	15	25	31		52	3	6	12	20	25
	36	2	9	16	25	32		01	3	5	12	19	24
	37	3	9	17	26	33		02	3	6	12	20	26
	38	2	8	16	24	31								

Table 27.2: Water to be Applied (lit/day) for One to Five Years Old Pomegranate Tree during Hasta Bahar

Months	MW	1st	2nd	3rd	4th	5th	Months	MW	1st	2nd	3rd	4th	5th
September	36	2	3	4	5	6	4th week of December to 2nd week of February	52	2	8	16	24	30
	37	2	3	4	6	9		01	2	8	15	23	29
	38	2	4	5	8	11		02	2	8	16	24	30
	39	2	4	6	10	13		03	3	9	16	25	31
	40	2	4	7	11	15		04	3	10	18	28	35
	41	2	5	8	13	18		05	3	10	19	29	37
1st week of October to 3rd week of December	42	2	6	9	15	20		06	3	10	20	31	39
	43	2	7	11	18	24		07	4	11	21	33	41
	44	2	7	12	19	25	3rd week of February to march	08	4	11	23	35	44
	45	2	8	13	21	28		09	4	12	24	37	47
	46	2	8	14	22	29		10	5	12	24	39	49
	47	2	8	14	23	30		11	5	12	25	39	49
	48	2	8	15	24	31		12	6	12	25	41	52
	49	2	9	16	25	32		13	6	12	25	41	52
	50	2	8	16	24	30		14	6	12	25	41	52
	51	2	8	16	24	30		15	6	12	26	43	54

Table 27.3: Water Applied (l/day) for One to Five Years Old Pomegranate Tree during Hasta Bahar

Months	MW	1st	2nd	3rd	4th	5th	Months	MW	1st	2nd	3rd	4th	5th
January	1	2	3	3	5	6	3rd week of April to 2nd week of june	17	5	19	35	54	68
	2	2	3	4	5	8		18	6	20	36	56	70
	3	2	4	5	8	11		19	6	19	36	56	70
	4	2	5	7	11	15		20	6	19	36	56	70
	5	2	6	8	13	18		21	6	19	35	54	67
February to March	6	3	7	10	17	23		22	5	18	33	51	65
	7	3	8	12	20	27		23	4	14	27	42	53
	8	3	9	15	25	33		24	4	12	24	37	47
	9	4	11	18	30	39	3rd week of June to 1st week of August	25	4	11	22	33	42
	10	4	12	21	34	45		26	4	10	19	30	38
	11	4	13	24	38	50		27	4	9	18	29	37
	12	4	15	28	44	57		28	4	8	17	27	34
	13	5	16	30	48	60		29	3	7	15	25	32
	14	5	17	32	49	62		30	3	6	13	22	27
	15	5	18	33	51	64		31	3	6	13	21	27
	16	5	19	35	54	68		32	3	6	13	21	27

2. Next day soil from this region in your fist and close your fist to compress the soil.

   A. If it remains loose not compress to form mould : water is deficient: irrigate the crop

   B. If forms a mould, throw it on the ground:

      a. If it loosen after falling- water is perfect; irrigate next day after checking

      b. If it remains in mould without much dispersing-water is excess: No irrigation is required, check regularly and irrigate accordingly.

Estimation of Crop Water Requirement

The daily water requirement for fully grown plants can be calculated as under

$V= Ep \times Kc \times Kp \times Wp \times Sp$

where,

   $V =$ Water requirement (litres/day/plant)

   $Ep =$ Evaporation data for open pan evaporimeter (mm/day)

   $Kc =$ Crop Coefficient

   $Kp =$ Pan coefficient

   $Wp =$ Wetted area (0.3 for fruit crops)

   $Sp =$ Spacing (m x m)

   $Re =$ Effective rainfall (mm)

   $A =$ Area of the plot (m²)

Estimation of Horse Power of Pumping Unit

Flow carried by each lateral line = discharge of dripper x No. of drippers per plant x No. of Plants along each lateral.

Flow carried by each submain = Flow carried by each lateral line x No. of latteral line per sub main

Flow carried by each main line= Flow carried by each submain x No. of submains

The friction head loss in mains can be estimated by Hazen-Williams formula

$Hf= 10.68 \times (Q/C) 1.852 \times D 4.87 \times (L + Le)$

where,

   $Hf =$ Friction head loss in pipe (m)

   $Q =$ Discharge (m³ sec)

   $C =$ Hazen-Wiliam constant (140 for PVC pipe)

D = Inner diameter of pipe (m)

L = Length of pipe (m)

Le = Equivalent length of pipe and accessories

Total pressure head drop in meters due to friction (Hf) is sum of fraction head loss of main, submains and laterals.

Operating pressure head required at the dripper= He

Total static head= Hs

Total pumping head= Hf + He + Hs

Discharge of main= dm (l/s)

Overall Efficiency= 60 per cent in the case of electric pump, 40 per cent in diesel pump

$$Hp = \frac{H \times dm}{75 \times e}$$

Table 27.4: Critical Period for Moisture in Various Fruit Crops

Crop	Sensitive Stage
Citrus	Flowering, fruit setting, fruit growth
Banana	Early vegetative period flowering and fruit formation
Mango	Start of fruiting to maturity
Pine apple	Vegetative growth
Grape	Vegetative growth. Frequent irrigation during vegetative stage may cause rotting of fruits
Guava	Period of fruit growth
Ber	A drought resistant plant ; irrigation is required during fruit growth

Table 27.5: Water Requirements of Fruit Crops

Crop	Water requirement (mm)
Banana	1200-2200
Citrus	900-1200
Grapes	500-1200
Pineapple	700-1000
Coconut	80-100 (lit/plant/day)
Mango	30-40 (lit/plant/day)
Guava	22-30 (lit/plant/day)
Banana in drip	8-12 (lit/plant/day)

Source: Precision Farming Development Centre, Division of Horticulture, UAS, Bangalore.

Worked Example for Setting Drip Fertigation Unit for Citrus Crop

1. No. of plants per hectare (100 m x 100 m)

Plants spacing 6 m x 6 m (36 m²)

= 10000/36 m²

= 277

2. Estimation of water requirement

Table 27.6: Normal Monthly Pan Evaporation Data (mm)

Month	Evaporation (mm)	Month	Evaporation (mm)
January	112.2	July	108.6
February	141.6	August	100.8
March	303.0	September	99.6
April	367.5	October	108.0
May	306.3	November	81.0
June	147.0	December	82.41

According to above data evaporation losses will be more in February, March, April, May and June months as a result water requirement will also be higher. Maximum daily evaporation losses during summer months were 9.89 mm with 355 litres per day per plant.

Pump rate per ha   = 98500 l/ha/day

If pump operate at 5 hrs per day

= 98500/5/60

= 5.47 lps

Alternatively a tank of 100 m³ capacity can be provided so that uninterrupted irrigation may continue for 5 hrs even in areas of power shut offs are frequent.

3. Selection of drippers

For a pressure head of 10 m and discharge at 5.47 l/hr number of drippers required is

No. of drippers per plant= Rate of pumping hr/plant

Average discharge of one dipper = 71/5.47/4

= 3.23 lph

4. Main line and laterals

It is showing in Table 27.7.

**Table 27.7: Water Saving and Productivity Gains under
Drip Method of Irrigation in Fruit Crops**

Crop	Water Consumption (mm/ha)		Yield (tons/ha)		Water Saving over FIM (per cent)	Yield Increase over FIM (per cent)	Water Use Efficiency (yield/ha)/(mm/ha)	
	FIM	DIM	FIM	DIM			FIM	DIM
Papaya	2285	734	13.0	23.0	68	77	0.006	0.031
Banana	1760	970	57.5	87.5	45	52	0.033	0.090
Grapes	532	278	26.4	32.5	48	23	0.050	0.117
Lemon	42	8	1.88	2.52	81	35	0.045	0.315
Watermelon	800	800	29.47	88.23	Nil	179	0.037	0.110
Mosambi*	1660	640	100.0	150.0	61	50	0.060	0.234
Pomegranate*	1440	785	55.0	109.0	45	98	0.038	0.139

Note: FIM- furrow irrigation; DIM- drip irrigation;* - yield in 1000 numbers:

Source: INCID (1994) and NCPA (1990).

28

Policy Issues for Protection of Weaker Sections of the Society in the Event of Water Use

D.M. Chandargi

The time when India got independence in 1947, major population was landless, illiterate, and poor livelihood. India adopted the Democratic form of government. The Constituent Assembly debates recognized that a section of people in Indian Society had been denied certain basic rights since ancient times and therefore remained economically, socially and educationally backward. As a result, this had created widespread disparities among various groups. This scenario of disparities leads to a situation that needs special measures to uplift the status of the marginalized and depressed groups. According to Government of India Act, 1935 "weaker section implies to those sections of society who are either because of traditional custom of practice of untouchability or because of tribal origin, tribal way of living or other backwardness have been suffering from educational and economic backwardness and some aspects of social life."

On the basis of their overall status in a view to their socio-economic and cultural life, for the convenience of effective administration of development and welfare programmes with special focus to their socio-economic stands, the weaker sections have been grouped basically into three distinct categories by different resolutions of Government of India. Those groups can be serially stated here on the basis of magnitude of their problems or vulnerability to various disabilities suffered by them.

1. Scheduled Caste
2. Scheduled Tribes and
3. Other Backward Classes

In addition to the above major groups, there are also some vulnerable groups like women, destitute children, and handicapped people *etc.* who are generally considered as weaker members of the society.

Strategies and Approaches for Planning and Development

1. Market Induced Plans
2. Trickledown Theory
3. Financial Assistance to Individuals
4. Promotion of Co-operatives
5. Encouraging Community Development
6. State Intervention at Community Level
7. Pro-Active Target Groups oriented Schemes.

Development Approach

1. Individual
2. Community Centered
3. Cluster approach
4. Saturation approach
5. Integration approach

Policies/Acts

1. Enforcing Equality and Removing Disability Untouchability Offences Act, 1955
2. Protection of Civil Rights Act, 1955
3. Bonded Labour System (Abolition), Act 1976
4. The Minimum Wages Act, 1948
5. Equal Remuneration Act, 1976
6. Child Labour (Prohibition and Regulation) Act, 1986
7. Land Reforms Laws
8. National Human Rights Commission
9. National Commission for Women

Schemes

1. 'Rajiv Gandhi National Fellowship':- for Providing Scholarships to Scheduled Caste Students to persue Programmes in Higher Education such as M.Phil and Ph.D (Effective from 01-04-2010)

2. Pilot Scheme of PRADHAN MANTRI ADARSH GRAM YOJANA (PMAGY):- State, District and Block wise abstract of villages selected under PMAGY

3. Babu Jagjivan Ram Chhatrawas Yojana (Letter, Annexures, and National Allocation for 2009-10)

4. Post-Matric Scholarship for SC Students

5. Pre-Matric Scholarships for the Children of those Engaged in Unclean Occupations

6. Central Sector Scholarship Scheme of Top Class Education for SC Students (Effective from June 2007)

7. Self Employment Scheme for Rehabilitation of Manual Scavengers

8. National Overseas Scholarships for Scheduled Castes (SC) *etc.* Candidates for Selection Year 2010-2011

9. Form for the Scheme of National Overseas Scholarship for SC *etc.* Candidates for the Selection Year 2010-11.

10. Special Educational Development Programme for Scheduled Castes Girls belonging to low Literacy Levels

11. Upgradation of Merit of SC Students

12. Scheme of free Coaching for SC and OBC Students

13. National Scheduled Castes Finance and Development Corporation (NSFDC)

14. National Safaikaramcharis Finance and Development Corporation (NSKFDC)

15. Assistance to Scheduled Castes Development Corporations (SCDCs)

16. National Comission for Safai Karamcharis

However, in the recent past and over the next decades, particularly those in developing countries, face shortages of water and food and greater risks to health and life as a result of climate change. Concerted global action is needed to enable developing countries to adapt to the effects of climate change that are happening now and will worsen in the future. Land and ecosystems are being degraded, threatening to undermine food security. In addition, water and air quality are deteriorating while continued increase in consumption and associated waste have contributed to the exponential growth in the region's existing environmental problems. Furthermore, the region is highly subject to natural hazards, such as the 2004 Indian Ocean Tsunami, There is evidence of prominent increases in the intensity and/or frequency of many extreme weather events such as heat waves, tropical cyclones, prolonged dry spells, intense rainfall, tornadoes, snow avalanches, thunderstorms, and severe dust storms in the region.

Impacts of such disasters range from hunger and susceptibility to disease, to loss of income and livelihoods, affecting human survival and well-being. Climate change will affect many sectors, including water resources, agriculture and food security, ecosystems and biodiversity, human health and coastal zones. Under climate change, crop yields are predicted to fall by up to 30 per cent, creating a very high risk of hunger in several countries.

Global warming is causing the melting of glaciers in the Himalayas. In the short term, this means increased risk of flooding, erosion, mudslides in north India during the wet season. Because the melting of snow coincides with the summer monsoon season, any intensification of the monsoon and/or increase in melting is likely to contribute to flood disasters in Himalayan catchments. In the longer term, global warming could lead to a rise in the snowline and disappearance of many glaciers causing serious impacts on the populations. The principal impacts of climate change on health will be on epidemics of malaria, dengue, and other vector-borne diseases.

Adaptation Strategies, Plans and Programmes

Adapting to climate change will entail adjustments and changes at every level – from community to national and international. Communities must build their resilience, including adopting appropriate technologies while making the most of traditional knowledge, and diversifying their livelihoods to cope with current and future climate stress. Local coping strategies and traditional knowledge need to be used in synergy with government and local interventions. The choice of adaptation interventions depends on national circumstances. To enable workable and effective adaptation measures, ministries and governments, as well as institutions and non-government organizations, must consider integrating climate change in their planning and budgeting in all levels of decision making.

One way of grouping adaptation options is to identify whether they are sectoral, crosssectoral or multi-sectoral.

Sectoral adaptation measures look at actions for individual sectors that could be affected by climate change. For example, in agriculture, reduced rainfall and higher evaporation may call for the extension of irrigation; and for coastal zones, sea level rise may necessitate improved coastal protection such as reforestation. Often adaptation measures in one sector will involve a strengthening of the policy that already exists, emphasizing the importance of including long term climate change considerations alongwith existing local coping mechanisms and integrating them into national development plans.

Multi-sectoral adaptation options relate to the management of natural resources which span sectors, for example, integrated management of water, river basins or coastal zones. Linking management measures for adaptation to climate change with management measures identified as necessary from the other Rio Conventions: the Convention on Biological Diversity and the United Nations Convention to Combat

Table 28.1: Regional Impacts and Vulnerabilities to Climate Change

Impacts	Sectoral Vulnerabilities	Adaptive Capacity
Temperature	**Water**	Adaptive capacity varies between countries depending on social structure, culture, economic capacity, geography and level of environmental degradation.
☆ Warming above the global mean in central Asia, the Tibetan Plateau, northern, eastern and southern Asia.	☆ Increasing water stress to over a hundred million people due to decrease of freshwater availability in Central, South, East and Southeast Asia, particularly in large river basins such as Changing.	Capacity is increasing in some parts of Asia, for example the success of early warning systems for extreme weather events in Bangladesh and the Philippines. However, capacity is still constrained due to poor resource bases, inequalities in income, weak institutions and limited technology.
☆ Warming similar to the global mean in Southeast Asia.		
☆ Fewer very cold days in East Asia and South Asia.	☆ Increase in the number and severity of glacial melt-related floods, slope destabilization followed by decrease in river flows as glaciers disappear.	
Precipitation, snow and ice	**Agriculture and food security**	Capacity is increasing in some parts of Asia, for example the success of early warning systems for extreme weather events in Bangladesh and the Philippines. However, capacity is still constrained due to poor resource bases, inequalities in income, weak institutions and limited technology.
☆ Increase in precipitation in most of Asia. Decrease in precipitation in central Asia in Summer.	☆ Decreases in crop yield for many parts of Asia putting many millions of people at risk from hunger.	
☆ Increase in the frequency of intense precipitation events in parts of South Asia, and in East Asia.	☆ Reduced soil moisture and evapotranspiration may increase land degradation and desertification.	
☆ Increasing reduction in snow and ice in Himalayan and Tibetan Plateau glaciers	☆ Agriculture may expand in productivity in northern areas.	
Extreme Events	**Health**	
Increasing frequency and intensity of extreme events particularly:	☆ Heat stress and changing patterns in the occurrence of disease vectors affecting health.	
☆ droughts during the summer months and El Niño events;	☆ Increases in endemic morbidity and mortality due to diarrhoeal disease in south and Southeast Asia.	
☆ increase in extreme rainfall and winds associated with tropical cyclones in East Asia, Southeast Asia and South Asia;		

Impacts	Sectoral Vulnerabilities	Adaptive Capacity
☆ intense rainfall events causing landslides and severe floods;	☆ Increase in the abundance and/or toxicity of cholera in south Asia.	
☆ heat waves/hot spells in summer of longer duration, more intense and more frequent, particularly in East Asia.	**Terrestrial Ecosystems**	
	☆ Increased risk of extinction for many species due to the synergistic effects of climate change and habitat fragmentation.	
	☆ Northward shift in the extent of boreal forest in north Asia, although likely increase in frequency and extent of forest fires could limit forest expansion.	
	Coastal Zones	
	☆ Tens of millions of people in low-lying coastal areas of south and Southeast Asia affected by sea level rise and an increase in the intensity of tropical cyclones.	
	☆ Coastal inundation is likely to seriously affect the aquaculture industry and infrastructure particularly in heavily-populated megadeltas.	
	☆ Stability of wetlands, mangroves, and coral reefs increasingly threatened.	

Table 28.2: Adaptation Measures in Key Vulnerable Sectors for Developing Countries

Vulnerable Sectors	Reactive Adaptation	Anticipatory Adaptation
Water Resources	Protection of groundwater resources	Better use of recycled water
	Improved management and maintenance of existing water supply systems	Conservation of water catchment areas
	Protection of water catchment areas	Improved system of water management
	Improved water supply	Water policy reform including pricing and irrigation policies
	Groundwater and rainwater harvesting and desalination	Development of flood controls and drought monitoring
Agriculture and food security	Erosion control	Development of tolerant/resistant crops (to drought, salt, insect/pests)
	Dam construction for irrigation	Research and development
	Changes in fertilizer use and application	Soil-water management
	Introduction of new crops	Diversification and intensification of food and plantation crops
	Soil fertility maintenance	Policy measures, tax incentives/subsidies, free market
	Changes in planting and harvesting times	Development of early warning systems
	Switch to different cultivars	
	Educational and outreach programmes on conservation and management of soil and water	
Human health	Public health management reform	Development of early warning system
	Improved housing and living conditions	Better and/or improved disease/vector surveillance and monitoring
	Improved emergency response	Improvement of environmental quality
		Changes in urban and housing design

Vulnerable Sectors	Reactive Adaptation	Anticipatory Adaptation
Terrestrial ecosystems	Improvement of management systems including control of deforestation, reforestation and afforestation	Creation of parks/reserves, protected areas and biodiversity corridors
	Promoting agroforestry to improve forest goods and services	Identification/development of species resistant to climate change
	Development/improvement of national forest fire management plans	Better assessment of the vulnerability of ecosystems
	Improvement of carbon storage in forests	Monitoring of species
		Development and maintenance of seed banks
		Including socioeconomic factors in management policy
Coastal zones and marine ecosystems	Protection of economic infrastructure	Integrated coastal zone management
	Public awareness to enhance protection of coastal and marine ecosystems	Better coastal planning and zoning
	Building sea walls and beach reinforcement	Development of legislation for coastal protection
	Protection and conservation of coral reefs, mangroves, sea grass and littoral vegetation	Research and monitoring of coasts and coastal ecosystems

Desertification; could be a useful multi-sectoral approach which addresses a range of environmental stresses.

Cross-sectoral measures also span several sectors and can include: improvements to systematic observation and communication systems; science, research and development and technological innovations such as the development of drought-resistant crop varieties or new technologies to combat saltwater intrusion; education and training to help build capacity among stakeholders; public awareness campaigns to improve stakeholder and public understanding on climate change and adaptation; strengthening or making changes in the fiscal sector such as new insurance options; and risk/disaster management measures such as emergency plans.

29

Machinery and Implements for Water Conservation

M. Anantachar, K.V. Praksh and Sushilendra

The climate change in different regions of Karnataka provided a way for different adoption of improved farm machinery/implements to suit the local conditions. The farm mechanisation has played a pivotal role in increasing the cropping intensity and timeliness of the operation. It has thereby resulted in increased productivity, efficient utilization of scarce resources, placement of expensive crop inputs by precision placement and thus overcoming shortage of labour during peak periods. Also the quantity of the operation improve and reduce loses during various post harvest operations. The farm mechanization has greatly helped to reduce drudgery involved in various agricultural operations.

In the present context, the global climatological changes caused the adoption of different technical inputs including the use of improved agricultural machinery for partial mechanization in agriculture. Karnataka state in different zones receives heavy rainfall and arid zones receive less rainfall (Mehata, 1989). The suitable equipment should be used based on climate, soil conditions and cropping intensity in order to mitigate the labour shortage which is a serious problem in agricultural sector. The mechanization is not just the introduction of the machines, but it is the use of right type of machines suitable under specific conditions and for given task. The most important factor which requires serious attention is the judicious judgement to identify the package of farm machinery to suite the farming situations based on the climate changing situations (Gajendra Singh, 1996).

The energy input in Indian Agriculture has increased largely resulting in increased tractor production from 50,000 per year by 1981 to 1,68,000 per year by 1995. The total number of tractors exceeded 1.7 million as against only 10,000 tractors available in the country in 1951. The farm electrification has brought tremendous increase in the use of machines like threshers and irrigation pump sets. In fact the irrigated areas has increased to 80 m ha from 22.6 m ha in recent past (Surendra Singh 2007). The adoption of suitable equipment under different climatic conditions in Karnataka based on rainfall situation and rainfed conditions has been suggested for carrying out farming operations to meet out partial mechanization.

I) Heavy Rainfall with High Infested Weed Condition

Under this condition, the tractor drawn rotavator is a promising solution to the farming community.

Tractor Drawn Rotavator

It is suitable for preparing seedbed in a single pass both in dry and wet land conditions. It also incorporates green manure crop in the field. It basically consists of steel frame, a rotary shaft on which blades are mounted, power transmission system and gear box. The blades are of L type made from medium carbon steel or alloy steel, hardened and tempered to suitable hardness. The PTO of tractor drives the rotavator. The working width of rotovator vary from 1000 – 2000 mm. The number of blades per flange is six. A good seed bed and pulverisation of the soil can be achieved in a single pass. The working of rotavator in the field is shown in Figure 29.1.

II) Combined Sowing and Land Development Equipment in Agriculture

The land preparation work as well as sowing work can be simultaneously done using tractor operated roto till drill and raised bed planter.

Roto Till Drill

It is tractor operated planter which is suitable for sowing of maize and wheat. The dimensions of the roto-till drill are 1800 x 1090 x 900 (LXWXH). It is run by 45 hp tractor. The sowing of maize is completed in a single operation leading to substantial savings on diesel consumption and time required for conventional seedbed preparation. The field capacity of the planter found to be 0.252 ha/h with field efficiency of 69 per cent. The cost of operation found to be Rs. 1655 per ha. The planter could save 68 per cent of time and 59 per cent of cost in sowing (Figure 29.2).

Raised Bed Planter

It is a tractor operated planter. The overall dimensions of the planter are 1800 x2100 x1400(LXWXH). Box section of 40 x 40 x 5 mm size made by welding 40 x40 x5 mm size angle iron. It requires power source of 45 -50 hp tractor. It is suitable for planting of small and bold size seeds like groundnut, bhendi, maize, gram. The effective field capacity 0.3ha/h with afield efficiency of 63.45 per cent. Cost of

Figure 29.1: Working of Rotavator.

Figure 29.2: Operation of Roto Till Drill.

operation is Rs. 1388 per ha for bhendi. There is a saving of 65 to 70 per cent in labour and saving of 30 per cent of irrigation water by adopting alternate row irrigation method. The working of raised bed planter is shown in Figure 29.3.

II) Land Levelling and Water Saving Device

Laser leveller plays an important role to meet out land levelling as well as act as water saving device.

Tractor Operated Laser Leveller

The land levelling work with prescribed slope of land can be maintained in field. The automatic land refilling in depressed positions can be done by laser controlled rays and operator can easily the tractor operated laser leveller and preferred for paddy fields. The water requirement for the crops can be easily saved and perfect levelling can be easily achieved (Figure 29.4).

Figure 29.3: Raised Bed Planter.

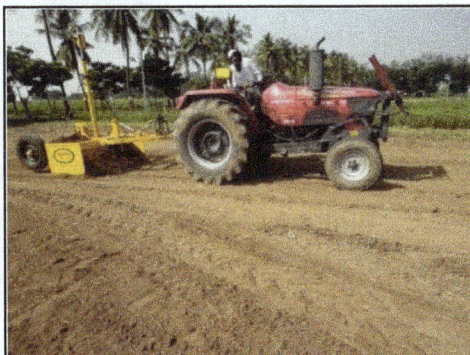

Figure 29.4: Tractor Operated Laser Leveller.

IV) No Rain Condition Equipment with Short Time available for Weeding

Under this climatic situation, the suitable equipment suggested for carrying out the works are zero till drill, three row rotary weeder and plastic mulching machine.

Zero Till Drill

It is a tractor operated planter. The overall dimensions of the planter are 2560 x 1370 x 1090 mm (LXWXH) with inclined plate seed metering mechanism. This is suitable for cotton. Timeliness of operation can be attained which is critical for sowing operation and this could solve problem of labour shortage during peak sowing season. The field capacity of the zero till planter 0.72 ha/hr with a field efficiency of 76.20 per cent. The fuel consumption was 3.2 l/h. The cost of operation in terms of Rs/h and Rs/ha is Rs.366.00 and Rs.508 respectively. Saving in time of 21.85 per cent was observed with zero till planter compared to sowing with conventional method. The field trials of zero till drill is shown in Figure 29.5.

Three Row Rotary Weeder

It is operated by a tractor and three rotary weeders are attached in the equipment. The weeding operation can be easily done in the field for weeding purpose. The rotary weeder is suitable for no rain conditions/dry land conditions (Figure 29.6).

Figure 29.5: Zero Till Drill.

Figure 29.6: Tractor Operated Three Row Rotary Weeder.

Tractor Operated Mulching Machine

The mulching machine can be attached to the tractor for spreading of plastic shhet over the entire field to control the weeds and moisture conservation. The black/blue coloured sheet is rolled and as the tractor moves it and covers the plastic sheet.

This machine is suggested to control the weed growth in open field. The operator can easily operate this machine and commercially it is available in the market based on width of plastic sheet.

Conclusion

The climate change scenario affects the suitable selection of farm equipment based on requirement in Karnataka. The farm mechanization for sustained growth in agricultural production is a necessity. The growth of mechanization demands for continuous upgradation of technological improvements in the production and operations of existing machines, climate change, cropping patterns, change of government policies and industrial advance of the country. In the mechanised states like Punjab the demand for quality machines for mechanization has increased tremandously for oil seeds, pulses, vegetable crops, forage grasses, cotton and for orchards. The use of rotavator, zero till drill, laser leveller, three row rotary weeder, mulching machine and improved version of tractor operated roto till drill depends on climatic and soil conditions in Karnataka.

References

Anantachar, M. 2005. Farm mechanization in Karnataka. Annual Report of All India Co-ordinated Research Project on Farm Implements and Machinery. pp : 10-14.

Gajendra Singh, 1996. Energy input in agricultural production of India. Paper presented in XI National Convention of Agricultural Engineers Sept 6-7, organised by Institution of engineers (Dharwad local centre)

Mehata, M.M. 1989. Farm mechanization growing recognition of role. Survey of Indian Agriculture *The Hindu* pp: 193-197

Surendra Singh, 2007. Farm Mechanization Scenario in India.Project Co-ordinator report of All India Co-ordinated Research Project on Farm Implements and Machinery. pp : 25-35.

30

Irrigation Water Legislation, Regulation and Distribution

Amrutha T. Joshi, Suresh S. Patil, G.B. Lokesh and H.M. Swamy

"Water is the softest thing, yet it can penetrate mountains and earth. This shows clearly the principle of softness overcoming hardness"

— *Lao Tzu*

Water is an essential resource for life. About 80 countries (40 per cent of the world's population) are suffering from water shortages and by the year 2025 two thirds of the world population may suffer from serious water problems. Agriculture accounts more than 70 per cent of the fresh water consumption inturn they give back 40 per cent of the world's food production. World agriculture faces an enormous challenge over the next 40 years to produce almost 50 per cent more food up to 2030 and double production by 2050 (Shilp Verma and Sanjiv Phansalkar, 2007). This will probably have to be achieved with less water, mainly because of pressure from growing urbanisation, industrialisation and climate change. India has many rivers whose total catchment area is estimated to be 252.8 mha out of which about 1 869 km³ of surface water resources, about 690 km³ of water is available for different uses. The ultimate irrigation potential of the country has been estimated to be 139.5 mha. India has achieved an irrigation potential of about 84.9 mha against the ultimate irrigation potential. About 360 km³ of groundwater is also available for irrigation. Water is widely shared among nations, states, religions, groups and communities and a total of 261 rivers are shared by 2 or more countries and which makes the management of water resources one of the most important.

Irrigation in India includes a network of major and minor canals, groundwater well based systems, tanks, and other rainwater harvesting projects for agricultural activities. Irrigation in India helps improve food security, reduce dependence on monsoons, improve agricultural productivity and create rural job opportunities. Dams used for irrigation projects help produce electricity and transport facilities, as well as provide drinking water supplies to a growing population, control floods and prevent droughts.

Keeping this in mind we need to give our more attention towards the issues of water policies such as water pricing, water legislation permits, inspection and enforcement, investment in infrastructure, scientific research, providing information to public and monitoring and evaluation and auditory issues of water such as water quality, rivers and lakes, flooding, drinking water and sanitation, water in relation to nature and biodiversity and marine environment. The problems of water management are multidimensional in nature and thus the approach to deal with it also needs to be multi-pronged and multi-faceted.

Water in Agriculture

India's food-grain production will report a decline of 3 per cent in 2014-15 accounting 257.07 million tons as compared to the highest ever food-grain production of 272.0 million tons in 2016-17. In India, some of the major challenges in agriculture water management relates to aging infrastructure and low water efficiency. Climatic changes will impact water availability and will pose a threat in times to come. Attaining efficiency in irrigation, developing ways to minimize losses and use of technology that uses less water to produce more per unit of land will be critical in meeting the increasing food demand simultaneously reducing the impact on environment. Water for agriculture has mainly been through major and minor irrigation projects. India's irrigation infrastructure is expanding by 1.8 M ha of irrigation potential with a public outlay of 7,000 crore per annum. Current annual expansion is one-third less than the maximum growth achieved in the past. The problems are due to poor implementation and the long gestation period of irrigation projects which results in spill over leading to more cost.

A study by the Water Resources Group has predicted that in 2030, the gap between demand and availability in India will be 50 per cent, with the demand touching 1,498 billion metre^3 and availability at mere 744 billion metre^3. It also states that a 58 per cent rise in demand from 2005 baseline in 2030, with demand almost doubling for the three sectors of agriculture, domestic and industry. The report cautions that the impact of the water crisis will be severe in the water rich basins and measures for water security will have to factor impacts of climate change into any planning for future. In case of agriculture, water demand is projected to rise from 656 Billion Cubic Metres (BCM) in 2005 to 979 BCM in 2020 and further raised to 1,195 BCM in 2030 (Kathpalia and Rakesh Kapoor, 2002). Availability of irrigation water to agriculture by source in India includes canals, Tanks, Tube-wells and other source. The irrigated area by canal has increased from 152 lakh hectares to 156.67 lakh hectare during 2001-02 to 2010-11. Irrigated area by tanks decreased from 21.96 lakh hectares to 20.04 lakh hectares during 2001-02 to 2010-11. The tube well irrigation has increased from 232.45 lakh hectares to 285.50 lakh hectares. Similarly the net irrigated area has also shown increasing trend from 56.93 lakh hectares to 636.01 lakh hectares.

Historic View of Irrigation in India

The history of the subject 'Irrigation and Power' dates back to 1855 when it was made the responsibility of the then newly created "Department of the Public Works" but not much impetus was given to irrigation work till the famine of 1858, when it was decided to take up canal construction work on an extensive scale and accordingly, an Inspector General of Canals was appointed. Under the Government of India Act 1919, irrigation became a Provincial subject and the Government of India's responsibility was confined to advice, co-ordination and settlement of disputes over right on the water of Inter-Provincial Rivers. In 1951, a new Ministry of National Resources and Scientific Research was set up and it took over the subject of 'Irrigation and Power' from the Ministry of Works, Mines and Power.

A separate Ministry of Irrigation and Power was set up in 1952 to look after the subject of irrigation. In the wake of unprecedented floods, a Flood Control Board was constituted to consider, flood control programme at the highest level. In 1969, an Irrigation Commission was set up to go into the matter of future irrigation development programme in the country in a comprehensive manner. To help in ensuring unified and coordinated programme for the speedy implementation of Irrigation and Command Area Development Projects, as well as for providing other inputs for maximizing agricultural produce, a separate Department of Irrigation was set up in November, 1974 under the reconstituted Ministry of Agriculture and Irrigation, consequently upon the bifurcation of erstwhile Ministry of Irrigation and Power. In January 1980, Department of Irrigation comes under the new Ministry of Energy and Irrigation. In 1980, the then Ministry of Energy and Irrigation was bifurcated and the erstwhile Department of Irrigation was raised to the level of Ministry with a view to having a coordinated and comprehensive view of the entire irrigation sector. In addition to major and medium irrigation, major irrigation sector, both surface and ground water as a Command Area Development Programme were brought within the purview of Ministry of Irrigation.

The irrigation for agricultural purpose, Minor and major irrigation and ground water exploration were transferred from the Ministry of Agriculture (Department of Agriculture and Cooperation) to the Ministry of Irrigation in 1980. In January 1985, the Ministry of Irrigation was once again combined under the Ministry of Irrigation and Power. However, in re-organization of the Ministries of the Central Government in September 1985, then Ministry of Irrigation and Power was bifurcated and the Department of Irrigation was re-constituted as the Ministry of Water Resources. This recognition of the necessity of planning for the development helps the country's water recourses in a coordinated manner.

With the nomenclature of the Ministry as the Ministry of Water Resources, perspective planning was taken up to fulfill the role expected of the Ministry. In this new perspective, calling for overall planning and coordination of all aspects of the development of the country's water resources, it was felt necessary to formulate a National Water Policy, laying down, *inter-alia*, priorities for various uses of water. National Water Resources Council was constituted under the Chairmanship of Hon'ble Prime Minister to look into this aspect. The National Water Resources Council (NWRC) adopted the National Water Policy in September 1987. National Water Board was constituted in September, 1990 to review the progress of implementation of the stipulations of the National Water Policy for reporting to the NWRC and also initiate effective measures for systematic development of the country's water resources.

The National Water Resources Council adopted the revised 'National Water Policy-2002' and further revised Draft National Water Policy 2012 adopted as recommended by the Drafting Committee, at its 14th Meeting of NWRC. The Council adopted the NWP 2012 during the India Water Week, 2013.

Water Legislation in India

The legislative and functional jurisdiction of the development and management of water lies with the state governments. The central government is required to step in only in the case of inter-state rivers. Panchayats, local governments, and municipalities fulfill a functionary role for several aspects related to water use as and when allocated by the respective state legislatures. Matters related to inter-state disputes need to be adjudicated under the Inter-States Disputes Act (1956) through the water Dispute Tribunal when the centre is of the opinion that the matter cannot be settled through negotiations. Of the 10 tribunals set up so far under the act, the final settlement under the Ravi–Beas Tribunal is still pending while the adjudication of four is currently under process. The River Boards Act (1956) enables the central government to control and regulate aspects relating to inter-state rivers. However, no boards have been formed so far to manage the basin as a whole.

The Second Irrigation Commission (1972) was constituted to review the development of irrigation and recommend essential irrigation works in chronically drought prone areas of the country in order to achieve food self-sufficiency. Among other suggestions, it proposed the adoption of benefit–cost ratio as a criterion for the sanction of irrigation projects, and the raising of water rates to cover O and M and other running works costs at a reasonable rate of interest charged on the capital cost. The Commission also recommended the protection of watershed areas through afforestation, pasture development, protection of riverbanks and shorelines, a participatory approach to water management, and the promotion of a special agency for the expeditious and coordinated development of command areas.

National Water Policy: Formulated in 1987 and updated in 2002 by the Ministry of Water Resources (MOWR), the NWP recognizes water as a prime natural resource, a basic human need and a precious national asset. The policy *inter-alia* prioritizes water use, stresses the promotion of the IWRM, and emphasizes conservation and efficient use of water. The relevant section of the NWP on private sector participation should be encouraged in planning, development and management of water resources projects for diverse uses, wherever feasible.

Private sector participation may help in introducing innovative ideas, generating financial resources and introducing corporate management and improving service efficiency and accountability to users. Depending upon the specific situations, various combinations of private sector participation, in building, owing, operating, leasing and transferring of water resources facilities, may be considered. Policy formulation setup a high political council called the National Water Resources Council (NWRC) has been set up to take all policy decisions in the country. This is headed by the Prime Minister and its members include the Chief Ministers of all states and union territories. In order to look into policy issues at an official level, a National Water Board has also been set up, which is headed by the Union Secretary of Water Resources alongwith the Chief Secretaries of all states and Union Territories. NGOs and other professional institutions also add to the insight as well as inputs on various policy matters with respect to development and management of water, for the consideration of the government.

Water Distribution

The water distribution rules of an irrigation system define a pattern of water delivery that does not match technically feasible irrigation services desired by the users, and then the users, often in cooperation with system managers, will modify or subvert the rules to bring water delivery into accord with their desires. Subversion of the water distribution rules will adversely affect water delivery performance, especially equity of distribution, and will raise the cost of irrigation to the users. Inconsistencies in the water distribution rules create difficulties in system operations that lead to inefficient and inequitable water distribution performance (Sakthivadivel and Raju, 1997). For the better operation and maintenance of irrigation system there is a need for improving cost recovery from all users including irrigators of the water, offers one of the most important avenues for raising financial resources. A study from Karnataka in rotational water distribution in Upper Krishna project area given below.

Rotational Water Distribution in Upper Krishna Project Area

The State government has adopted rotational distribution of water to improve the irrigation potential besides providing a thrust to horticultural and high value crops in Upper Krishna Project (UKP) area. Based on the recommendation of committee headed by Dr. S A. Patil rotational distribution of water with 14 days interval was introduced to facilitate availability of water to tail end farmers and also provide time to them for land preparation and sowing. For efficient utilisation of water and maximise production, the committee has also suggested promotion of horticultural and high value crops like mango, guava, fig, lime, papaya, pomegranate, custard apple, drumstick, amla *etc.* as these were found to be most suitable for the area. The committee favoured implementation of a cropping pattern involving crops which require less water and discoraged raising paddy crops, a practice which leads to increased degradation of land. It has also favoured offering of a special package for development of horticulture crops in UKP area as in Bijapur by providing 100 per cent subsidy for irrigation equipment and other infrastructure.

Issues and Challenges in Water Distribution

1) Mismatch between Rules and Farmers' Desires

Farmers' desires for irrigation services are based on the crops that they want to grow within the recognized limits of water availability in the irrigation system. Water distribution rules limit the quantity to be supplied or the period of supply. Farmers' crop choices are restricted to those crops that can be grown within the water quantity or schedule defined by the rules. If farmers perceive an opportunity to grow a more profitable crop outside the choices permitted by the water distribution rules but within the technical possibilities of the irrigation system, they can (a) submit to the limitations imposed by the distribution rules, (b) evade or subvert the rules, or (c) attempt to change the rules

2) Farmers' Desires and the Limitations of the Irrigation System

The overall water availability and the technical features of an irrigation system limit the possibilities of water distribution. There is a need to ensure that farmers understand the limits so that they do not demand more than the system can feasibly provide. Conflicts among farmers and between farmers and system managers may be exacerbated by farmers' failure to understand the limitations of the irrigation system.

Inter-States Water Disputes in India

The Interstate River Water Disputes Act, 1956 (IRWD Act) is an Act of the Parliament of India enacted under Article 262 of Constitution of India on the eve of reorganization on linguistic basis to resolve the water disputes that would arise in the use, control and distribution of an interstate river or river valley. Article 262 of the Indian Constitution provides a role for the Central government in adjudicating conflicts surrounding inter-state rivers that arise among the state/regional governments. The existing inter-states water disputes in India given are given below.

A) *Cauvery Water Disputes*

The Cauvery Water Disputes Tribunal (CWDT) was constituted by the Government of India on 2nd June 1990 to adjudicate the water dispute regarding inter-state river Cauvery and the river valley thereof. The Tribunal had also passed an Interim Order in June, 1991 and further Clarificatory Orders on the Interim Order in April, 1992 and December, 1995. The Cauvery Water Disputes Tribunal has submitted its reports and decision under Section 5 (2) of Inter-State River Water Disputes Act, 1956 to Government on 5th February, 2007. The party states and the Central Govt. have sought clarification and guidelines under Section 5(3) of the Act. The terms of the tribunal has been extended upto 2/11/2008 as per provisions of ISRWD Act, 1956. Further, the party states have also filled SLPs in the Hon'ble Supreme Court against Cauvery tribunals report and Hon'ble Supreme Court has granted leave. The problem of water sharing till persists between the states of Karnataka and Tamil Nadu.

B) *Ravi and Beas Waters Tribunal*

Surplus Ravi-Beas waters refer to available Ravi-Beas waters excluding the pre-partition utilization of 3.13 MAF by Rajasthan, the then Punjab and J and K. The surplus Ravi-Beas waters were first allocated in a Conference of the Chief Ministers held in January 1955, then by a Govt. of India Notification dated 24/3/1976 subsequent to the reorganization of Punjab in Nov. 1966 and later, in an agreement among the Chief Ministers of Punjab, Haryana and Rajasthan. In 1985 the issues got re-opened, there were prolonged negotiations which culminated in signing of the Punjab Memorandum of Settlement (Rajiv-Longowal Accord) between the then Prime Minister of India, Shri Rajiv Gandhi and the then President of Shiromani Akali Dal, Sant Harchand Singh Longowal.

C) *Vansadhara River Water Dispute*

The State of Orissa in February 2006 sent a complaint to the Central Government under Section 3 of the Inter-State River Water Disputes (ISRWD) Act, 1956 regarding water disputes between the Government of Orissa and Government of Andhra Pradesh pertaining to Inter-State River Vansadhara for constitution of a Inter-State Water Disputes Tribunal for adjudication. The main grievance of the State of Orissa in the complaint sent to the Central Government is basically adverse effect of the executive action of Govt. of Andhra Pradesh in undertaking the construction of a canal taking off from the river Vansadhara called as flood flow canal at Katragada and failure of Govt. of Andhra Pradesh to implement the terms of inter-State agreement understanding *etc.* relating to use, distribution and control of waters of inter-State river Vansadhara and its valley. Basic contention of State of Orissa in the complaint is that the flood flow canal would result in drying up the existing river bed and consequent shifting of the river affecting ground water table. It has also raised the issue of scientific assessment of available water in Vansadhara at Katragada and Gotta Barrage and the basis for sharing the available water.

D) *Mahadayi/Mandovi River*

In July, 2002, the State of Goa made a request under Section 3 of the Inter-State River Water Disputes Act, 1956 (as amended) for constitution of the Tribunal under the said Act and refers the matter for adjudication and decision of dispute relating to Mandovi River. The issues mentioned in the request included the assessment of available utilisable water resources in the basin at various points and allocation of this water to the 3 basin States keeping in view priority of the use of water within basin as also to decide the machinery to implement the decision of the tribunal *etc.* The Act requires that Central Government shall constitutes a tribunal if it is of the opinion that water dispute cannot be settled by negotiation.

Therefore, actions and efforts of Central Government in MoWR since July, 2002 were basically guided by the aforesaid provision of the Act. In continuation of this process, Hon'ble Union Minister for Water Resources convened an inter-State meeting on 4.4.2006 at the level of Chief Ministers of the States of Goa, Karnataka and Maharashtra. Subsequent actions of Government of Goa with regard to follow up action on decisions taken in the inter-State meeting gave impression that State of Goa is not ready to pursue the negotiation process further and wants constitution of tribunal and reference of the dispute to the Tribunal immediately. Accordingly, the Central Government in the MOWR concluded that the dispute contained in the request of State of Goa of July, 2002 cannot be resolved by negotiation and initiated further action in the matter as per the provisions of Inter-State River Water Disputes Act, 1956 and rules made there under.

Meanwhile the Govt. of Goa filed a suit in the Hon'ble Supreme court in Sept, 06 for setting up of a water dispute tribunal for adjudication of the above river water dispute and an interlocutory application (IA) for stay in construction activities. The Writ Petition with the application has been listed on a number of occasions before the Hon'ble Supreme Court. Meanwhile, the Cabinet considered and approved the proposal of constitution of Mahadayi Tribunal. Central government has constituted Mahadayi Water Disputes Tribunal (MWDT) 2010 and the issue is still not been solved.

E) *Krishna River Water Dispute Tribunal*

River Krishna is an inter-state river traversing states of Maharashtra, Karnataka and Andhra Pradesh. Krishna Water Disputes Tribunal (KWDT) was set up under Inter-State River Water Disputes Act, 1956 vide notification No. S.O. 1419 dated 10th April, 1969 (hereinafter referred to as KWDT-I) to adjudicate upon the water dispute regarding the Inter-State river Krishna and the river valley thereof.

Water Disputes between the Countries

A) *Indus Waters Treaty*

Water-sharing treaty between India and Pakistan, brokered by the World Bank was signed in Karachion September 19, 1960 by Indian Prime Minister Jawaharlal Nehru and President of Pakistan Ayub Khan. The treaty was a result of Pakistani fear that since the source rivers of the Indus basin were in India, it could potentially create droughts and famines in Pakistan, especially at times of war. Since the ratification of the treaty in 1960, India and Pakistan have not engaged in any water wars. Disagreements and disputes have been settled via legal procedures, provided for within the framework of the treaty. The treaty is considered to be one of the most successful water sharing endeavours in the world today even though analysts acknowledge the need to update certain technical specifications and expand the scope of the document to include climate change. As per the provisions in the treaty, India can use only 20 per cent of the total water carried by the Indus river.

B) *The India-Pakistan Water Dispute*

The water dispute between India and Pakistan is serious not only because of water, but also due to the political rivalry between the two countries. Their rivalry made things more complicated than they really are. The water dispute between them started soon after the partition of the subcontinent in 1947. Until the Indus

Waters Treaty, arrangements to share east and west flowing rivers were ad hoc. Unfortunately, over the years the IWT, too, has failed to pacify the water conflict. Hindu right-wing groups in India call on [the government] to stop flow of water to Pakistan or flood it. In the meantime, Islamic radicals in Pakistan call for water jihad against India. Moreover, the water dispute between the two countries is embedded in their political relationship.

C) Indo-China Water Dispute

The dispute between India and china is mainly regarding the Brahmaputra river flowing through the two countries.The search for water resources in China and India has persistently been a source of tension between the two countries. Chinese efforts to divert the water resources of the Brahmaputra river away from India will worsen a situation that has remained tense since the 1962 Indo-China war. The melting glaciers in the Himalayas as a result of accelerating global climate change will have a dramatic effect on this river's water supply. This will increase water scarcity as well as the likelihood of floods, impact agrarian livelihoods and strain the fragile equilibrium between the two Asian giants.

D) Indo-Bangla Water Disputes

The dispute between India and Bangladesh (East Pakistan till 1971) over the sharing of the Ganga water arose when in 1960s. The Indian government decided to construct a barrage at Farakka, close to Indo-Bangladesh border. The objective behind the construction of the Farakka Barrage was to increase the lean period flow of the Bhagirathi-Hooghly branch of Ganga to increase the water depth at the Kolkata port which was threatened by siltation. As irrigation withdrawals increased in Bangladesh, dispute arose between India and Bangladesh over the sharing of the lean season flow at Farakka. The inadequacy of water during the lean season to meet the assessed demands in the two countries is the root cause of the conflict.

F) Indo-Bangla Water Disputes

India and Bangladesh share 54 rivers between them. Despite setting up a Joint River Commission for water management as early in 1972, tensions between the countries on how to share resources recently came to a head in a dispute over the Teesta River. At stake are the lives of countless people from West Bengal and Bangladesh who depend upon the river for survival.

Conclusion

Given the anticipated growth in demand for food and water and increasing pressures from climate, agriculture will be a key target for policy makers as it consumes about 70 per cent of the world's freshwater withdrawals. The groundwater policies usually involve licenses and other regulatory instruments. Farmers are only paying the operation and maintenance costs for water supplied, with little or no recovery of agriculture's share of capital costs for water infrastructure. For the better operation and maintenance of irrigation system there is a need for improving cost recovery from all users including irrigators of the water, offers one of the most important avenues for raising financial resources. But illegal groundwater pumping is difficult to observe or control and remains a major challenge for the sustainability of farming and the policy makers should address the issue of water management. Given the competing water demands by other sectors, it is important that, the water allocation priorities be rationalized. But before getting into this, there is a need to enhance efficiency in irrigation water use. In order to facilitate this, WUAs should play a key role. This should not remain a onetime initiative, but be institutionalized within the existing systems and processes of the Ministry.

References

Sakthivadivel, J.D.R., Raju, K.V. 1997, Water distribution rules and water distribution performance: A case study in the Tambraparani Irrigation System. *Research Report, International Irrigation Management Institute*, Colombo, Sri Lanka.

Seminar on reforms in management of public irrigation system. *Information Bulletin* 30-31st October, 2014, Bangalore.

Verma, S., Phansalkar, S., 2007. India's water future 2050: Potential deviations from 'Business-as-Usual', *IWMI-TATA Water Policy Program*, India.

The report of the Cauvery water disputes tribunal with the decision, *volume I* background of the dispute and framing of issues, New Delhi, 2007.

www.oecd.org/agriculture/water.

Ministry of Water Resources, Government of India.

Ministry of Water Resources, Government of Karnataka.

Conversion Factors Used in Irrigation Water Management

1 m³= 1,000 litres	1 acre-foot= 43560 cubic feet
1 litre= 1,000 cm³	1 acre-foot = 1,233 m³
1 cm³= 1 ml	1 acre-inch= 3630 cubic feet
1 m³= 35,314 cubic feet	1 acre-inch = 102.8 m³
1 litre= 0.0353 cubic feet	1 ha- mm = 10,000, 000 cm³
1 litre= 0.2201 Imperial gallon	1 ha- =mm= 10,000 litres
1 U.S. Gallon= 3.7854 litres	1 ha-cm = 100 m³
1 imperial gallon= 4.5436 litres	1 ha-cm = 100,000,000 cm³
1 TNC= 1,000,000,000 cubic feet	1 ha-cm 100,000 l
1 TMC= 28,300,000 m³	1 ha-m = 10,000 m³
1 km³= 1,000,000,000 m³	1 cubic-feet= 28.32 litres
	1 cubic-feet= 0.0283 m³

Rate of Flow Measurement

1 m³/sec= 35.314 cubic feet per second
1 m³/h = 0.278 l/s
1 l/s= 0.0353 cubic feet per second
1 l/s= 3.6 m³/h
1 cubic feet/s = 0.028 m³/s
1 cusec= 28.32 l/s
1 cusec= 1 acre-inch/h

Metric Units Prefixes

exa - one trillion (10^{18})
peta - one thousan million million (10^{15})
tera - one million million (10^{12})
giga - one thousand millin (10^{9})
mega - one million (10^{6})
kilo - one thousand (10^{3})
hecto - one hundred (10^{2})
deca - ten (10^{1})
deci - one tenth- (10^{-1})
centi - one hundredth (10^{-2})
milli - one thousandth (10^{-3})
micro - one millionth (10^{-6})
rano - one thousand millionth (10^{-9})
pico - one million millionth (10^{-12})
femto - one thousand million millionth (10^{-15})
atto - one trillionth (10^{-18})

Index

www.ingramcontent.com/pod-product-compliance
Lightning Source LLC
Chambersburg PA
CBHW050508190326
41458CB00005B/1472